UCLA Symposia on Molecular and Cellular Biology, New Series

Series Editor, C. Fred Fox

RECENT TITLES

Volume 50
Interferons as Cell Growth Inhibitors and Antitumor Factors, Robert M. Friedman, Thomas Merigan, and T. Sreevalsan, *Editors*

Volume 51
Molecular Approaches to Developmental Biology, Richard A. Firtel and Eric H. Davidson, *Editors*

Volume 52
Transcriptional Control Mechanisms, Daryl Granner, Michael G. Rosenfeld, and Shing Chang, *Editors*

Volume 53
Progress in Bone Marrow Transplantation, Robert Peter Gale and Richard Champlin, *Editors*

Volume 54
Positive Strand RNA Viruses, Margo A. Brinton and Roland R. Rueckert, *Editors*

Volume 55
Amino Acids in Health and Disease: New Perspectives, Seymour Kaufman, *Editor*

Volume 56
Cellular and Molecular Biology of Tumors and Potential Clinical Applications, John Minna and W. Michael Kuehl, *Editors*

Volume 57
Proteases in Biological Control and Biotechnology, Dennis D. Cunningham and George L. Long, *Editors*

Volume 58
Growth Factors, Tumor Promoters, and Cancer Genes, Nancy H. Colburn, Harold L. Moses, and Eric J. Stanbridge, *Editors*

Volume 59
Chronic Lymphocytic Leukemia: Recent Progress and Future Direction, Robert Peter Gale and Kanti R. Rai, *Editors*

Volume 60
Molecular Paradigms for Eradicating Helminthic Parasites, Austin J. MacInnis, *Editor*

Volume 61
Recent Advances in Leukemia and Lymphoma, Robert Peter Gale and David W. Golde, *Editors*

Volume 62
Plant Gene Systems and Their Biology, Joe L. Key and Lee McIntosh, *Editors*

Volume 63
Plant Membranes: Structure, Function, Biogenesis, Christopher Leaver and Heven Sze, *Editors*

Volume 64
Bacteria–Host Cell Interactions, Marcus A. Horwitz, *Editor*

Volume 65
The Pharmacology and Toxicology of Proteins, John S. Holcenberg and Jeffrey L. Winkelhake, *Editors*

Volume 66
Molecular Biology of Invertebrate Development, John D. O'Connor, *Editor*

Volume 67
Mechanisms of Control of Gene Expression, Bryan Cullen, L. Patrick Gage, M.A.Q. Siddiqui, Anna Marie Skalka, and Herbert Weissbach, *Editors*

Volume 68
Protein Purification: Micro to Macro, Richard Burgess, *Editor*

Volume 69
Protein Structure, Folding, and Design 2, Dale L. Oxender, *Editor*

Volume 70
Hepadna Viruses, William Robinson, Katsuro Koike, and Hans Will, *Editors*

Volume 71
Human Retroviruses, Cancer, and AIDS: Approaches to Prevention and Therapy, Dani Bolognesi, *Editor*

Please contact the publisher for information about previous titles in this series.

UCLA Symposia Board

Plant Gene Systems and Their Biology

Plant Gene Systems and Their Biology

Proceedings of a CIBA-Geigy–UCLA Symposium
Held at Tamarron, Colorado, February 2–8, 1987

Editors

Joe L. Key
Department of Botany
University of Georgia
Athens, Georgia

Lee McIntosh
MSU-DOE Plant Research Laboratory
Michigan State University
East Lansing, Michigan

Alan R. Liss, Inc. • New York

Address all Inquiries to the Publisher
Alan R. Liss, Inc., 41 East 11th Street, New York, NY 10003

Printed in the United States of America

Library of Congress Cataloging-in-Publication Data

CIBA-Geigy-UCLA Symposium (1987 : Tamarron, Colo.)
Plant gene systems and their biology.

(UCLA symposia on molecular and cellular biology ; new ser., v. 62)
Includes index.
1. Plant genetics—Congresses. 2. Plant molecular genetics—Congresses. 3. Plant molecular biology—Congresses. I. Key, Joe L. II. McIntosh, Lee. III. Title. IV. Series.
QK981.C53 1987 581.1′5 87-22791
ISBN 0-8451-2661-X

Contents

Contributors

W. Michael Ainley, Botany Department, University of Georgia, Athens, GA 30602 **[153]**

D.C. Alexander, ARCO Plant Cell Research Institute, Dublin, CA 94568 **[121]**

M.A. Anderson, Plant Cell Biology Research Centre, School of Botany, University of Melbourne, Parkville, Victoria 3052, Australia **[53]**

A. Bacic, Plant Cell Biology Research Centre, School of Botany, University of Melbourne, Parkville, Victoria 3052, Australia **[53]**

Ram Bandyopadhyay, Department of Microbiology and Cell Science, University of Florida, Gainesville, FL 32611 **[279]**

Cynthia Bauerle, Department of Botany, University of Wisconsin, Madison, WI 53706 **[363]**

John Bedbrook, Advanced Genetic Sciences, Inc., Oakland, CA 94608 **[289]**

Jeffrey L. Bennetzen, Department of Biological Sciences, Purdue University, West Lafayette, IN 47907 **[183]**

Mark Bloom, Department of Biochemistry, Michigan State University, East Lansing, MI 48824 **[227]**

J. Botterman, Plant Genetic Systems N.V., 9000 Gent, Belgium **[171]**

A.H.D. Brown, CSIRO Division of Plant Industry, Canberra, ACT 2601, Australia **[121]**

Willis E. Brown, Department of Biological Sciences, Purdue University, West Lafayette, IN 47907 **[183]**

Wesley B. Bruce, Department of Microbiology and Cell Science, University of Florida, Gainesville, FL 32611 **[279]**

John J. Burke, Department of Plant and Soil Science, Texas Tech University, Lubbock, TX 79409; USDA-ARS, Plant Stress and Water Conservation Unit, Lubbock, TX 79401 **[109]**

Thomas J. Burke, Department of Horticulture, University of Wisconsin-Madison, Madison, WI 53706 **[251]**

Nam-Hai Chua, Laboratory of Plant Molecular Biology, The Rockefeller University, New York, NY 10021 **[321]**

The numbers in brackets are the opening page numbers of the contributors' articles.

A.E. Clarke, Plant Cell Biology Research Centre, School of Botany, University of Melbourne, Parkville, Victoria 3052, Australia **[53]**

Y.S. Cody, Department of Plant Pathology, Washington State University, Pullman, WA 99164-6430 **[161]**

Pam Cooper, Department of Plant Biology, USDA-ARS, University of Illinois, Urbana, IL 61801 **[353]**

E.C. Cornish, Plant Cell Biology Research Centre, School of Botany, University of Melbourne, Parkville, Victoria 3052, Australia **[53]**

Janelle Cortner, Department of Horticulture, University of Wisconsin-Madison, Madison, WI 53706 **[251]**

Gloria M. Coruzzi, Laboratory of Plant Molecular Biology, The Rockefeller University, New York, NY 10021-6399 **[217]**

April Cresse, Department of Biological Sciences, Purdue University, West Lafayette, IN 47907 **[183]**

Eva Czarnecka, Department of Microbiology and Cell Science, University of Florida, Gainesville, FL 32611 **[279]**

Jeffery L. Dangl, Abteilung Biochemie, Abteilung Genetische Grundlagen der Pflanzenzüchtung, Max-Planck Institut für Züchtungsforschung, D-5000 Köln 30, Federal Republic of Germany **[139]**

C.H. Daniels, Department of Plant Pathology, Washington State University, Pullman, WA 99164-6430 **[161]**

Caroline Dean, Advanced Genetic Sciences, Inc., Oakland, CA 94608 **[289]**

M. De Block, Plant Genetic Systems N.V., 9000 Gent, Belgium **[171]**

Richard J. Debus, Plant Research Laboratory and Department of Biochemistry, Michigan State University, East Lansing, MI 48824 **[371]**

Liz Dennis, Division of Plant Industry, CSIRO, Canberra, ACT 2601, Australia **[263]**

R.E. Dewey, Department of Crop Science and Genetics, North Carolina State University, Raleigh, NC 27695-7614 **[65]**

Carl J. Douglas, Abteilung Biochemie, Abteilung Genetische Grundlagen der Pflanzenzüchtung, Max-Planck Institut für Züchtungsforschung, D-5000 Köln 30, Federal Republic of Germany **[139]**

Pamela Dunsmuir, Advanced Genetic Sciences, Inc., Oakland, CA 94608 **[289]**

Janice W. Edwards, Laboratory of Plant Molecular Biology, The Rockefeller University, New York, NY 10021-6399 **[217]**

Jeff Ellis, Division of Plant Industry, CSIRO, Canberra ACT 2601, Australia **[263]**

Thomas E. Elthon, MSU-DOE Plant Research Laboratory, Biochemistry Department, Michigan State University, East Lansing, MI 48824 **[243]**

Mitchell Favreau, Advanced Genetic Sciences, Inc., Oakland, CA 94608 **[289]**

R. Finkelstein, MSU-DOE Plant Research Laboratory, Michigan State University, East Lansing, MI 48824 **[311]**

Joan S. Gebhardt, Plant Sciences Group, Department of Biology, Rensselaer Polytechnic Institute, Troy, NY 12180-3590 **[39]**

John W. Grula, Phytogen, Pasadena, CA 91105 **[207]**

Michael Gurevitz, Plant Research Laboratory and Department of Biochemistry, Michigan State University, East Lansing, MI 48824 **[371]**

William B. Gurley, Department of Microbiology and Cell Science, University of Florida, Gainesville, FL 32611 **[279]**

L.A. Hadwiger, Department of Plant Pathology, Washington State University, Pullman, WA 99164-6430 **[161]**

Klaus Hahlbrock, Abteilung Biochemie, Abteilung Genetische Grundlagen der Pflanzenzüchtung, Max-Planck Institut für Züchtungsforschung, D-5000 Köln 30, Federal Republic of Germany **[139]**

Wayne W. Hanna, USDA-ARS, Coastal Plain Experiment Station, Department of Agronomy, Tifton, GA 31793 **[75]**

A.D. Hanson, MSU-DOE Plant Research Laboratory, Michigan State University, East Lansing, MI 48824 **[121]**

P.J. Harris, Plant Cell Biology Research Centre, School of Botany, University of Melbourne, Parkville, Victoria 3052, Australia **[53]**

Peggy H. Hatfield, Department of Horticulture, University of Wisconsin-Madison, Madison, WI 53706 **[251]**

Thomas K. Hodges, Department of Botany and Plant Pathology, Purdue University, West Lafayette, IN 47907 **[339]**

Heidi Hoffmann, Abteilung Biochemie, Abteilung Genetische Grundlagen der Pflanzenzüchtung, Max-Planck Institut fur Züchtungsforschung, D-5000 Köln 30, Federal Republic of Germany **[139]**

N.E. Hoffman, MSU-DOE Plant Research Laboratory, Michigan State University, East Lansing, MI 48824; present address: Plant Science Institute, University of Pennsylvania, Philadelphia, PA 19104 **[121]**

H. Hofte, Plant Genetic Systems N.V., 9000 Gent, Belgium **[171]**

D. Hondred, MSU-DOE Plant Research Laboratory, Michigan State University, East Lansing, MI 48824 **[121]**

W.W.L. Hu, Department of Crop Science and Genetics, North Carolina State University, Raleigh, NC 27695-7614 **[65]**

Richard L. Hudspeth, Phytogen, Pasadena, CA 91105 **[207]**

Merten Jabben, Department of Horticulture, University of Wisconsin-Madison, Madison, WI 53706 **[251]**

Christer Jansson, Plant Research Laboratory and Department of Biochemistry, Michigan State University, East Lansing, MI 48824 **[371]**

Jonathan Jones, Advanced Genetic Sciences, Inc., Oakland, CA 94608 **[289]**

Kenneth Keegstra, Department of Botany, University of Wisconsin, Madison, WI 53706 **[363]**

M. Keil, Institut für Genbiologische Forschung Berlin, 1000 Berlin 33, Federal Republic of Germany **[331]**

D.F. Kendra, Department of Plant Pathology, Washington State University, Pullman, WA 99164-6430 **[161]**

Joe L. Key, Department of Botany, University of Georgia, Athens, GA 30602 **[xvii,87,279]**

Janice Kimpel, Agrigenetics Advanced Science Co., Madison, WI 53716 **[87]**

M. Krishnan, Department of Plant and Soil Science, Texas Tech University, Lubbock, TX 79409; USDA-ARS Plant Stress and Water Conservation Unit, Lubbock, TX 79401 **[109]**

A.R. Lachmansingh, Stauffer Chemical Co., Richmond, CA 94804 **[301]**

Anton Lang, MSU-DOE Plant Research Laboratory, Michigan State University, East Lansing, MI 48824-1312 **[3]**

Ray Larsen, Institute of Biological Chemistry, Washington State University, Pullman, WA 99164 **[227]**

Lisa Lee, Department of Biological Sciences, Purdue University, West Lafayette, IN 47907 **[183]**

J. Leemans, Plant Genetic Systems N.V., 9000 Gent, Belgium **[171]**

C.S. Levings III, Department of Crop Science and Genetics, North Carolina State University, Raleigh, NC 27695-7614 **[65]**

Susanne Lipphardt, Abteilung Biochemie, Abteilung Genetische Grundlagen der Pflanzenzüchtung, Max-Planck Institut für Züchtungsforschung, D-5000 Köln 30, Federal Republic of Germany **[139]**

Danny Llewellyn, Division of Plant Industry, CSIRO, Canberra ACT 2601, Australia **[263]**

Horst Lorz, Max-Planck Institut für Züchtungsforschung, D-5000 Köln, Federal Republic of Germany **[339]**

Thomas Lubben, Department of Botany, University of Wisconsin, Madison, WI 53706 **[363]**

J.M. Martinez-Zapater, MSU-DOE Plant Research Laboratory, Michigan State University, East Lansing, MI 48824 **[311]**

D. Mascarenhas, Stauffer Chemical Co., Richmond, CA 94804 **[301]**

S.-L. Mau, Plant Cell Biology Research Centre, School of Botany, University of Melbourne, Parkville, Victoria 3052, Australia **[53]**

Carl N. McDaniel, Plant Science Group, Department of Biology, Rensselaer Polytechnic Institute, Troy, NY 12180-3590 **[39]**

Lee McIntosh, MSU-DOE Plant Research Laboratory, Biochemistry Department, Michigan State University, East Lansing, MI 48824 **[xvii,243,371]**

I.J. Mettler, Stauffer Chemical Co., Richmond, CA 94804 **[301]**

Matthew Morell, Department of Biochemistry, Michigan State University, East Lansing, MI 48824 **[227]**

Ronald T. Nagao, Botany Department, University of Georgia, Athens, GA 30602**[87,279]**

Ferenc Nagy, Laboratory of Plant Molecular Biology, The Rockefeller University, New York, NY 10021 **[321]**

Henry T. Nguyen, Department of Plant and Soil Science, Texas Tech University, Lubbock, TX 79409; USDA-ARS, Plant Stress and Water Conservation Unit, Lubbock, TX 79401 **[109]**

Roxy L. Nickels, MSU-DOE Plant Research Laboratory, Biochemistry Department, Michigan State University, East Lansing, MI 48824 **[243]**

Joan T. Odell, Laboratory of Plant Molecular Biology, The Rockefeller University, New York, NY 10021 **[321]**

Thomas W. Okita, Institute of Biological Chemistry, Washington State University, Pullman, WA 99164 **[227]**

Donald R. Ort, Department of Plant Biology, USDA-ARS, University of Illinois, Urbana, IL 61801 **[353]**

Heinz D. Osiewacz, Plant Research Laboratory and Department of Biochemistry, Michigan State University, East Lansing, MI 48824 **[371]**

Jim Peacock, Division of Plant Industry, CSIRO, Canberra, ACT 2601, Australia **[263]**

H. Pena-Cortes, Institut für Genbiologische Forschung Berlin, 1000 Berlin 33, Federal Republic of Germany **[331]**

D.A. Pierce, Stauffer Chemical Co., Richmond, CA 94804 **[301]**

L.M. Pomeroy, Stauffer Chemical Co., Richmond, CA 94804 **[301]**

Jack Preiss, Department of Biochemistry, Michigan State University, East Lansing, MI 48824 **[227]**

Natasha V. Raikhel, MSU-DOE Plant Research Laboratory, Michigan State University, East Lansing, MI 48824-1312 **[153]**

Subbanaidu Ramagopal, USDA-ARS Experiment Station, Hawaiian Sugar Planters'Association, P.O. Box 1057, Aiea, HI 96701 **[131]**

A. Reynaerts, Plant Genetic Systems N.V., 9000 Gent, Belgium **[171]**

M. Rocha-Sosa, Institut für Genbiologische Forschung Berlin, 1000 Berlin 33, Federal Republic of Germany **[331]**

J. Sanchez-Serrano, Institut für Genbiologische Forschung Berlin, 1000 Berlin 33, Federal Republic of Germany **[331]**

Pablo A. Scolnik, E.I. DuPont de Nemours, Plant Science Group, Experimental Station, 402/4252, Wilmington, DE 19898 **[383]**

John Shanklin, Department of Horticulture, University of Wisconsin-Madison, Madison, WI 53706 **[251]**

Susan R. Singer, Plant Science Group, Department of Biology, Rensselaer Polytechnic Institute, Troy, NY 12180-3590; present address: Department of Biology, Carleton College, Northfield, MN 55057-4025 **[39]**

Karam Singh, Division of Plant Industry, CSIRO, Canberra ACT 2601, Australia **[263]**

Sjef Smeekens, Department of Molecular Cell Biology, State University of Utrecht, Utrecht 3584 CH, The Netherlands **[363]**

C.R. Somerville, MSU-DOE Plant Research Laboratory, Michigan State University, East Lansing, MI 48824 **[311]**

Stanley Tamaki, Advanced Genetic Sciences, Inc., Oakland, CA 94608 **[289]**

Scott V. Tingey, Laboratory of Plant Molecular Biology, The Rockefeller University, New York, NY 10021-6399; present address: E.I. DuPont de Nemours Co., Central Research and Development, Wilmington, DE 19898 **[217]**

Fong-Ying Tsai, Laboratory of Plant Molecular Biology, The Rockefeller University, New York, NY 10021-6399 **[217]**

M. Vaeck, Plant Genetic Systems N.V., 9000 Gent, Belgium **[171]**

Elizabeth Vierling, Department of Biochemistry, University of Arizona, Tucson, AZ 85721 **[99]**

Richard D. Vierstra, Department of Horticulture, University of Wisconsin-Madison, Madison, WI 53706 **[251]**

Daphne Vince-Prue, Department of Botany, University of Reading, Reading, England **[25]**

Elsbeth L. Walker, Laboratory of Plant Molecular Biology, The Rockefeller University, New York, NY 10021-6399 **[217]**

John Walker, Division of Plant Industry, CSIRO, Canberra ACT 2601, Australia **[263]**

E.A. Weck, Stauffer Chemical Co., Richmond, CA 94804 **[301]**

Peter Weisbeek, Department of Molecular Cell Biology, State University of Utrecht, Utrecht 3584 CH, The Netherlands **[363]**

Thea A. Wilkins, MSU-DOE Plant Research Laboratory, Michigan State University, East Lansing, MI 48824-1312 **[153]**

L. Willmitzer, Institut für Genbiologische Forschung Berlin, 1000 Berlin 33, Federal Republic of Germany **[331]**

David Wolstenholme, Division of Plant Industry, CSIRO, Canberra ACT 2601, Australia; present address: Department of Biology, University of Utah, Salt Lake City, UT 84112 **[263]**

J.R. Woodward, Plant Cell Biology Research Centre, School of Botany, University of Melbourne, Parkville, Victoria 3052, Australia **[53]**

Preface

The papers included in this volume are representative of topics presented at the CIBA-Geigy–UCLA Symposium on Plant Gene Systems and Their Biology held at Tamarron, Colorado, February 2–8, 1987. These papers review most, but not all of the oral presentations. Some articles included here were presented in poster sessions, and the organizers felt these would add significantly to the volume.

Following the onset of the "boom" of research in plant molecular biology, the field has undergone a number of transitions. A great increase in relative interest in the area and also in the number of researchers working in it has led to applications of molecular genetic techniques to many facets of plant research. As with most fields where an abundance of technology has arrived within a short time period, many of the primary applications were based on what could be accomplished immediately. The tone of this meeting indicated another transition in plant molecular biology—the change from doing what we can do now to that of accomplishing what is necessary to solve long-range problems in basic plant biology. In order to achieve that goal, it is necessary to select relevant plant genetic systems, and at the same time, work to comprehend their biology. It became apparent at this meeting that one of the greatest challenges in applying molecular technology to plant biology is at the level of filling in large gaps in our knowledge of plant biochemistry. We feel that the meeting illustrated this premise and that the presentations were a first step toward meeting this challenge.

We gratefully acknowledge CIBA-Geigy Corporation for their generous sponsorship of this meeting. Additional gifts were received from Zoecon Research Institute of Sandoz Crop Protection Corporation; Northrup King, Co.; Plant Genetic Systems N.V.; Monsanto; Calgene, Inc.; FMC Corporation; Agricultural Chemical Group, and Phytogen.

Joe L. Key
Lee McIntosh

I. REPRODUCTIVE BIOLOGY: FLOWERING

Plant Gene Systems and Their Biology, pages 3–24

PERSPECTIVES IN FLOWERING RESEARCH

Anton Lang

MSU-DOE Plant Research Laboratory, Michigan State University, East Lansing, MI 48824-1312

ABSTRACT. The first part of this paper will be a survey of certain common features in the transition from vegetative growth to flower formation in different plants, with emphasis on the changes in the shoot meristem undergoing this transition and on graft-transmissible materials, presumably hormones, which promote or inhibit that transition - "florigen" and "antiflorigen". Both florigen and antiflorigen are formed in the leaves, are freely interchangeable between different species and genera, and are also freely interchangeable between different physiological plant types, namely long-day, short-day and day-neutral plants. Although chemically not identified they can be assumed to be identical in different plants.

The second part of the paper will consist of a review of some molecular approaches to flower formation, pointing out premises that should be observed in such work, and emphasizing the need of controls necessary to ascertain the specificity of the results obtained.

In the third and last part of the paper, the existence and role of physical factors in flower formation will be considered. These factors, which are being studied particularly by P.B. Green, are inherent in the plant as a whole. They are derived from cellulosic reinforcement patterns in the cell walls and can guide development in a selfperpetuating or - as in the case of flower formation - cascade-like manner. While dependent for their "substrate" on cellular, biochemical and molecular processes they operate at a different, higher hierarchical level and are an essential factor particularly in large-scale developmental processes in plants, including flowering.

INTRODUCTORY REMARKS

Interest in flowering needs hardly to be justified. From the fundamental viewpoint, flower formation means a profound change in the development of the higher plant, yet this change is not accomplished by the production of any new structures but by a thorough modification of existing ones. The axis of the flower (the receptacle) is homologous to the stem of a vegetative shoot, the individual flower parts (sepals, petals, anthers, carpels) are homologous to the leaves of that shoot. Yet the flower is so different from the vegetative shoot that anybody not familiar with comparative plant morphology will quite understandably consider it a new, fundamentally different organ. Flower formation may be called the metamorphosis of the higher plant.

From the commercial viewpoint, flowers are of course widely used for ornamental purposes, but most importantly, most agricultural and horticultural crops are grown for fruits or seeds, and fruit and seed production depend on the production and functioning of flowers. Thus, understanding the regulation of flowering may help us, directly or indirectly, in manipulating crop performance.

In this paper I shall address three topics: features in the transition from vegetative growth to flower formation that are common to many and possibly all plants; some recent molecular approaches to flower formation; and some physical forces inherent in the plant that need to be considered when trying to understand flower formation and other major developmental processes in plants. The second topic depends, to some extent, on the information summarized in the first part of the paper, and the third one adds a different, but very important perspective to the understanding of flower formation.

COMMON DENOMINATORS OF FLOWER INITIATION IN DIFFERENT PLANTS

In her chapter in this volume, Daphne Vince-Prue has drawn attention to the enormous variation in the flowering process and the diversity of conditions that can influence this process in different plants and sometimes even in one and the same plant. If one looks at the flowering process as a whole, beginning with flower initiation, i.e. the production of a shoot meristem committed to flower formation (a flower primordium) and the preparatory physiological processes preceding this event, and ending with the presence of a mature flower one is indeed faced with a great diversity in the process itself as well as in the conditions that may be required for or may affect the process. This heterogeneity of the flowering process in different plants has lead some plant biologists to argue that there are different routes to flowering in different plants, i.e. that there is no common "mechanism" of flower formation and development. However, if one focuses on the first morphological changes in a shoot meristem that is becoming a flower primordium, and the preceding physiological changes in the plant, particularly in the leaves, which are necessary for this transition although not apparent morphologically, one recognizes important common denominators in quite different plants.

The focus on flower initiation is justified as this is the actual turning point in the development of the plant, or at least of the shoot meristem that is switching from vegetative growth to flower formation. Further development of the flower primordium can be suppressed by unfavorable conditions but the meristem that has reached that turning point cannot revert to vegetative growth. On the other hand, if the flower primordium continues to develop this is where the great diversity in the flowering process comes into play. Flower initiation is the theme of the flowering process; the following stages are the variations on this theme.

Morphology of Flower Initiation

The changes which constitute the common morphological denominator of flower initiation are quite inconspicuous. In most plants they consist in an increase in both diameter and height of the shoot meristem and are usually accompanied by a reduction in growth in length of the subtending young internodes. In some plants, e.g. grasses, the meristem first elongates and floral meristems are formed on its flanks. However, once these changes have reached a certain point, but well before the initiation of any of the flower parts, the meristem or meristems have become flower primordia, irreversibly committed to flower formation. This is not to say that size is the only difference between a vegetative meristem and a flower primoridum. The transition of the former to the latter usually involves numerous histological and biochemical changes, including increases, redistribution and synchronization of mitotic activity; changes in the DNA, RNA and protein levels; and other changes. It is however not clear whether and to what extent these changes are essential for that transition. Thus, it has been shown that flower initiation can occur without the synchronization of mitoses (1). Many of the internal changes in the meristem may moreover already relate to the later stages of flower development, which as just pointed out can indeed be quite different in different plants and moreover can obey different factors and forces. It is therefore very important when studying the regulation of flower initiation not to confound this stage and the subsequent ones. In fact, failure to observe this distinction has contributed to the impression of the great heterogeneity of the flowering process in different plants.

Florigen and Antiflorigen

The other major common denominator in flower initiation are hormone-like materials that arise in the leaves and regulate the behavior of the shoot meristems. The first evidence for such agents came from experiments with photoperiodic plants - plants in which flower formation depends on daylength or photoperiod - in which the leaves, or part of the leaves, of such a plant were kept in one and the rest of the plant, particularly the shoot apex, in the opposite photoperiod. The evidence was greatly strengthened and expanded by grafting experiments between plants that were capable of flower formation, either autonomously or because they were held on the flower-inducing photoperiod, and plants that were kept on the non-inductive photoperiod or had not yet reached the stage at which they flower autonomously. Evidence for a graft-transmissible flower-promoting agent was obtained 50 years ago by Chailakhyan (2) and Moshkov (3) (and Melchers, 4, using a cold-requiring plant); the first-named author coined for it the term "florigen". Evidence for the inhibitory graft-transmissible agent was obtained 10 years ago (5) and this agent will be called "antiflorigen".

I shall illustrate the current status of florigen as well as antiflorigen by grafting experiments with several Solanaceous plants since for florigen it is the most comprehensive, and for antiflorigen almost the sole existing evidence based on such experiments. The plants used were several photoperiod-insensitive (day-neutral) cultivars of tobacco (*Nicotiana tabacum* L.), the short-day (SD) tobacco cultivar Maryland Mammoth (MM), and the long-day (LD) plants *Nicotiana silvestris* Spegg. et Comes (NS) and *Hyoscyamus niger* L., the black henbane (HYO). I shall begin with florigen and, following well-established practice, shall call the grafting partner capable of flowering, the donor; and the one not capable of flowering, the receptor. Work of several authors (2,6,7,8,9), extending over many years, has shown that --

a) NS and HYO donors induce flower formation in NS, HYO and MM receptors;
b) MM donors induce flower formation in MM, NS and HYO receptors;
c) Day-neutral tobacco cultivars as donors induce flower formation in NS, HYO and MM receptors;
d) NS, HYO and MM donors accelerate flower formation in day-neutral tobaccos. With NS donors, the vegetative phase of Trapezond tobacco has been shortened by approximately two thirds, the receptors forming a total of 13 nodes before initiating flowers while ungrafted individuals of this cultivar form at least 40 nodes before flowering (9). The effect of HYO and MM donors is less dramatic but significant;
e) In some, although not all combinations, a single leaf is sufficient as the donor.

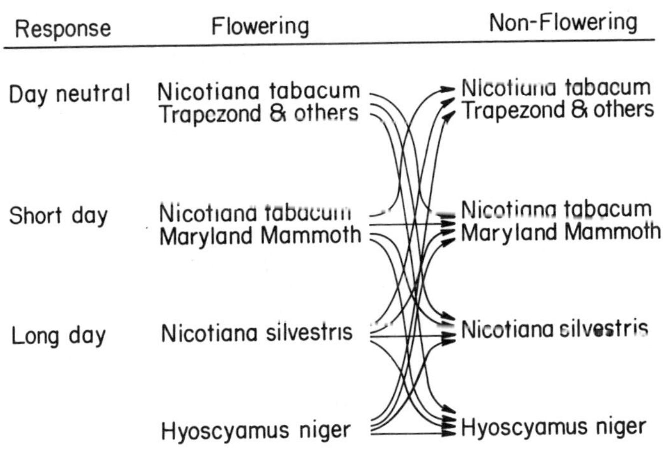

FIGURE 1. Diagrammatic representation of florigen transmission from donors ("Flowering") to receptors ("Non-Flowering") in various Solanaceae.

These results are diagrammatically illustrated in Fig. 1, and the following conclusions may be drawn from them:

(i) Florigen is formed in the leaves of the plants and is transported to the shoot meristems, in this process being capable of crossing graft unions.

(ii) Florigen is not specific in the taxonomic sense as it is readily interchangeable between grafting partners belonging to the same species, to different species of the same genus, and to species of different genera. The only limitation appears to be graft compatibility, particularly connection of the vascular tissues since florigen is moving in the phloem.

(iii) Florigen is not specific for the different photoperiodic types nor the autonomous (day-neutral) plants as it can be readily interchanged between LD, SD and day-neutral plants.

The most economic interpretation is that the florigen of the different plants is identical, but that in photoperiodic plants it is formed only under inductive photoperiods while in day-neutral ones it is formed independent of the photoperiod (autonomously) once the plant has reached a certain age.

Analogous, although mostly less entended grafting experiments have been done, with quite analogous results, in the families listed below (for details, see 10,11):

Asteraceae (Compositae)
Caryophyllaceae
Chenopodiaceae
Crassulaceae
Fabaceae (Leguminosae)
Malvaceae
Menthaceae (Labiatae)

Two results from these experiments need to be mentioned, as particularly important in our context:

(1) The work with Crassulaceae included, in addition to LD and SD plants, long-short- and short-long-day plants, showing that the conclusions derived from work with LD, SD and day-neutral plants are also valid for these less common photoperiodic response types.

(2) In the LD plant _Silene armeria_ flower formation can be induced not only by LD, but also under SD conditions by either high or low temperatures, and in some genotypes by treatment with gibberellic acid (GA_3). However, regardless of how a plant has been induced, it can serve as a donor for inducing flower formation in a vegetative receptor (12,13). This shows that even if induction can be consummated by different conditions it ends in all cases with the

production of florigen.

Literature contains several reports according to which no induction of flower formation at all could be obtained in some donor-receptor combinations, or could be obtained in one direction more effectively than in the other. Various reasons are conceivable for these apparent failures. The grafting set-up may not have been optimal. The "florigenic" efficiency of the grafting partners can be different. There may have been differences in tissue union and/or in the efficiency of florigen translocation. These matters need to be explored before those reports can be adduced as arguments against the florigen concept in general. The most thoroughly investigated case is that of the day-neutral tobacco cultivar Delcrest which is ineffective as florigen donor to the short-day cultivar Maryland Mammoth (MM) (7). However, Delcrest is an effective florigen donor for _Nicotiana silvestris_ (7) while other day-neutral tobacco cultivars are effective donors for MM. Thus, the behavior of the Delcrest-MM grafts is something peculiar to this particular combination but of no general nature.

Now turning to _antiflorigen_, the main evidence for this agent and its physiological properties derives from grafting experiments between, on the one hand, the LD plants NS and HYO, and on the other, the day-neutral tobacco cultivar Trapezond and the SD tobacco MM. When these grafts were maintained on SD, neither the Trapezond nor the MM partner flowered (5,9,14). The growth habit of the day-neutral and the SD partners was also markedly modified; the internodes were greatly thickened and shortened so that the growth of these partners to some extent approached that of noninduced NS and HYO plants which, like most LD plants in this state, grow in the rosette habit, with a greatly shortened stem. A single leaf of the LD partner was sufficient to cause these effects. However, when shoots of young plants of the day-neutral Samsun tobacco were grafted onto MM plants and the grafts maintained on LD, flower formation in Samsun was _not_ affected, and in similar grafts with Trapezond, NS and _HYO_, any pronounced and consistent inhibition of flower formation was not observed (9). Thus, whereas NS and HYO produce florigen in inductive photoperiods and antiflorigen in noninductive ones, MM does not seem to produce antiflorigen on LD, or at most produces it in much lesser quantities.

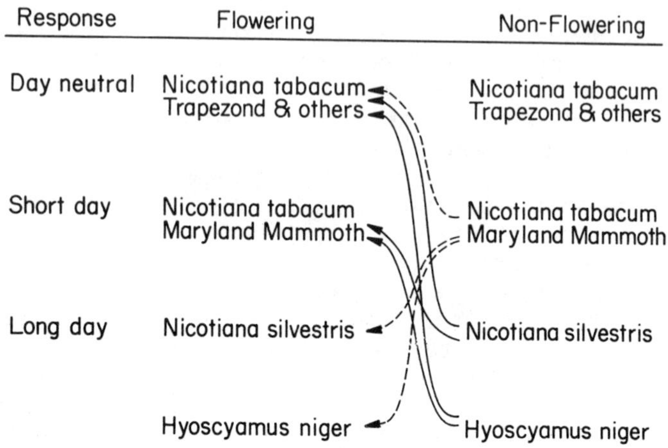

FIGURE 2. Diagrammatic representation of grafting experiments demonstrating the existence of antiflorigen in some plants. Solid arrows = flower formation was inhibited in partners capable of flower formation ("Flowering") by partners maintained on noninductive photoperiods and hence not capable of flower formation ("Non-Flowering"). Broken arrows = no inhibitory effect of the non-flowering graft partner on the one capable of flowering.

In analogy with the florigen grafting experiments the antiflorigen experiments can be summarized as shown in Fig. 2, and although because of the different behavior of MM the pattern is somewhat different, exactly the same conclusions can be drawn for antiflorigen as had been drawn for florigen:

(i) antiflorigen is produced in the leaves and acts in the meristems (the latter conclusion being based on the fact that the inhibition of flower formation by florigen is associated with the change in growth habit which can result only from a change in the function of the meristem);

(ii) it is not specific in a taxonomic sense as antiflorigen from _Nicotiana silvestris_ and _Hyoscyamus_ is effective in _N. tabacum_;

(iii) it is not specific in a physiological sense as antiflorigen from two LD plants affects flower formation in a day-neutral and a SD plant.

Some evidence from grafting experiments for an antiflorigen has so far been obtained in only one other case, and in this case in a short-day plant, namely, a species of *Coleus* (15). However, in the long-day plant *Lolium temulentum* Evans (16) has shown that whereas leaves kept in long days promote flower formation leaves kept in short days inhibit it. This was in fact the first clear demonstration of a transmissible flower-inhibitory agent, and in the short-day plant *Salvia occidentalis* analogous results have been obtained (17). Thus, while it would be premature to consider antiflorigen as ubiquitous as florigen it seems that, like florigen, it can be produced in quite different plants, and that one and the same plant can produce florigen *and* antiflorigen, the former under the inductive, the latter under the noninductive photoperiods. Failure to flower under the latter conditions may thus not be the consequence only of absence of florigen, but also of the presence of antiflorigen. Experiments in which a scion of MM and one of NS were grafted onto a MM stock, the number of leaves on the scions was varied (2-5-10), and the grafts were kept on short days, a simple quantitative relation between florigen and antiflorigen was found. The more MM leaves (producing florigen) were present, the earlier was flower formation; the more NS leaves (producing antiflorigen), the later it was. Ultimately, however, all grafts formed flowers, even if there were only two promotive (MM) and ten inhibitory (NS) leaves (see 14).

The obvious way to prove the reality of florigen as well as antiflorigen is to isolate the compounds and determine their chemical nature. For florigen, such experiments were started soon after the florigen concept arose but have remained notable by their lack of success. At some times, some investigators seemed to have obtained promising results but further work failed to support and extend these. The most recent case is work by Chailakhyan and associates (18,19) who obtained flower formation in seedlings of the SD plant *Chenopodium rubrum* with extracts from induced Maryland Mammoth tobacco. The assay plants were either given threshold inductive treatment, or kept on long days entirely. Judging by the authors' illustrations the effect was quite striking. Extracts from noninduced MM also gave some response but a much lesser one. However, no follow-up experiments seem to have been published. Attempts at extracting antiflorigen, carried out during the past 5 years (9), have so far been unsuccessful, too. Thus, both florigen and antiflorigen are presently physiological rather than chemical entities.

The failure to extract florigen has been at least one reason why some investigators have abandoned the florigen concept and tried to replace it with other hypotheses. The farthest-reaching one is the nutrient diversion hypothesis (21,22), according to which flower formation is a matter of proper assimilate supply to the meristem, and photoperiod and other environmental conditions that regulate flower formation do so by modifying assimilate partitioning within the plant. It is difficult to envisage how a non-specific and quantitative process like assimilate partitioning can be responsible for a highly specific and often, in plants with an absolute photoperiod (or temperature) requirement, qualitative process, in the sense that the plant either forms flowers or does not form any, and does so, as long as the proper photoperiod is maintained, under a wide variety of light, temperature and nutritional conditions. However and obviously, as long as florigen has not been shown to be a distinct chemical entity, neither the nutrient diversion hypothesis, not any other alternative hypothesis, can be rejected out of hand. But what can be expected from any such hypothesis is that it provides at least as complete and consistent an explanation for the results which have been used to support the florigen and antiflorigen concepts, such as e.g. those summarized in Figs. 1 and 2. For example, adherents of the nutrient diversion hypothesis should explain, and do so by precise measurements of nutrient production and distribution, why e.g. a single leaf of NS promotes flower formation in MM on long days and inhibits it (and greatly modifies growth) on SD; or why that single NS leaf has a powerful flower inducing action in MM on LD while all the leaves normally present on a MM plant - under favorable growing conditions 20 and more - are not capable of inducing flower formation. The evidence so far accumulated for the nutrient diversion hypothesis and other alternative explanations is very far from meeting this premise; in fact, the "pro-florigen" (and "pro-antiflorigen") evidence is largely ignored.

MOLECULAR APPROACHES TO FLOWER FORMATION

Several authors have approached flower formation - specifically, photoperiodic flower induction - using concepts and methods of molecular biology. Several of these studies are vitiated by the use of techniques which by now are outdated or were too crude in the first place (use of onedimensional chromatography) or too unspecific (inhibitors of RNA and protein synthesis) to provide meaningful results. Two recent reports are however free from such gross deficiencies.

Warm (22), working with the LD plant *Hyoscyamus niger*, examined the poly(A)RNA patterns of leaves of plants that had been either exposed to three days of continuous light of high irradiance (in this plant this treatment is sufficient to induce flower initiation; CL) or given three short days of 10 h high-irradiance light and 14 h of darkness each (in which the plants remain vegetative; SD). Lay-Yee et al. (23) worked with *Pharbitis nil*, the Japanese morning-glory, a SD plant which can be photoinduced in the cotyledonary stage, and examined the poly(A)RNA pattern in the cotyledons of seedlings raised in continuous light and exposed to one of three treatments:

CL = 16 more hours of high-irradiance light (noninductive);
SD = 16 h of darkness (inductive)
NB = 16 h of darkness but with a "night break" of 10 min of low-energy red light (non-inductive)

Such night breaks are known to nullify the effect of long dark periods, inducing flower formation in LD plants and inhibiting it in SD plants. In either case, the isolated poly(A)RNA was in-vitro translated and the translation products separated twodimensionally. In *Hyoscyamus*, quantitative differences were found in several polypeptide "spots", some increasing and others decreasing in intensity in the CL plants, indicating analogous changes in the responsible mRNAs. In *Pharbitis*, when the CL and the SD plants were compared, there were also changes in several polypeptides, but when SD and NB plants were compared, only one consistent change was observed, this polypeptide being highest in the SD-, lower in the NB- and non-detectable, at least within the limits of the technique used, in the CL-treated plants.

These two reports can serve as a convenient basis for discussing some premises that are essential in this kind of work. Three points can be made.

First, the authors of both reports made a wise choice in directing their studies at the inductive changes taking place in the leaves. These changes - in essence, production of florigen and perhaps antiflorigen - must be much simpler and much less numerous than those occurring in the meristem developing into a flower primordium; moreover, as indicated earlier it is not certain whether all of the latter changes are essential for flower initiation or are not already related to somewhat later stages of flower development.

With the current state of our physiological and biochemical information the smaller the differences between flower-induced and non-induced material, or between different stages in the flowering process, the higher the prospects of identifying molecular changes unequivocally related to flowering.

Second, Warm compared plants that had been given either 72 h or 3 x 10 h of high light, the latter interrupted by 16-h dark periods. These are quite different regimes and will undoubtedly result in different activation of "light dependent" enzymes other than any that may be related to flower induction. The author tried to reduce this problem by keeping the plants, induced and noninduced alike, for 24 h in darkness before analyzing the poly(A)RNA patterns, in the hope that the "nonspecific" light-activated enzymes would be deactivated in this time, but he has to admit that the results remain difficult to interpret. By using the night-break technique Lay-Yee and coworkers were able to reduce the difference in the treatment of induced and noninduced plants very substantially, to 10 min of low light, and found much lesser differences in the poly(A)RNA patter in these two treatments[1]. Obviously, the lesser the differences between the inductive and the non-inductive treatment, the greater the probability that the molecular differences found have something to do with flower induction.

Third, in both reports a very important control is missing, namely to apply the same treatments to a related day-neutral plant. The importance of this control is illustrated by what is probably the first "molecular" approach to flowering, done at a time when molecular biology had not yet become the catchword it is today. In 1957 Aach and Melchers (25) undertook a serological comparison of cold-treated and non-cold-treated plants of a cold-requiring strain in *Hyoscyamus niger*. This species occurs in

[1]Warm says he did not use the night-break technique as it is not effective in inducing flower formation in *Hyoscyamus*. However, *Hyoscyamus* plants can be quite effectively induced by daily cycles consisting of 7 h of light and another 1-h light period approximately in the middle of the dark period while they remain strictly vegetative on 9-h short days (24). While perhaps not ideal it seems preferable to compare plants receiving 9 h of light daily, as a single period, or 7+1 h, rather than plants receiving daily either 24 h or 10 h of light.

cold-requiring (biennial) and non-cold requiring (annual) strains; it is the latter that were used in all grafting work summarized in Fig. 1 and also in the work of Warm (22). Aach and Melchers found that cold treatment of biennial Hyoscyamus did result in the disappearance of a soluble protein, a result that on its face could be interpreted as removal of a material inhibiting flower formation. However, annual Hyoscyamus was found to contain the same protein, regardless of whether or not it was cold-treated, although it is of course quite capable of flowering in either case. Thus, that protein is a response of the biennial strain to low temperature, but is unrelated to flower induction. The genus Hyoscyamus includes a quantitative long-day plant (H. albus). I am not aware whether there are also fully day-neutral Hyoscyamus species, nor whether there are day-neutral strains of Pharbitis. However, there are species and genera that do contain photoperiodic and day-neutral plants, such as Nicotiana with day-neutral and SD cultivars of N. tabacum and the LD plant N. silvestris, or Xanthium with short-day and day-neutral species. Optimally, one should have isogenic lines differing only in one "flowering" gene, but as long as these are not available we should at least make use of the controls with which Nature has provided us. Without such controls it is not possible to be certain whether any molecular differences between induced and noninduced plants are specific for flowering.

Leaves of photoinduced and noninduced photoperiodic plants (and of related day-neutral controls) are not the only material potentially suitable for molecular approaches to flower formation. Another "system" worth exploring are the thin tissue layers of day-neutral tobacco which, depending on their origin on the plant and on the culture medium regenerate flower or vegetative buds (see paper by D. Meeks-Wagner in this volume). A caveat that needs to be raised here is that the tissues capable of flower regeneration may have been "programmed" a good deal earlier in their development; they may consummate the program or may be inhibited to do so, but they may not be suitable to study those processes which have instituted the flowering program. From this viewpoint another system may be preferable, namely leaf discs from the SD plant Streptocarpus nobilis first studied by Rossini and Nitsch (26). When taken from a LD-grown plant and cultured on LD or SD these discs will regenerate vegetative shoots or flower buds, respectively; they are clearly uncommitted. However, with any system to be used for molecular studies of flowering, the three points discussed above should be satisfied as far as possible. Otherwise there is no assurance that one has to do with processes indeed dealing with flowering, and we may find ourselves drowning in data but lacking meaningful information. On the other hand, if such a specificity can be reasonably established, molecular studies should offer great new potentials for research on flowering. To name only one example, using the technique of gene fusion, i.e. placing under the control of the genetic locus to be studied another gene coding for a protein with an easily measurable activity, e.g. an enzyme, it might even be possible to develop an alternative approach to the isolation and identification of florigen and antiflorigen. Using this kind of approach Peters et al. (27) have recently succeeded in isolating and identifing the inducer, luteolin, by monitoring the expression of _Rhizobium_ nodulin gene in terms of β-galactosidase activity.

INTERNAL PHYSICAL FORCES INVOLVED IN FLOWER FORMATION AND DEVELOPMENT

In this part of the paper I would like to draw attention to restraining physical forces that are inherent in the plant and can exercise profound effects in development, including flower formation. The presence of these forces can be discerned in the earliest phases of the development of a higher plant. For example, the seeds in a pea pod are all arranged so that the cotyledons are parallel to the pod walls and the axis in line with the axis of the pod; the vascular tissue, which in the pea seedling root is triarch, i.e. consists of three phloem and three xylem strands, when beginning to develop, is also oriented in a precise and consistent manner in relation to the pod. Furthermore, as Torrey (28) has shown, a 0.5 mm long piece of the tip of a pea seedling root, excised and cultured in a suitable medium, may grow into a root that at first has only two xylem and two phloem strands, or even only one of each (is diarch or monarch), but that it always reverts to the triarch pattern. Thus, this pattern has evidently been imprinted on the root of the embryo, early in the latter's development, by its position in the pod, and imprinted in a highly persistent manner. The fruit, by means of its structural characteristics, imposes spatial physical restraints on the embryo and these result in developmental patterns at least some of which, once established, seem to have a selfperpetuating character.

The role of such restraining forces in plant development has recently been studied by Green (29,30), with main emphasis on phyllotaxis and flower development. The starting point for this work is a very elementary fact of plant structure, namely, that plant cells are surrounded by cell walls in which, at least in their primary state, the main material giving them strength is cellulose. Cellulose occurs in microfibrils consisting of cellulose molecules arranged in parallel. In cells of a cylindrical organ like the stem the cellulose microfibrils are arranged at right angles to the long axis of the constituent cells so that each cell is surrounded by cellulose hoops and the organ as a whole is thus also encased in such hoops. These hoops permit growth in length but greatly restrict growth in girth. When lateral organs (leaves) are produced or when the entire growth pattern of the organ changes, as in flower formation, its reinforcement pattern also changes. These changes can, because of the birefringent nature of the cellulose fibrils, be seen in polarized light; the pattern of reinforcement in the surface plane, which opposes the strong tangential stresses there, can be studied in paradermal (hand) sections.

I shall describe here briefly only two "cases" studied by Green, namely, alternate or distichous phyllotaxis in which leaves are inserted, in an alternating manner, "right" and "left" on a stem (as in grasses, irises, gladiolus; see Fig. 3A); and initiation of stamens in a typical dicotyledonous flower. The first case is the simplest one to illustrate the major aspects of the role of the reinforcement pattern of an organ in its development; the second is an example for a more complex situation and is of course directly related to the topic of this paper.

On the organ level (Fig. 3B) a new leaf appears as a small, arc-like bulge on the side of the apical meristem, slightly above the preceding leaf. This primordium grows in a pincer-like fashion around the meristem and at the same time initiates from its central portion (the portion that was first visible) prominent cell fields, the central distal files (CDF), running inwards, i.e. towards the center of the meristem. When the "pincers" of the leaf primordium meet the CDF of the preceding leaf their circumferential growth stops and the next leaf is initiated at this point, growing in the opposite direction. On the tissue level (Fig. 3C), each leaf develops, in the region of the meristem above it, a reinforcement field with lines of cellulosic reinforcement roughly tangential to the inner face of the leaf base. In

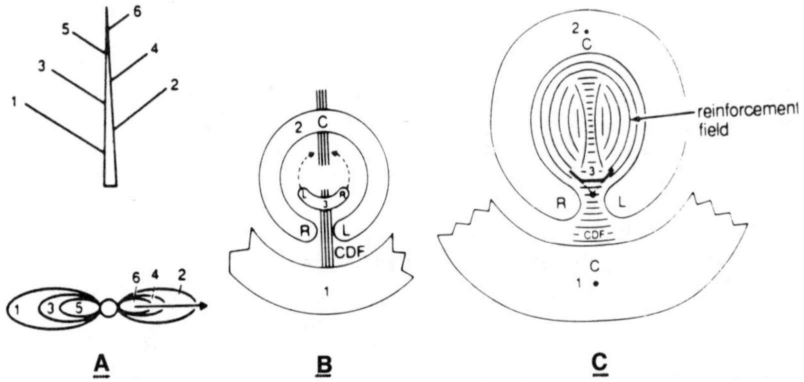

Figure 3. Distichous phyllotaxis, diagrammatic. Leaves are numbered in the order of their origin on the shoot meristem. A Side view and top view. B Top view of the meristem, showing the pincer-like growth of the leaf primordia and the central distal cell files (CDF) formed inward to the central region of the leaf (C). L and R, left and right tips of the "pincer". C Top view, showing the reinforcement patterns (lines) on an apical meristem. For details, see text. From Green (30), by permission.

the meristem of a distichous shoot nearby portions of the left and right parts of the reinforcement field of the youngest leaf (leaf 2) combine with a portion of the CDF of the next-older leaf (1) to give a three-membered reinforcement pattern as the site for the initiation of the next leaf (3). The bulging that initiates the new leaf is presumable fostered by the relatively high curvature at this site. It occurs along the reinforcement lines (lines of least resistance) and transforms the three-membered reinforcement irregularity into a smooth arc, the new leaf, and the subsequent pincer-like growth of this leaf is based on the bulging continuing along the lines of reinforcement of the preceding leaf. In summary, a leaf is produced on a site of the meristem with a modified cellulose reinforcement pattern; the leaf in turn reestablishes the reinforcement pattern for the next leaf, and so on, so that altogether we have a selfperpetuating, cyclic alternation in the reinforcement pattern of the meristem resulting in the continuous production of leaves at precisely determined

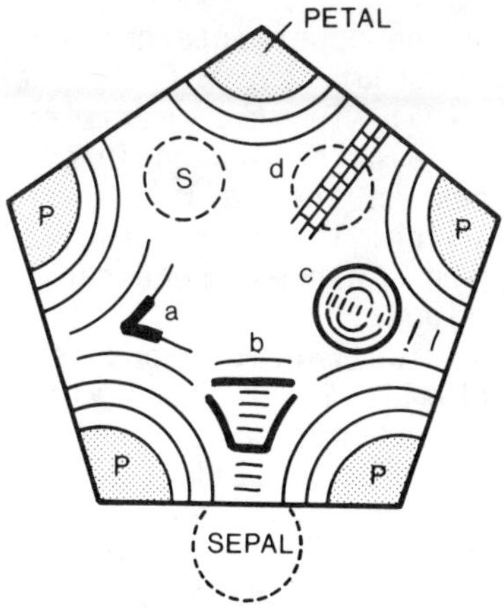

Figure 4. The meristem of a pentameric flower (i.e., a flower having whorls of 5 sepals, 5 petals, 5 anthers) viewed from above. Each petal primordium (P) has made a circumferential field of reinforcements (curved parallel lines). If these merely contact (a) no new reinforcement pattern results, but if these fields combine with a field from a sepal a rough new reinforcement field results (b). This region smooths out and becomes a stamen field (c) which has reinforcement fields over its center (shown in d). From Green (29), by permission.

sites on the meristem. Regarding the very first leaf of the shoot, and recalling that the position of the vascular pattern of the root of the embryo is imprinted on it by the seed coats, it seems logical to assume that its site of initiation is also determined by the constraining influence of the seed coats on the embryo.

The changes in the reinforcement pattern of the flower meristem which result in the formation of a stamen are illustrated in Fig. 4. The two cases differ insofar as stamen formation involves the interaction of reinforcement fields from three organs - two petals, a sepal - and as the result is formation not of the same but a different type of organ. However, they share the self-regulatory, cyclic or, in the stamen case, cascade-like nature.

Green does not seem to have considered the transition of the shoot meristem from vegetative growth to flower formation, i.e. flower initiation. However, if we recall that the first morphological change in this transition is an increase in the size of the meristem and a reduction in elongation growth in the upper internodes of the shoot it is tempting to propose that one of the first effects of florigen, upon arrival in the meristem, is to reduce elongation growth in favor of growth in girth, and that this effect causes changes in the cellulose microfibril pattern in the meristem which in turn lead to a cascade of reinforcement pattern changes required for the further development of the flower.

According to a widely held hypothesis the orientation of cellulose microfibrils in the cell wall is determined by the orientation of the cortical microtubules located near the plasma membrane of the cell. It is therefore very likely that further understanding of the changes in the reinforcement patterns in the meristem will come from studies of the behavior of these microtubules and perhaps other elements of the cytoskeleton. Such studies during the transition of a meristem from vegetative growth to flower initiation may be more informative than the studies of changes in mitotic activity, DNA and RNA contents, etc., whose essentiality for and precise relation to flower formation are not understood.

Drawing attention to the role of physical forces acting in the shoot meristem, and perhaps other developing regions of the plant, does not mean to question or downgrade the role of cellular, biochemical, and molecular processes in flowering and in developmental processes in plants in general. In order to become operative the physical forces need substrate. Thus, the cyclic alternation of the reinforcement pattern in the meristem, which results in the continuous production of leaves, depends on the supply of new cells and of more cellulose, microtubules and other materials all of which are produced by cellular and molecular processes. However, those physical forces are acting after the cellular and molecular ones have largely completed, at any given time point of the development, their own action. They represent another hierarchy in developmental regulation. Moreover, as illustrated by the imposition of the exact arrangement of the vascular pattern in the embryonic root by the seed coats, they can exert their influence from one generation to the next and may be an essential factor in the initiation of developmental regulation in the new organism. Clearly, they need to be included in our analyses if we are to achieve a complete, integrated understanding of flower formation and other developmental processes in plants, particularly large-scale ones which lead to major changes in form or pattern.

CONCLUDING REMARKS

The title of this paper is "Perspectives in Flowering Research", and its three major sections reflect my views of this problem.

Flowering as a whole is a complex process, consisting of several stages which may be very different in different plants and may depend on different processes and conditions. It is therefore very important precisely to define the stage which one may want to study, and not to confound it with other stages.

Despite the diversity of the process as a whole, its first stage, flower initiation, and the preparatory processes in the leaves preceding it, exhibit important common features in quite different plants. This stage is also the actual turning point in the development of the plant.

As to further research, there is still great need for physiological information. As shown by the work of Carl McDaniel and associates (see their chapter in this volume) or by the finding of a hormone-like, graft-transmissible flower inhibitor (antiflorigen) new and very important findings can be made even at the whole-plant level. As long as an at least equally well-supported alternative hypothesis is not available, efforts at isolation and identification of florigen and antiflorigen are entirely justified and should be vigorously pursued in cooperation with bio-organic chemists. Research at the cellular level and biochemical research are equally needed.

Molecular approaches should be a powerful new tool for studies of the flowering process. However, at present it seems most profitable to select "simple" systems, i.e. systems where the difference between the flower-induced and the non-flowering material can be expected to be limited to few processes or changes - preferably of course, but this is presently difficult to ascertain, one single change. It is also advisable to keep the conditions which determine flowering or no flowering as comparable as possible, e.g. by using not long days versus short days but short days versus short days with night breaks. Last not least, it is most important to conduct parallel studies with as closely related plants as possible which however do not depend on the flower-inducing conditions used, e.g. to subject day-neutral plants to the same "inductive" treatments as photoperiodic plants. Without this one cannot be certain that the molecular changes one may find are related to flowering, or are a more general response to the inductive conditions applied.

Finally, attention needs to be directed at biophysical forces that are inherent in the structure of plants. They derive from the cellulosic reinforcements present in cells and organs, and change in a selfperpetuating or cascade-like manner when the developmental pattern changes, keeping development "on track." Emanating from the mother plant they may be instrumental in initiating regulated development in the young daughter plant. While dependent on biochemical and molecular processes for "substrate", these forces operate at a higher hierarchical level, playing an essential part particularly in large-scale developmental processes including flower formation.

ACKNOWLEDGEMENTS

My own work, referred to in this paper, has been supported by the U.S. Department of Energy under Contract DE-AC02-76ER0-1338 and the National Science Foundation (grant PMC-7910791). I am very grateful to Professor Paul B. Green, Stanford University for checking the section dealing with the physical forces involved in development.

REFERENCES

1. Grose, S, Lyndon, RD (1984) Inhibition of growth and synchronization of cell division in the shoot apex in relation to flowering in Silene. Planta 161: 289-294.
2. Chailakhyan, MKh (1036) New facts in support of the hormonal theory of plant development. CR (Dokl) Acad Sci URSS 13 (1936 4): 79-83.
3. Moshkov, BS (1937) Flowering of short-day plants in continuous light as a result of grafting (in Russ). Bull Appl Bot Genet Plant Breed Ser A Plant Industry in USSR 21:145-156.
4. Melchers, G (1937) Die Wirkung von Genen, tiefen Temperaturen und bluhenden Pfropfpartnern auf die Bluhreife von Hyoscyamus niger L. Biol Zentralbl 57:568-614.
5. Lang, A, Chailakhyan, MKh, Frolova IA (1977) Promotion and inhibition of flower formation in a day-neutral plant. Proc Natl Acad Sci USA 74:2412-2416.
6. Melchers G, Lang A (1941) Weitere Untersuchungen zur Frage der Bluhhormone. Biol Zentralbl 61:16-39.

7. Zeevaart, JAD (1958) Flower formation as studied by grafting. Meded Landbouwhogeschool Wageningen 58, No 3.
8. Chailakhyan, MKh, Yanina, LI, Khazhakyan, KhK (1977) Flowering of plants in reciprocal grafts of short-day and long-day species (in Russ). Dokl Akad Nauk SSSR 237:1248-1251.
9. Lang, A. Unpublished data.
10. Lang, A (1965) Physiology of flower initiation. In Ruhland, W (ed) "Encyclopedia of Plant Physiol", vol XV pt 1, Berlin, Heidelberg, New York: Springer-Verlag, pp. 1380-1536.
11. Zeevaart, JAD (1979) Physiology of flower formation. Annu Rev Plant Physiol 27:321-348.
12. Wellensiek, SJ (1966) The flower forming stimulus in Silene armeria. Ztschr Pflanzenphysiol 55:1-10.
13. Van de Pol, PA (1972) Floral initiation, floral hormones and flowering. Meded Landbouwhogeschool Wageningen 72, No 9.
14. Lang, A (1980) Inhibition of flowering in long-day plants. In Skoog, F (ed) "Plant Growth Substances 1979," Berlin, Heidelberg, New York: Springer-Verlag, pp. 310-322.
15. Jacobs, WP (1980) Inhibition of flowering in short-day plants. In Skoog, F (ed) "Plant Growth Substances 1979," Berlin, Heidelberg, New York: Springer-Verlag, pp. 301-309.
16. Evans, LT (1960) Inflorescence initiation in Lolium temulentum L. II. Evidence for inhibitory and promotive photoperiodic processes involving transmissible products. Aust J Biol Sci 13:429-440.
17. Bhargava, SC (1964) Photoperiodism, floral induction and floral inhibition in Salvia occidentalis. Meded Landbouwhogeschool Wageningen 64, No 12.
18. Chailakhyan, MKh, Grigoryeva, NYa, Lozhnikova, VN (1977) Influence of extracts from leaves of flowering tobacco plants on flowering in seedlings and seedling plants of Chenopodium rubrum L. (in Russ). Dokl Akad Nauk SSSR 236:773-776.
19. Chailakhyan, MKh, Lozhnikova, VN (1985) The florigen hypothesis and its substantiation by extracting substances which cause flowering in plants (in Russ). Fiziol Rast 32:1172-1181.

20. Sachs, RM, Hackett, WP (1983) Source-sink relationships and flowering. In Meudt, WJ (ed), "Strategies of Plant Reproduction" (Beltsville Symp in Agric Res No 6), Totowa, NJ: Allenhead, Osmun & Co/St Albans, Herts: Granada Publishing, pp. 263-272.
21. Sachs, RM, Roles of photosynthesis and assimilate partitioning in flower initation. In Atherton, J (ed) "The Manipulation of Flowering" (45th Easter School in Agric Sci) London: Butterworth Scient Publ, in press.
22. Warm, E (1984) Changes in the composition of in vitro translated leaf mRNA caused by photoperiodic flower induction in _Hyoscyamus niger_. Physiol Plant 61:344-350.
23. Lay-Yee, M, Sachs, RM, Reid MS (1987) Changes in cotyledon mRNA during floral induction of _Pharbitis nil_ cv. Violet. Planta 170: in press.
24. Claes, H, Lang, A (1947) Die Blütenbildung von _Hyoscyamus niger_ in 48-stündigen Licht-Dunkel-Zyklen und in Zyklen mit aufgeteilten Lichtphasen. Ztschr f Naturforsch 2b:56-63.
25. Aach, HG, Melchers, G (1957) Serologische Untersuchungen an kältebedürftigen Pflanzen. Biol Zentralbl 78:466-475.
26. Rossini, LME, Nitsch, JP (1966) Induction de la floraison in vitro chez une plante de jours courtes, _Streptocarpus nobilis_. CR Acad Sci Paris 263:1379-1382.
27. Peters, NK, Frost, JW, Long, SR (1986) A plant flavone, luteolin, induces expression of _Rhizobium meliloti_ nodulation genes. Science 233:977-980.
28. Torrey, JG (1955) On the determination of vascular patterns during tissue differentiation in excised pea roots. Am J Bot 42:183-198.
29. Green, PB (1986) Form and pattern: Linkage through cellulose reinforcement as seen in flowering. In "Progress in Developmental Biology, Part A," New York: Alan Liss, pp. 195-198.
30. Green PB, Inheritance of pattern: analysis from phenotype to gene. In Moore, JA (ed) "Science as a Way of Knowing - Developmental Biology," Allen Press, in press.

Plant Gene Systems and Their Biology, pages 25–37

THE INDUCTION AND INITIATION OF FLOWERS

Daphne Vince-Prue

Department of Botany, University of Reading, U.K.

ABSTRACT Although there may be no general mechanism for the control of flowering, it is argued that induction in the leaves is a common feature. Floral induction is often under daylength control and this offers a useful probe for studying its molecular basis. The photoperiodic process itself appears to be highly conserved between species and a model for its operation in short-day plants is described. Long-day processes may have additional or different components. Problems in the use of daylength for studying molecular events are discussed.

INTRODUCTION

Flowering is an important stage in the growing of most, if not all crops. Consequently the control of flowering has received much attention from agronomists and horticulturists, and there is a wealth of empirical information on the requirements for flowering in a wide range of plants (1). The variation among species and even among cultivars of the same species is enormous. Many physiologists have looked for a general mechanism of action in the control of flowering and attempted to integrate the data into a unifying theory. However, such attempts have usually concentrated on results obtained with a small number of 'model' plants and largely ignored the enormous diversity of behaviour that is apparent when a wider range of plant species is considered. Recently physiologists have begun to place more emphasis on this diversity. For example, Bernier (2) has argued that there are several alternative routes to flowering, even within the same plant, and that the availability of these alternative pathways may be important for survival under natural conditions. To a number of workers, the complexity of the responses and the number of factors involved seem to indicate that there are no specific 'floral' mechanisms and that the formation of flowers at the apex is merely a nutritional event, involving no special 'florigenic' substances (3).

An important question for molecular approaches, therefore, is the extent to which there are general mechanisms operating in the control of flowering. If there are many different pathways to flowering, are there any common features which might point to highly conserved components of the overall processes? Despite the enormous range of effective triggers, inhibitors and modifiers of flowering in different plants (4), it is evident that a number of common features can be identified. For example, the many recorded cases where flowering has been effected by grafting between plants of different species and between plants responding to different environmental triggers (5) supports the idea that the end product of the floral pathway(s) may be the same. From this it is reasoned that the pathway leading to the end product may also be the same. Moreover, grafting experiments and other physiological approaches have shown that essential events occur in the leaves. From this has evolved the classical view that there is a specific pathway to flowering which involves a process of *induction* in the leaves. This leads to the production of a transmissible *floral stimulus* which, on arrival at a competent shoot apex, brings about *evocation* (the events preceding morphological changes) and the *initiation* of floral primordia. As indicated above, this classical view is currently challenged by a number of workers. This paper will largely concentrate on induction in the leaf and its relationships with other components of the flowering process. The floral stimulus is considered elsewhere in this volume and has recently been reviewed (6,7) as have the events of evocation (2,8).

CONTROLS FOR FLOWERING

The switch to flowering appears to be under two major types of control. In some cases, flowering is determined entirely endogenously and without reference to specific environments; this is commonly termed *autonomous induction*. More often environmental signals such as daylength and temperature are important, if not crucial, for the transition to reproductive growth. Even in autonomous induction, the course of flower development is often modified by environmental conditions. An important question in relation to fundamental mechanism(s), therefore, concerns the relationships between autonomous and environmental induction. Because daylength is such a useful non-invasive probe for triggering the switch to flowering and has been used in almost all biochemical studies, it is particularly important to determine the relationship between autonomous induction and photoperiodically-controlled flowering. One of the well established facts about the photoperiod process is that it takes place in the leaves. This has been shown by localised treatments and, even more stringently in the short-day plant (SDP) *Perilla*, by experiments in which an excised leaf was exposed to SD and subsequently grafted to a plant maintained in non-inductive long days (LD) where flowering was evoked (9). Daylength treatment of the apex normally has little effect and there is no evidence for photoperiodic perception by the roots. The site of autonomous induction is less easy to identify. However, autonomously-induced plants without roots have been shown to generate a transmissible floral stimulus (10) and grafting a single leaf of the day-neutral soybean, Agate, was found to evoke flowering of the SD cultivar, Biloxi, in LD (11). Consequently, it seems likely that, in intact plants, autonomous induction also occurs in the leaves and in this respect, at least, is the same as photoperiodic induction.

A second common feature of these two pathways is the apparent identity of the end product(s) exported to the shoot apex. For example day-neutral cultivars of tobacco are able to cause flowering in both the SD Maryland Mammoth and the LD species *N. sylvestris* under non-inductive conditions. Similarly, flowering in the day-neutral, Trapezond, was accelerated when grafted to the LD-induced donor, *N. sylvestris* or to the SD-induced donor Maryland Mammoth, whereas grafting to a non-induced donor was ineffective or even inhibitory (12). Moreover, although responding to different daylength signals, grafting experiments indicate that the end product of induction in plants of different photoperiodic response categories is also the same. However, the two-step nature of induction in plants with dual photoperiodic requirements indicates that, here, SD and LD must lead to different products in the leaf; the relationship between the two inductive treatments is clearly sequential since the order of exposure to SD and LD is crucial. Moreover, not all grafting experiments support the concept of a universal floral stimulus and, in some cases, transmissible inhibitors of flowering also seem to be involved. Finally, detailed studies of events at the apex have indicated that sequential stimuli may be generated by the leaves (13) and that different components in the overall evocation process may be triggered by different stimuli (2). Thus the physiological evidence does not prove that a single pathway in the leaves leads to induction and to a single end product; multiple pathways and multiple products are not excluded. Certainly no unique florigenic substance has ever been extracted from induced plants. However, a potent inhibitor of flowering (a lipoidal compound called BEHD) has recently been extracted from plants of several different families (14). It is interesting that BEHD disappeared with induction in three SDP and one LDP, but not in a day-neutral plant nor in a second LDP.

Vernalization, like daylength, is an inductive treatment and leads to a flowering response some time after the exposure to low temperature is given. However, vernalization appears to take place primarily at the shoot apex, although all dividing cells, including those in the leaves, may be potential sites of perception. Interactions between photoperiod and vernalization are many and complex (15). In relation to the position of vernalization in the chain of events leading to flowering, two contrasting types of interaction are particularly interesting. The biennial *Hyoscyamus niger* can only be photoperiodically induced (by LD) after it has received at least 7 days of vernalization. Thus, here, vernalization at the shoot apex appears to be an earlier step than photoperiodic induction in the leaf on a sequential pathway to flowering. In contrast, in *Campanula medium*, SD (acting in the leaf) can completely substitute for vernalization (at the apex) and it has been concluded that, here, the two processes are different, alternative pathways to flowering. There are strong arguments, therefore, for the conclusion that the vernalization process may be different in different plants. Where low temperature can substitute for photoperiod, how does it act? Does vernalization at the apex lead to the development of the induced state in the leaves without exposure to an inductive daylength? Could floral initiation take place at the apex in the absence of the product(s) of induction in the leaves? It is evident that much further work is needed before relationships between vernalization and the photoperiodic induction mechanism can be understood in molecular terms.

PHOTOPERIODIC INDUCTION

Almost all studies of floral induction in the leaves have involved photoperiodically-sensitive plants and to some extent this complicates the interpretation of experimental results. On the one hand, daylength controls many processes in addition to flowering, so that it is difficult to be certain that any observed changes necessarily relate to floral induction. On the other hand, treatments may be affecting the operation of the photoperiodic process itself rather than directly influencing floral induction. It is known, for example, that some actions of light are on the circadian oscillator which controls photoperiodic time-measurement, while others are more directly involved in the induction process itself. It is important, therefore, to understand how the photoperiodic mechanism operates, especially when using light as a probe. The general model for photoperiodic induction shown in Fig 1 and described below is largely based on experiments with the SDP, *Pharbitis nil*, using a single cycle (16), but data supporting the individual components of the model have also been obtained from experiments with other plants and it appears to be widely applicable.

The essence of photoperiodism is that it achieves seasonal flowering, or other displays, through a mechanism whereby the plant responds to the duration of the daily light/dark periods. Under natural conditions, the durations of light and darkness are absolutely related and the critical daylength could, therefore, be detected by measuring the duration of either light or darkness. From several early experiments in which the lengths of the light and dark periods were varied independently, it has generally been concluded that time is primarily measured in darkness and that floral induction in SDP requires exposure to a longer than critical duration of darkness. A variety of experimental approaches have indicated that the mechanism for photoperiodic timekeeping is based on a circadian rhythm, i.e. a rhythm which continues to oscillate under constant environmental conditions with a periodicity of about 24 hours (17). For example, in many plants, there is a circadian rhythm in the response to a pulse of light interrupting a long dark period. Even when not evident under normal conditions, an underlying rhythm of sensitivity to light may be uncovered when the conditions are changed. This is seen, for example, in *Pharbitis*, where the circadian rhythm is only evident in dark-grown seedlings which are exposed to a relatively short photoperiod. The rhythmic nature of the time-measuring system can also be demonstrated by so-called resonance experiments. In these, the flowering response changes rhythmically with increase in the duration of the dark period which follows a short photoperiod.

Although several lines of evidence suggest that photoperiodic time-measurement in higher plants is based on a circadian oscillator, the precise way in which this is coupled to achieve measurement of the critical nightlength in SDP is less well established. One approach has been to investigate the effect of the photoperiod on the phasing (entrainment) of the photoperiodic rhythm, using the time of maximum inhibition of flowering by a night-break as an indication of rhythm phasing (16,18,25). The results suggest that photoperiodic time-measurement involves a single circadian rhythm in the response to light. This rhythm starts to run at light-on (dawn) but is suspended at a constant phase-point after several hours in continuous light until released by a light-off signal (dusk) to free-run during the dark period (Fig 1). Floral induction is thought to take place only when a particular phase of this photoperiodic rhythm-called the inducible phase -- is reached in darkness and in this way the critical night-length is measured. Induction is prevented when light (night-break or dawn) is given during this phase. In this model, therefore, the basis of photoperiodic timekeeping is thought to be a rhythm in an inducible phase, which is required for floral induction. It is envisaged that the rhythm is coupled to an underlying oscillator and that it is the action of light on the oscillator which initiates, suspends, or releases the photoperiodic rhythm (Fig 2). Long exposures to light, therefore, act on the oscillator to entrain the rhythm, but they also have other more direct effects on floral induction in the Pfr-requiring reaction discussed below.

A characteristic feature of overt rhythms is that they can be phase-shifted by a brief exposure to light given during the subjective dark period. The rhythm can be moved either forwards or backwards in real time, the precise

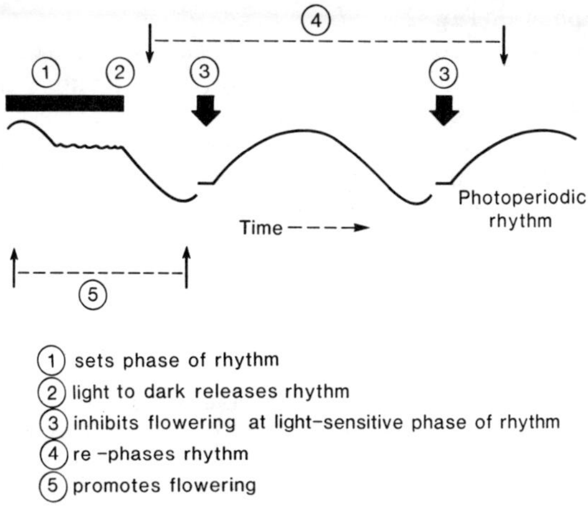

FIGURE 1. A model for the photoperiodic mechanism in SDP

light (entrainment)
↓
oscillator
↓
night-break ⟶ photoperiodic rhythm
↓
floral induction

FIGURE 2. Proposed sites of light action in SDP

response depending on the phase of the rhythm when light is presented. The photoperiodic rhythm has been shown to be phase-shifted by light in a similar way (19) and one model for the control of floral induction proposes that it is the phase-shifting action of light which is responsible for the night-break effect. The night-break response in SDP would then result from the fact that, at certain times in the dark period, the rhythm is phase-shifted in such a way as to prevent flowering because the inducible phase is not reached before dawn. In this model, therefore, the *only* action of light is on the oscillator to control the phase of the rhythm. However, a more direct action of light on floral induction has also been proposed and, in this scheme, the night-break acts at a different site -- the photoperiodic rhythm -- to inhibit flowering at the inducible phase (Fig 2).

One approach to distinguish between the two hypotheses has been to compare the sensitivity of the night-break response with phase-shifting on the assumption that, if the night-break effect is simply a consequence of re-phasing, the two responses will have the same dose-relationship. In *Pharbitis*, the relative sensitivities of the night-break and phase-shift have been compared at different times after a 24 h photoperiod; the two responses were found to have different dose-response relationships at both 6 and 8 h into the dark period, with the night-break showing an increase in sensitivity at 8 h (compared with that at 6 h) while the sensitivity for a phase-shift decreased (20). Similar results were obtained with dark-grown seedlings following a 10 min photoperiod (21). Thus, although the phase of the photoperiodic rhythm can be shifted by a R pulse, a separate night-break effect can be distinguished and must result from another action of light, most probably a direct interaction with the circadian rhythm as proposed in Fig 2. Thus, if night-break light is used to study the control of floral induction in the leaf, it should be borne in mind that such a light pulse probably has more than one site of action (Fig 1).

In addition to the light actions already discussed, it is known that light promotes flowering in SDP in a reaction requiring the Pfr-form of phytochrome (17). This reaction also occurs in the leaf (22) but its contribution to the overall mechanism for floral induction is not understood. Pfr is required during and sometimes after the photoperiod for flowering to occur (Fig 1); if this Pfr-requiring reaction is not completed during the day, it will continue during the night, indicating that Pfr does not decay rapidly following transfer to darkness. In contrast, the night-break effect appears to involve a different pool or type of Pfr, which decays more rapidly in darkness (17,23). Thus light not only has multiple effects in the control of floral induction in SDP but also appears to operate through different forms of phytochrome or possibly through different modes of action of this photoreceptor (24,25). Thus, when using light to study the control of floral induction, the complexity of its actions should always be borne in mind. It is also important to recognise that, in LDP, the operation of the photoperiodic mechanism to control of induction is much less well understood, both with respect to the timekeeping mechanism and the identity and mode of action of the photoreceptor pigment(s). In LDP, the mechanism does not appear to be simply the mirror image of that in SDP (27). Long exposures to light are often necessary to effect induction and the duration of darkness does not seem to be the overriding factor. Long-day plants, therefore, pose special problems in distinguishing between true photoperiodic effects and direct responses to light (see below).

WHAT EVENTS OCCUR IN THE LEAVES DURING INDUCTION?

It is often assumed that floral induction is achieved by switching on, or off, the transcription of specific floral genes in the leaf. However, the evidence for such an assumption is limited. It is supported indirectly by grafting experiments in which it has been shown that, in some SDP (e.g., *Perilla*) the induced state is extremely long-lived (9). Such a persistent developmental change would be expected to involve the expression of new genetic information. The involvement of protein synthesis in photoperiodic floral induction might also be deduced from experiments using high temperature. Flowering is inhibited in SDP, but promoted in LDP by exposure to temperatures near 40°C, but the effect is strictly limited to a particular time in the night. In the SDP, *Xanthium*, this is 8-10 h from the beginning of darkness (26) and in the LDP, *Silene*, *Lunaria* and *Samolus*, high temperature between the 5th and 9th hour promoted flowering (27). A possible explanation is that high temperature switches on the synthesis of heat-shock protein and prevents the formation of others. If this interpretation is correct, it indicates the involvement of protein synthesis in photoperiodic induction -- particularly in *Xanthium*, where the effect of high temperature coincides with the time of maximum sensitivity to a night-break and the presumptive time of the inducible phase in the photoperiodic rhythm.

There is, as yet, little direct evidence to support the hypothesis that floral induction involves the synthesis of specific proteins in the leaf. Much of the early work concentrated on the use of inhibitors but the results were often conflicting (28). A more direct approach was to look for the formation of new proteins in the leaf during induction but once again the results were conflicting; electrophoretic studies revealed new protein bands in some cases but not in others. From published information, it would appear that few attempts have so far been made to use modern molecular biological approaches in the study of floral induction or the photoperiodic mechanism, although I am aware that this is currently being attempted in several laboratories. It may, therefore, be appropriate to conclude with a brief consideration of some of the problems of interpretation -- as I see them -- in a few of the more recently published investigations into the biochemical aspects of floral induction and/or photoperiodism.

Differences in the products of *in vitro* translation have been found in the leaves of the LDP, *Hyoscyamus niger*, after photoperiodic floral induction (29). However, these proteins -- some of which increase and some decrease compared with those in non-induced leaves -- have not yet been shown to be associated with floral induction. In *Hyoscyamus*, as in many other rosette LDP, long days bring about elongation of the stem as well as flowering. Moreover, as in other LDP, long daily exposures to light are required to induce flowering and it is not necessarily clear that the observed changes in the leaf are related to the photoperiodic mechanism (i.e. to the *timing* of the light). This particular problem has been avoided in a study with *Pharbitis* which has recently been reported briefly (30). There were several differences in the translation products obtained from poly(A)$^+$RNA from induced and non-induced cotyledons but one product in particular was claimed to be reproducibly observed following SD induction, which was not found in cotyledons given a 10 min night-break 8 h into the dark period. Although both inhibited flowering, there were a number of differences between the night-break treatment and continuous light, emphasizing the problems inherent in using prolonged exposures. The significance for floral induction of these apparent changes in translatable mRNAs has yet, however, to be determined.

Perhaps the best approach to studying the photoperiodic mechanism itself, at the present time, is to use a chemically identified marker known to be under daylength control. Several lines of evidence have shown that, in spinach, the formation of GA_{20} from GA_{19} requires long days. Using cell-free extracts, the percentage conversion of (^{14}C)GA_{19} was found to be much lower in leaves maintained in SD than after exposure to 3 LD (31). However, the activity of the enzyme decreased rapidly during the first few hours of darkness and the published time-course does not suggest that the GA_{19} to GA_{20}

conversion depends on a critical night-length. Neither did there seem to be a critical duration of light. Thus further physiological investigations are required to establish whether the GA_{19} to GA_{20} conversion involves a true photoperiodic mechanism of the kind shown in Fig 1, with timekeeping coupled to a circadian rhythm. The development of the CAM pathway in *Kalanchoe blossfeldiana* has, on the other hand, been unequivocally shown to be under true photoperiodic control (32). Using a cDNA probe from maize, it has been established that the mRNA for PEPcarboxylase is increased in SD and that this increase is prevented by a classic night-break. The kinetic and other characteristics of this enzyme differ from those of the PEPc present in long days. Although this particular system is somewhat unwieldy in view of the large number of SD required to effect the response, one particularly interesting feature is that the photoperiodic control of CAM seems to be very closely coupled to the photoperiodic control of floral induction in this plant.

The inhibition (SDP) of promotion (LDP) of flowering by a night-break is consistent with the action of phytochrome to repress or increase mRNA synthesis in a number of direct responses to light (33). However, phytochrome also appears to modify the properties of cell membranes. Since Pfr action in a night-break can be extremely rapid, it has been argued that this may also be dependent on a modification of membrane properties. An extension of this hypothesis is that floral induction itself is a consequence of changes in membrane properties brought about by the operation of the photoperiodic clock. However, there is little direct evidence for this hypothesis (28).

REFERENCES

1. Halevy AH (1985/6). "Handbook of Flowering" Boca Raton: CRC Press.
2. Bernier G (1986). Symp Soc Exp Biol 40:257.
3. Sachs RM (1977). HortScience 12:220.
4. Vince-Prue D (1985). Nordbrugsf Foren 22:16.
5. Vince-Prue D (1975). "Photoperiodism in Plants." London: McGraw Hill.
6. Zeevaart JAD (1984). In Vince-Prue D, Thomas B, Cockshull KEC (eds): "Light and the Flowering Process," London: Academic Press, p 137.

7. Vince-Prue D (1985). Encyc Pl Phys NS 11:308.
8. Francis D, Lyndon RF (1985). In Bryant J, Francis D (eds): "The Cell Division Cycle in Plants," Cambridge: Cambridge University Press, p 199.
9. Zeevaart JAD (1958). Meded LandbHoesch Wageningen 58:1.
10. Chailakhyan MKh, Yanina LI, Khazhakyan Kh (1977). Dokl Akad Nauk SSSR 237:1248.
11. Heinze PH, Parker MW, Borthwick HA (1942). Bot Gaz 103:518.
12. Lang A, Chailakhyan MKh, Frolova IA (1977). Proc Nat Acad Sci USA 74:2412.
13. Vince-Prue D, Gressel J (1985). In Halevy AH (ed) "Handbook of Flowering vol iv," Boca Raton: CRC Press, p 47.
14. Jaffe MJ, Bridle K, Kopcewicz J (1986). Flowering Newsletter 1:6.
15. Napp-Zinn, K (1984). In Vince-Prue D, Thomas B, Cockshull KEC (eds): "Light and the Flowering Process," London: Academic Press, p.75.
16. Vince-Prue D, Lumsden PJ (1987). In Atherton J (ed): "The Manipulation of Flowering," Univ of Nottingham 45th Easter School (in press).
17. Vince-Prue D (1983). Encyc Pl Phys NS 16:457.
18. Lumsden PJ, Thomas B, Vince-Prue D (1982). Pl Physiol 70:277.
19. King RW, Cumming B (1972). Planta 103:281.
20. Lumsden PJ, Furuya M (1987). Pl & Cell Physiol (in press).
21. Lee HSS, Vince-Prue D, Kendrick RE (1987). Pl & Cell Physiol (in press).
22. Knapp PH, Sawhney S, Grimmet M, Vince-Prue D (1986). Pl & Cell Physiol 27:1147.
23. Takimoto A, Saji H (1984). Physiol Pl 61:675.
24. Thomas B, Lumsden PJ (1984). In Vince-Prue D, Thomas B, Cockshull KEC (eds): "Light and the Flowering Process," London: Academic Press, p. 107.
25. Vince-Prue D (1986). In Kendrick RE, Kronenberg GHM (eds): "Photomorphogenesis in Plants," Dordrecht: Martinus Nijhoff, p 269.
26. Salisbury FB (1963). Planta 59:518.
27. Wellensiek SJ (1986). Flowering Newsletter 1:4.
28. Cleland CF (1984). In Vince-Prue D, Thomas B, Cockshull KEC (eds): "Light and the Flowering Process," London: Academic Press, p 123.
29. Warm E (1984). Physiol Plant 61:344.

30. Lay-Yee M, Sachs RM, Reid MS (1986). Flowering Newsletter 2:6.
31. Gilmour SJ, Zeevaart JAD, Schwenen L, Graebe JE (1986). Pl Phys 82:190.
32. Brulfert J, Vidal J, Keryer E, Thomas M, Gadal P, Queiroz O (1985). Physiol Veg 23:921.
33. Jordan BJ, Thomas B, Partis MD (1986). BCPC Monograph 34 "Biotechnology and Crop Improvement and Protection.

Plant Gene Systems and Their Biology, pages 39–50

DEVELOPMENTAL PATTERNS OF FLORAL INDUCTION IN TOBACCO[1]

Carl N. McDaniel, Susan R. Singer,[2] and Joan S. Gebhardt

Plant Science Group, Department of Biology, Rensselaer Polytechnic Institute, Troy, NY 12180-3590

ABSTRACT Floral induction has been extensively studied in photoperiodic plants but not in dayneutral plants. We have compared floral induction in short-day and dayneutral tobaccos by measuring the result of the induction process, floral determination. That is, an inductive signal acts on competent cells to commit them to a new developmental fate (e.g., floral determination) which can be expressed *in situ* as well as in different environments (e.g., in isolation or in another place on the organism). In dayneutral *Nicotiana tabacum* cv. Wisconsin 38 both shoot apical meristems and internode tissues become stably committed to form flowers prior to the morphological appearance of flowers. Studies of *N. tabacum* cv. Maryland Mammoth, a short-day plant which differs from dayneutral tobacco in its photoperiodic response because of a single gene, indicate that the terminal meristem becomes florally determined before the initiation of the terminal flower. Floral determination occurs in both dayneutral and short-day tobaccos after considerable vegetative growth. This significant amount of vegetative growth appears to be regulated by the roots as well as the leaves.

[1]This work was supported by grants from USDA Competitive Grants Program (84-CRCR-1-1490) and NSF (DCB8 4-09708, PCM83-20407) to CNM and an NSF predoctoral fellowship to SRS.

[2]Present address: Department of Biology, Carleton College, Northfield, MN 55057-4025.

INTRODUCTION

The flowering process, which is of central importance in the biology of angiosperms, has been extensively studied (1,2,3). We have focused our efforts on the early developmental events which commit cells and meristems to a program of floral development. Our approach as well as the results obtained have been described in several recent reviews (4,5,6,7). In this paper we will consider floral induction and floral determination in dayneutral and short-day tobaccos. Dayneutral varieties like *Nicotiana tabacum* cv. Wisconsin 38 will flower under essentially all photoperiods while the *N. tabacum* Maryland Mammoth types require short-day photoperiods. This difference is controlled by the recessive Maryland Mammoth allele which makes plants carrying these alleles responsive to the photoperiod (8).

We have shown that the terminal bud of *N. tabacum* cv. Wisconsin 38 becomes determined for floral development prior to the morphological initiation of floral structures (9). It has been known for over two decades that internode tissues of dayneutral tobacco can form *de novo* floral shoots *in vivo* (10) and *in vitro* (11) indicating that the developmental state, floral determination, can be acquired by tissues other than organized meristems. We have shown that floral determination in internode tissues first occurs in a plant at about the same time that the terminal bud becomes florally determined (12).

Numerous experiments with photoperiodic plants have shown that exposure to some minimal number of inductive photoperiods will enable the plants to produce flowers when they are placed in a non-inductive photoperiod (1,2,13). The inductive photoperiod has changed the plants' developmental state from being vegetative to being floral: the plant has become determined for floral development. We have shown that the terminal bud of *N. tabacum* cv. Maryland Mammoth becomes florally determined at the same time that the whole plant becomes florally determined (14). However, cells and tissues do not express floral determination in determination in culture since they only produce vegetative shoots (11,15,16). This paper compares floral induction and floral determination in dayneutral tobacco with that in short-day Maryland Mammoth tobacco.

METHODS

N. tabacum cv. Wisconsin 38 (seed source, Ian Sussex, Yale) plants were grown from seed and raised to maturity in the field as previously described (9). *N. tabacum* cv. Hicks Maryland Mammoth (seed source, Verne Sisson, USDA *Nicotiana* germplasm collection, Oxford, NC) and *N. tabacum* cv. Maryland Mammoth (seed source, Anton Lang, MSU) plants were grown in growth rooms as described elsewhere (14).

N. tabacum cv. Wisconsin 38 plants were assigned "ages" based on the number of leaves greater than or equal to 3.0 cm in length. The first leaf greater than 15 cm in length was designated leaf number one. *N. tabacum* cv. Hicks Maryland Mammoth and cv. Maryland Mammoth plants were assigned "ages" based on the number of leaves greater than or equal to 1.0 cm in length. The first leaf greater than 10 cm in length was designated leaf number one.

The developmental state of terminal buds on plants at various ages was assessed by rooting the bud and counting the number of nodes below the terminal flower on the resulting plant. When rooted, florally-determined buds produced few nodes along with a terminal flower. This number of nodes was comparable to that produced by equivalent buds *in situ* (9,14).

RESULTS

Floral determination in dayneutral tobacco.

Floral determination in terminal buds. Floral determination was studied in the terminal bud of a population of field-grown plants (9). Terminal buds were removed from plants at various ages and rooted (Figure 1). At ages 10 to 20, terminal buds were not determined for floral development since they produced a number of nodes similar to that of untreated plants grown from seed. At age 21, half of the rooted terminal buds were florally determined since they produced a number of nodes equivalent to that which they would have produced had they remained *in situ*. Terminal buds from all plants at ages 22 and older were florally determined. Histological examination of equivalent meristems showed that the terminal flower had not yet formed at the time of floral determination.

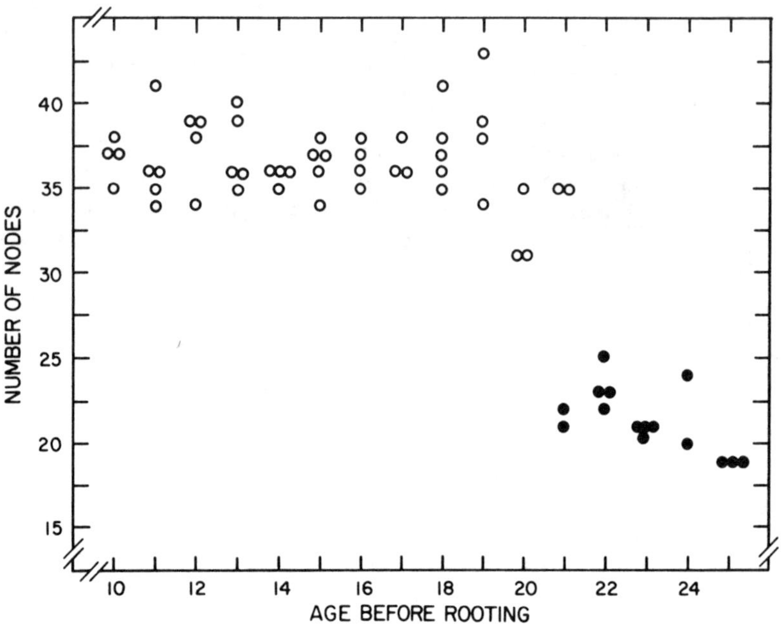

Figure 1. Timing of floral determination in terminal buds of N. tabacum cv. Wisconsin 38 plants. The number of nodes present on the plant resulting from the rooted bud is represented by O (bud not florally determined at time of rooting) or ● (bud florally determined at time of rooting). Taken from Singer and McDaniel (9) with permission.

Floral determination in short-day tobacco.

Floral determination in the terminal bud. Employing the same rooting assay used for establishing the time of floral determination in buds of dayneutral tobacco, the time of floral determination in short-day tobacco was established. N. tabacum cv. Hicks Maryland Mammoth plants were grown under inductive conditions and terminal buds from plants at various ages were rooted (Figure 2). At ages 5 thru 8, rooted buds were not florally determined.

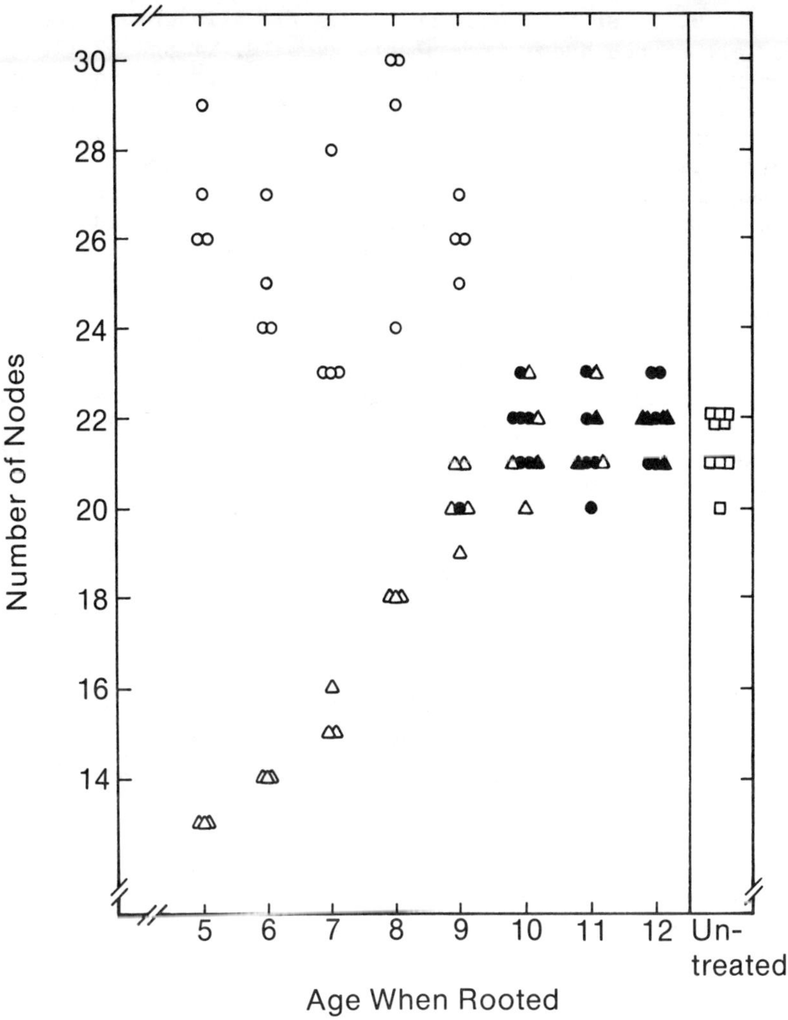

Figure 2. Timing of floral determination in terminal buds of N. tabacum cv. Hicks Maryland Mammoth. The total number of nodes produced by a bud before and after rooting is plotted: O (bud not florally determined at the time of rooting) or ● (bud florally determined at the time of rooting). Triangles indicate number of nodes present on untreated plants of a similar age: △ no terminal flower present, ▲ terminal flower being formed. □ represents nodes produced by untreated plants.

At age 9, 1 of 5 rooted buds was florally determined as were all rooted buds from plants ages 10, 11 and 12 (Table 1). Similar results were obtained for rooted terminal buds of *N. tabacum* cv. Maryland Mammoth (14). In this cultivar the first florally-determined buds were observed at age 15 where 1 of 3 rooted buds was florally determined. Also in these experiments terminal buds of Maryland Mammoth plants grown under inductive conditions were rooted under non-inductive conditions. Floral determination was first observed in buds taken from plants at age 15. Whole plants were also shifted from inductive to non-inductive conditions at various ages. Again, floral determination was first observed for plants shifted at age 15. Histological examination of equivalent meristems of *N. tabacum* cv. Hicks Maryland Mammoth and cv. Maryland Mammoth showed that the terminal flower had not yet formed at the time of floral determination.

TABLE 1

NODES PRODUCED BY ROOTED TERMINAL BUDS OF *NICOTIANA TABACUM* CV. HICKS MARYLAND MAMMOTH[a]

Developmental State at Time of Rooting	Number of Nodes Produced±S.D.(n)(range) Rooted Shoot Only	Before and After Rooting
Not florally determined[b]	22.2±2.6(20)(19-27)	26.1±2.3(20)(23-30)
Florally determined[b]	15.0±1.1(17)(13-17)	21.7±1.0(17)(20-23)
Untreated	-----------	21.4±0.7(9)(20-22)

[a]This is a numerical analysis of the data presented in Figure 2. For experimental details see Methods and legend for Figure 2.

[b]A one-sided Student's t test of means indicated that the number of nodes produced by determined and undetermined terminal buds are different with $p < 0.001$.

DISCUSSION

The developmental state of the terminal buds of dayneutral and short-day tobacco plants of various ages were established by assessing the developmental fate of terminal buds _in situ_ and in isolation (i.e., rooted). Terminal buds exhibited two distinctly different patterns of development after being rooted: they produced a number of nodes equivalent to that produced by an untreated plant grown from seed or they produced the same number of nodes as they would have produced had they remained _in situ_. The former pattern was replaced by the latter pattern at a specific age establishing that the terminal bud became florally determined at a specific time in the ontogeny of the plant. In short-day tobacco, shifting whole plants from inductive to non-inductive conditions demonstrated that the whole plant exhibited a florally-determined state at the same time that the terminal bud became florally determined. Histological analysis of both types of tobacco indicated that floral determination occurred prior to the initiation of sepals and that one to several leaf primordia were formed after the bud became florally determined. Floral determination in the terminal bud of _N. silvestris_, a long-day tobacco, is similar to that reported here for dayneutral and short-day tobaccos (6).

The above observations are important because they firmly establish that the developmental pathways which lead to flowering in three photoperiodic types of tobacco have common features. We have measured the end product of floral induction, floral determination, in the three photoperiodic types of tobacco. Since the final outcome is similar, at least at the organizational level of the terminal bud, it is reasonable to think that the induction process may be similar. However, if floral induction and the resulting floral determination are similar, why are there different photoperiodic requirements?

To discuss this question, we need to employ several fundamental concepts in developmental biology (Figure 3). We have placed the terms for these concepts within what we call a developmental "black box" because our knowledge of the molecular mechanisms involved is at best rudimentary. In regulating a process like flowering it is possible to imagine that each arrow within the "black box" in figure 3 might be a point at which control could be exerted. For simplicity let us consider what we know about dayneutral and Maryland Mammoth tobaccos where a single recessive allele, when in a homozygous condition, can convert dayneutral types into short-day types (8).

The fundamental difference in controlling the expression of flowering appears to be at the level of induction. Induction does not occur in the Maryland Mammoth types unless the leaves are exposed to short days (1,2,13). The terminal bud is apparently competent to respond to an inductive signal and the leaves do not appear to produce compounds which interfer with induction (17). Thus the Maryland Mammoth gene makes the leaves dependent upon the appropriate photoperiod in order for them to produce and transmit the inductive signal.

Dayneutral tobaccos are not dependent upon a specific photoperiod for floral induction but they do not flower

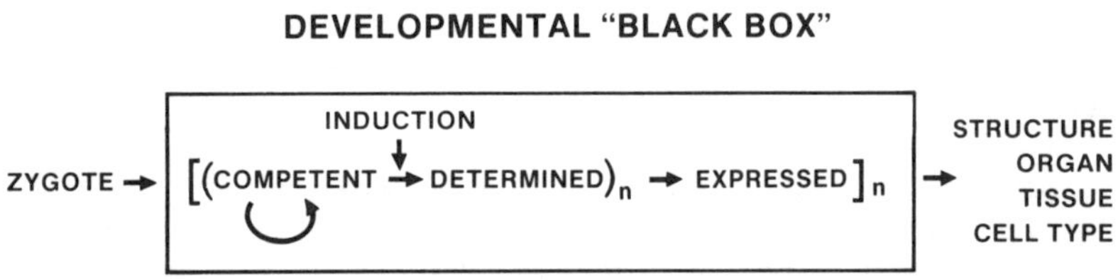

Figure 3. This developmental "black box" illustrates some of the possible events which may occur during the ontogeny of an individual. Arrows indicate the normal direction of events but do not imply irreversibility. Competent cells/tissues are those which have the ability to respond to a developmental signal in a specific way. Cells and tissues can acquire different competences either via endogenous or exogenous means. Induction occurs when a developmental signal acts on competent cells/tissues to determine them for a specific developmental fate. The determined state is subsequently expressed. The "n" indicates that the bracketed sequence may be reiterated. Modified from McDaniel et al. (7) with permission.

immediately upon germination. In the case of N. tabacum cv. Wisconsin 38 about 35 nodes are produced before the formation of a terminal flower. By inducing roots to form on the stem of a growing plant and by rooting the terminal portion of a plant each time a small number of leaves were produced, it was shown that the presence of roots near the terminal bud prevents the bud from forming flowers (18). Leaf removal experiments have established that several leaves in any position on the main axis (base, middle, apex) are sufficient to induce flowering (18,19). These observations indicate that all leaves are similar in their inductive capacity and that the roots somehow prevent induction when they are in proximity to an apical bud. Roots could exert their influence by regulating either competence or determination. Lang (20) has shown that when young apical buds of dayneutral tobacco are removed from the root influence, they can be induced to flower. However, the appropriate experiments for sorting out how the roots influence floral determination have not been conducted.

It is well known that Maryland Mammoth types do not flower immediately when raised under inductive conditions. Two observations support the contention that the roots in short-day tobacco play a role similar to that discussed above for dayneutral tobacco. First, when terminal buds of Hicks Maryland Mammoth plants at ages prior to floral determination are rooted, these buds produce more nodes than they would have produced had they remained _in situ_ (see Figure 2 and Table 1). Thus, the presence of roots near the terminal bud appears to permit vegetative growth and prevent flowering. Second, reciprocal grafts between _N. tabacum_ cv. Wisconsin 38 and _N. tabacum_ cv. Maryland Mammoth have shown that seedling scions produce the same number of nodes independent of the root-stock genotype (21). These observations indicate that the root influence appears to be similar in dayneutral and short-day tobaccos.

Leaf removal experiments with _N. tabacum_ cv. Maryland Mammoth have established that basal and apical leaves possess different inductive capacities (19). As many as six basal leaves (10 cm or greater in length) are not capable of inducing flowering while as few as two apical leaves (10 cm or greater in length) can induce flowering. The Maryland Mammoth gene thus appears not only to make leaves sensitive to the photoperiod but also to alter the inductive capacity of leaves as a function of their position on the main axis.

The Maryland Mammoth allele appears to have at least one more effect on the flowering process. Floral determination is expressed by tissues of dayneutral tobacco (11) and we have shown that floral determination of internode tissues commences at about the same time that the terminal bud becomes florally determined (12). Since both florally-determined terminal buds and internode tissues exhibit the same ontogeny (i.e., production of several nodes and internodes followed by the formation of a terminal flower), it is reasonable to assume that they express the same developmental program which leads to the formation of a terminal flower. If this assumption is true, then induction and floral determination are similar processes at the tissue and shoot-apical meristem levels of organization. Many laboratories have shown that tissues from the inflorescence of Maryland Mammoth types produce only vegetative shoots *in vitro* (15,16). It is clear that inflorescence tissues have been exposed to the inductive signal but they do not become stably determined for floral development since they cannot express this state *in vitro*. The presence of the Maryland Mammoth allele may reduce tissue level competence or make floral determination at the tissue level less stable than in dayneutral tobacco. Results from an investigation of floral determination in axillary buds of *N. silvestris* (long-day tobacco) have been interpreted as showing that floral determination is less stable in photoperiodic tobaccos than in dayneutral tobacco (22).

The comparison of induction and floral determination in dayneutral and short-day tobaccos has indicated how a single gene can modify several aspects of a developmental pathway. Our current understanding of an important developmental gene like the Maryland Mammoth gene has provided an example of how little we know about the developmental processes that are influenced by the presence of such a gene. This situation exemplifies the field of developmental botany, where for most of the important problems our undergirding information base at the level of developmental physiology is severely deficient.

REFERENCES

1. Bernier G, Kinet JM, Sachs RM (1981a). "The Physiology of Flowering, Vol. 1. The Initiation of Flowers." Boca Raton: CRC Press.
2. Bernier G, Kinet JM, Sachs RM (1981b). "The Physiology of Flowering, Vol. 2. Transition to Reproductive Growth." Boca Raton: CRC Press.

3. Halevy AH (ed.) (1985). "Handbook of Flowering." Boca Raton: CRC Press.
4. McDaniel CN (1984a). Shoot meristem development. In Barlow PW, Carr DJ (eds): "Positional Controls in Plant Development," Cambridge: Cambridge University Press, p 319.
5. McDaniel CN (1984b). Competence, determination and induction in plant development. In Malacinski G (ed): "Pattern Formation: A Primer in Developmental Biology," New York: Macmillan, p 393.
6. McDaniel CN, Singer SR, Dennin KA, Gebhardt JS (1985). Floral determination: Timing, stability, and root influence. In Freeling M (ed): "Plant Genetics," New York: Alan R. Liss, p 73.
7. McDaniel CN, Singer SR, Gebhardt JS, Dennin KA (1987). Floral determination: A critical process in meristem ontogeny. In Atherton JG (ed): "The Manipulation of Flowering," London: Butterworth, p 109.
8. Allard HA (1919). Gigantism in Nicotiana tabacum and its alternative inheritance. Amer Naturalist 53:218.
9. Singer SR, McDaniel CN (1986). Floral determination in the terminal and axillary buds of Nicotiana tabacum L. Dev Biol 118:587.
10. Aghion-Prat D (1965a). Floral meristem-organizing gradient in tobacco stems. Nature 207:1211.
11. Aghion-Prat D (1965b). Neoformation de fleurs in vitro chez Nicotiana tabacum L. Physiol Veg 3:229.
12. Singer SR, McDaniel CN (1987). Floral determination in internode tissues of dayneutral tobacco first occurs many nodes below the apex. Proc Natl Acad Sci: In press.
13. Lang A (1965). Physiology of flower initiation. In Ruhland W (ed): "Encyclopedia of Plant Physiology." Vol15/1, Berlin: Springer-Verlag, p 1380.
14. Gebhardt JS, McDaniel CN (1987) Induction and floral determination in the terminal bud of Nicotiana tabacum L. cv. Maryland Mammoth. Planta: Submitted.
15. Chailakhyan MD, Aksenova NP, Konstantinova TN, Bavrina TV (1975). The callus model of plant flowering. Proc Royal Soc Lond B 190:333.
16. Bridgen MP, Veilleux RE (1985) Studies of de novo flower initiation from thin cell layers of tobacco. J Amer Soc Hort Sci 110:233.
17. Lang A (1980). Inhibition of flowering in long-day plants. In Skoog F (ed): "Plant Growth Substances

1979," Berlin: Springer-Verlag, p 310.
18. McDaniel CN (1980). Influence of leaves and roots on meristem development in *Nicotiana tabacum* L. cv. Wisconsin 38. Planta 148:462.
19. Gebhardt JS (1987). "Floral Determination in *Nicotiana tabacum* L. cv. Maryland Mammoth." M.S. Thesis: Rensselaer Polytechnic Institute, Troy, NY 12180.
20. Lang A (1984) Die photoperiodische Regulation von Forderung und Hemmung der Blutenbildung. Ber Deutsch Bot Ges 97:293.
21. Bartkiewicz-Krasodomski J (1986) "The Influence of Plant Growth Regulators and Roots on Flowering in *Nicotiana tabacum*." M.S. Thesis: Rensselaer Polytechnic Institute, Troy, NY 12180.
22. Dennin KA, McDaniel CN (1985) Floral determination in axillary buds of *Nicotiana silvestris*. Dev Biol 112:377.

II. REPRODUCTIVE BIOLOGY: FERTILIZATION

Plant Gene Systems and Their Biology, pages 53–64

MOLECULAR ASPECTS OF SELF-INCOMPATIBILITY IN FLOWERING PLANTS

A.E. Clarke, M.A. Anderson, A. Bacic, E.C. Cornish, P.J. Harris, S-L. Mau and J.R. Woodward

Plant Cell Biology Research Centre, School of Botany, University of Melbourne, Parkville, Victoria 3052, Australia

ABSTRACT The study of self-incompatibility in flowering plants provides a model for approaching a more general understanding of cell-cell recognition in plants. The genetics and biology of the pollination process has been well described for several species. Application of molecular cloning techniques has laid the foundation for a new level of understanding of the process.

In this paper we review the contribution from our laboratory to understanding molecular aspects of gametophytic self incompatibility in two members of the Solanaceae, *Nicotiana alata* and *Lycopersicon peruvianum*, and briefly summarize work from two other laboratories on sporophytic self-incompatibility in *Brassica oleracea* and *B. campestris* (Cruciferae).

INTRODUCTION

Many flowering plant families have specific self-incompatibility genes (*S*-genes) which prevent inbreeding (for reviews see refs 1-6). Of the two types of self-incompatibility (gametophytic and sporophytic), gametophytic is the more common and, in many cases, is controlled by a single nuclear gene locus (*S*-locus) having multiple alleles. Pollen expresses its haploid *S*-genotype and matings are incompatible if the *S*-allele of the pollen is matched by one of the two *S*-alleles expressed in the diploid tissue of the pistil. During both incompatible and compatible matings, pollen germinates and the pollen tubes grow through the stigma into the transmitting tissue of the style. Tube growth from incompatible pollen grains is arrested in the upper part of the style; the tip of the tube becomes thickened, there is often a characteristic deposit of callose close to the tip, and the tip may burst. In sporophytic incompatibility, pollen behaviour is determined by the genotype of the pollen-producing plant; inhibition usually occurs on the stigma surface and is often associated with deposition of callose in the stigma cells. The gametophytic system is generally considered to be the ancestral form of self-incompatibility in flowering plants, with the sporophytic system being derived from it.

The genetics of self-incompatibility systems has been studied extensively (see, for example, refs 1, 7-11). Progress towards understanding the molecular basis of self-incompatibility was, until recently, limited to observation of *S*-allele-associated proteins or glycoproteins in pistil extracts. In gametophytic systems, *S*-genotype-associated style glycoproteins (*S*-glycoproteins) have been demonstrated in *Nicotiana alata* (5, 8, 12), *Lycopersicon peruvianum* (13), *Petunia hybrida* (14), *Prunus avium* (15, 16), and *Lilium longiflorum* (17). In sporophytic systems, many of the observations have been made with *Brassica* species (18-21).

In this paper, we summarize the progress our own group has made towards understanding the molecular basis of self-incompatibility in *N. alata* and *L. peruvianum* (13, 22-25) and briefly summarize the published data on molecular aspects of sporophytic self-incompatibility (26-28).

MOLECULAR BASIS OF SELF-INCOMPATIBILITY IN *N. ALATA* AND *L. PERUVIANUM*

Pollen tubes grow extracellularly between the files of the transmitting tissue cells of the style. There is the possibility therefore that the interaction leading to arrest or continued growth of the pollen tubes is between secreted components of the transmitting tissue and some component of the pollen tubes. The transmitting tissue cells in the style are separated by a "mucilage" which contains proteins, glycoproteins and proteoglycans as well as low molecular weight material. Our research program has been directed to characterizing individual components of the transmitting tissue extracellular "mucilage" and pollen tubes (see Fig. 1).

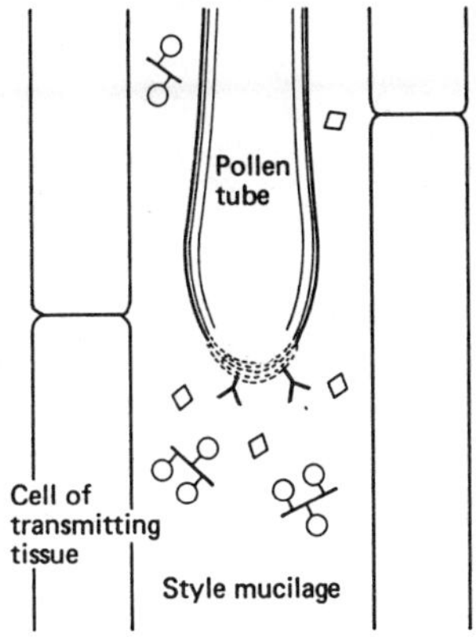

FIGURE 1. Diagram illustrating major goals of our current research program.

Style Components

Style glycoproteins which segregate with particular *S*-alleles. Various combinations of *S*-alleles were generated in plants of *N. alata* by reciprocal crossing of the genotypes S_2S_2, S_3S_3, S_2S_3 and S_1S_3. Analysis of stylar proteins by gel electrophoresis demonstrated that a 32K glycoprotein segregated with the S_2-allele. This 32K glycoprotein was termed the S_2-glycoprotein; it appears concomitantly with S_2-function in maturing styles and occurs in highest concentration in the part of the style in which inhibition of incompatible pollen tube growth occurs. These observations indicated that this glycoprotein was the product of either the S_2-allele or a gene closely linked to it. The glycoprotein was isolated and a full length cDNA encoding the mature protein and a putative signal sequence was obtained. Details of the cloning strategy and sequence data have been published (22).

A hydrophobicity plot of the deduced amino acid sequence is shown in Fig. 2. The putative signal sequence is typical of eukaryotic signal peptides with a relatively hydrophilic sequence of five amino acids at the C-terminal end, a relatively hydrophobic sequence (-16 to -14) and a more hydrophilic sequence at the N-terminal end. The mature protein is essentially hydrophilic except for a relatively hydrophobic stretch of 15 amino acids at the N-terminus.

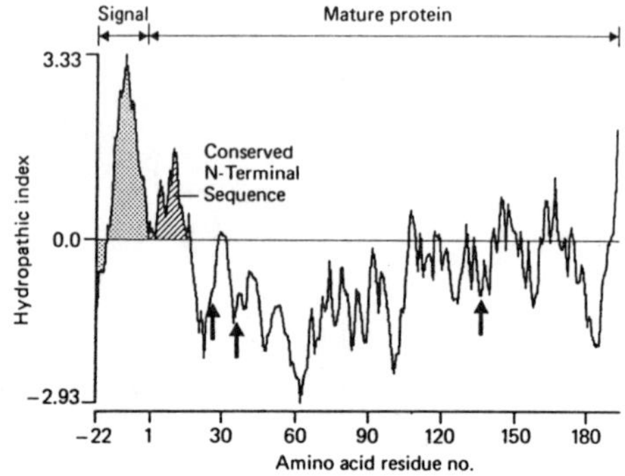

FIGURE 2. A hydrophobicity plot of the deduced amino-acid sequence of the unprocessed S_2-glycoprotein. The region of the profile corresponding to the putative signal peptide (amino acids -22 to -1) is stippled. The hydrophobic N-terminal sequence, which exhibits striking homology to the *S*-allele-associated style components from *L. peruvianum*, is shaded. The positions of the three putative *N*-glycosylation sites are indicated by arrows.

This N-terminal sequence shows striking homology to that of other *S*-allele associated proteins from *N. alata* and *L. peruvianum* (13) (Fig. 3). The finding that these proteins share homology at the N-terminus, although differing in pI and molecular weight, gives indirect support that they are allelic products of a single locus. Definitive proof will require demonstration that they have specific biological activity and that the *S*-allele specificity of individual plants can be altered by transformation experiments using cloned genes encoding the *S*-proteins.

A major proportion of the carbohydrate on the S_2-associated glycoprotein occurs as N-linked chains, as it was susceptible to degradation with endoglycosidase F (endo F) and peptide:N-glycosidase ("N-glycanase;"from *Flavobacterium meningosepticum*, designated PNGase F(29)). Hydrolysis of different *S*-allele associated glycoproteins from *N. alata* demonstrates that there are different numbers of N-linked carbohydrate chains associated with different *S*-allele associated glycoproteins. This might indicate that the carbohydrate is functionally significant or alternatively that the allelic differences in the protein sequence preclude N-glycosylation.

	pI	Mr	Amino-terminal sequence														
L. peruvianum			1	2	3	4	5	6	7	8	9	10	11	12	13	14	15
S_1	7.5	28,000	Y	F	E	Y	L	Q	L	V	L	Q	X	P	T	T	F
S_3	>9.5	28,000	D	F	D	Y	L	Q	L	V	L	Q	X	P	R	S	F
N. alata																	
S_2	>9.5	32,000	A	F	E	Y	M	Q	L	V	L	T	W	P	I	T	F
S_6	>9.5	31,000	A	F	E	Y	M	Q	L	V	L	Q	W	P	T	A	F
S_z	9.0	30,000	D	F	D	Y	M	Q	L	V	L	T	X	P	A	S	F
S_{f11}	9.5	27,000	D	F	E	Y	L	Q	L	V	L	T	W	P	A	S	F

FIGURE 3. Comparison of the pI, M_r and amino-terminal sequences (residues 1-15) of *L. peruvianum* and *N. alata S*-allele-associated proteins. Amino acids are identified by their single letter code, with X indicating residues which could not be clearly assigned. Residues which are identical in all six proteins are boxed.

In all mature styles bearing the S_2-allele, a predominant mRNA species of approximately 900 nucleotides hybridizes to the S_2-cDNA. Homologous mRNA species of identical size were also detected in S_3S_3, S_1S_3 and S_6S_6 styles of *N. alata*; the extent of hybridization being significantly lower than the level obtained in styles bearing the S_2-allele (see Fig. 4). This result indicates either lower expression from these alleles or a major sequence change or a combination of both. We are currently cloning and sequencing other *S*-alleles from *N. alata* and *L. peruvianum* to provide some insight into the nature of the allelic differences.

Hybridization of the S_2-cDNA to sections of the mature pistil showed that in the style the *S*-gene is expressed throughout the transmitting tract, but not in the epidermal or cortical cells(25). It is also expressed in the epidermis of the placenta. Parallel immunocytochemical studies using an S_2-specific antibody raised to a synthetic peptide show that the protein is present in all tissues where the gene is expressed (Bonig, I., unpublished observations). In the immature style, hybridization of the S_2-cDNA was restricted

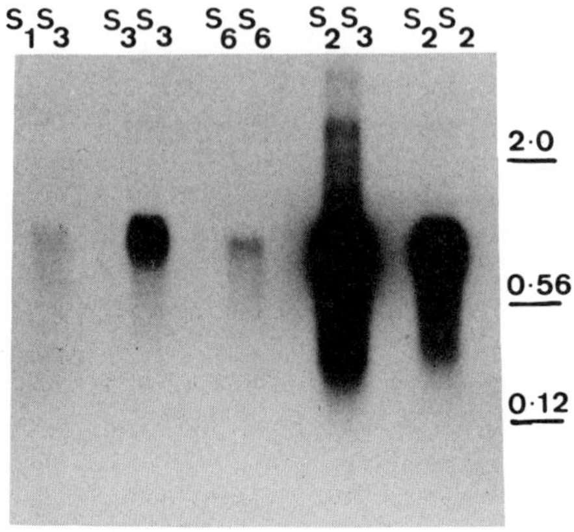

FIGURE 4. Northern blot analysis of total RNA isolated from styles of different *S*-genotypes of *N. alata*.

5μg of unfractioned RNA was used from genotypes S_1S_3, S_3S_3 and S_6S_6 and 0.1μg from genotypes S_2S_3 and S_2S_2.

to two sites corresponding to the proliferated stigma epidermis; there was no hybridization to either the transmitting tract or the ovary of immature pistils. The pattern of S_2-gene expression, first in the immature stigma but not in the style transmitting tissue or the placenta, and later in all three, reflects the development of this tissue which proceeds from the top to the base of the style. Our observations have shown that from the time it penetrates the stigma surface until the time it enters the ovule the pollen tube is growing through or over tissues in which, in mature pistils, the *S*-gene is expressed (Fig. 5).

The S_2-cDNA was used to analyse the organization of homologous sequences in the genomes of a number of different *S*-genotypes of *N. alata*. Restriction fragment length polymorphisms characteristic of other *S*-alleles were identified. As with the Northern analyses, the extent of hybridization to the genomes of the other alleles indicates that there are marked differences in sequence between the different alleles.

Style proteoglycans of the arabinogalactan-protein (AGP) class. AGPs are components of styles of many flowering plants (30). In *N. alata*, AGPs account for at least 60% of the total high molecular weight soluble style carbohydrate (31). The AGPs are localized in the extracellular mucilage

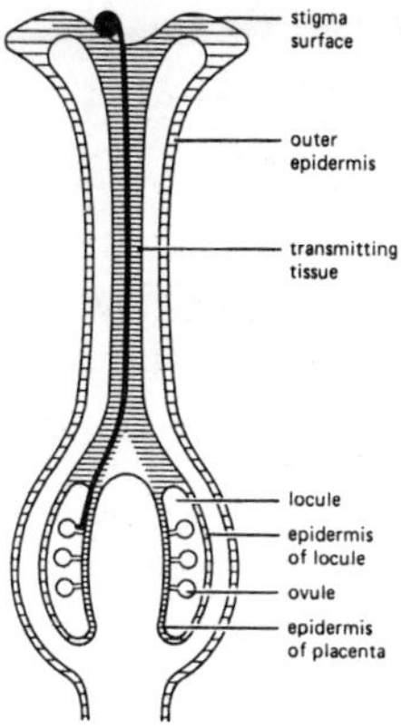

FIGURE 5. Diagram illustrating pathway of pollen tube through the pistil and region of hybridization with the S_2-cDNA (close cross hatching).

of the style transmitting tract tissue (32); and in the mucilage coating the inner epidermis of the ovary. Stigma, style and ovary contain AGPs which are developmentally regulated. Both the total amount and concentration of AGP in the stigma increase during flower development, whereas the AGP concentration in the style remains constant throughout the maturation period. The charge characteristics of AGPs present in the stigma, style and ovary are different and change during development (24; A. Gane unpublished observations). Pollination of flowers of *N. alata* with compatible or incompatible pollen results in a significant increase in the amount of AGPs in the stigma, but not the style, compared with unpollinated controls; the increases are unrelated to the incompatibility phenotype of the pollen. AGPs, isolated from stigma extracts of *N. alata* genotypes S_2S_2 and S_3S_3, do not show any significant differences in amino acid, monosaccharide (neutral and acidic) or glycosyl linkage composition (31).

It seems unlikely that AGPs are directly involved in the expression of self-incompatibility. On the stigma surface, they may be responsible for generating conditions which are conducive for the adhesion and germination of pollen. In addition they may provide nutrients for the growing pollen tube.

Pollen Tube Components

Chemical studies on the cell-wall polysaccharides of pollen tubes have been published for seven species (for review see 33). In each study, glucose was the most abundant monosaccharide, with the second most abundant being either arabinose or galactose. The monosaccharide analyses of cell walls isolated from pollen of *N. alata* germinated *in vitro* also contain glucose and arabinose as the dominant monosaccharides. Methylation analysis and cytochemical studies are consistent with the major polysaccharides being a (1→3)-β-D-glucan (callose) and an arabinan, together with small amounts of cellulose. The cell walls also contain 2.8% uronic acids which are apparently concentrated in the tip (23). Immunocytochemical studies using a monoclonal antibody directed to terminal α-L-arabinofuranoside residues show that the arabinosyl residues are located in the outer part of the pollen tube wall (34). The (1→3)-β-glucan is located in the inner wall which does not extend to the tip of the tube (33). We are currently undertaking a cytochemical study of formation of these two wall components using probes specific for terminal arabinofuranosyl residues and for (1→3)-β-glucosidic linkages. This information is an important basis for our ultimate understanding of how the style *S*-gene products interact with pollen tubes to signal arrest or continued growth of the tube.

MOLECULAR BASIS OF SELF-INCOMPATIBILITY IN *BRASSICA*

Progress in our understanding of the molecular basis of sporophytic self-incompatibility has recently been advanced in studies on two *Brassica* species. A cDNA clone encoding part of the S_6-allele-associated glycoprotein of *Brassica oleracea* has been described by Nasrallah *et al.* (26). This clone corresponds to a 2Kb mRNA species which is expressed exclusively in the surface papillae cells of the stigma. Unlike *N. alata*, the *S*-alleles in *B. oleracea* have a high degree of homology. This implies that allelic specificity in this system is either a consequence of quite small differences in the amino acid sequence or is determined by the carbohydrate component of the glycoproteins. Recently,

Takayama *et al.* (27, 28) isolated glycoproteins corresponding to the S_8, S_9 and S_{12} alleles of *B. campestris* and obtained N-terminal amino acid sequences. As for the different *S*-glycoproteins of *N. alata*, there was a high degree of homology between these N-terminal sequences. They also determined the sequence of the oligosaccharide side chains, but found no major differences attributable to allelic specificity.

Interestingly, there is no homology between the *S*-associated proteins of *N. alata* and *Brassica* characterized to date. This raises the possibility that the sporophytic and gametophytic systems of self-incompatibility evolved separately.

FUTURE DIRECTIONS

The study of self-incompatibility not only provides information on the mechanism of control of fertilization in plants, but it is also a relatively simple model for understanding the basis of cell-cell recognition in plants. The description of cDNA clones encoding style proteins which segregate with particular alleles and chemical and physical characterization of the structure of the *S*-allele associated glycoproteins opens the way for study of:

- transgenic plants, to determine whether the style glycoproteins are indeed the *S*-gene products
- sequences controlling tissue specific expression of the gene
- expression of the gene in pollen
- interaction of products of the *S*-gene from pollen and style
- how this interaction leads to arrest of pollen tube growth in incompatible matings.

REFERENCES

1. de Nettancourt D (1977). "Incompatibility in Angiosperms." Berlin: Springer.
2. Heslop-Harrison J (1978). Genetics and physiology of angiospermsincompatibility systems. Proc Roy Soc Lond B 202: 73-92.
3. Lewis D (1979). Genetic versatility of incompatibility in plants. New Zealand J Bot 17: 637-644.

4. Pandey KK (1979). Overcoming incompatibility and promoting genetic recombination in flowering plants. New Zealand J Bot 17: 645-663.
5. Clarke AE, Anderson MA, Bacic T, Harris PJ, Mau S-L (1985). Molecular basis of cell recognition during fertilization in higher plants. J Cell Sci Suppl 2: 261-285.
6. Knox RB (1984). Pollen-pistil interactions. In Linskens HF, Heslop-Harrison J (eds): "Encyclopaedia of Plant Physiology", Vol. 17, Berlin: Springer-Verlag, p 508.
7. Larsen K (1977). Self-incompatibility in *Beta vulgaris* L. I. Four gametophytic, complementary *S*-loci in sugar beet. Hereditas 85: 227-248.
8. Pandey KK (1967). Elements of the *S*-gene complex. II. Mutation and complementation at the SI locus in *Nicotiana alata*. Heredity 22: 255-283.
9. Lawrence MJ (1975). The genetics of self-incompatibility in *Papaver rhoeas*. Proc Roy Soc Lond B 188: 275-285.
10. Lawrence MJ, Afzal M, Kenrick J (1978). The genetical control of self-incompatibility in *Papaver rhoeas*. Heredity 40: 239-253.
11. Linskens HF (1975). Incompatibility in *Petunia*. Proc Roy Soc Lond B 188: 299-311.
12. Bredemeijer GMM, Blaas J (1981). *S*-specific proteins in styles of self-incompatible *Nicotiana alata*. Theor Appl Genet 59: 185-190.
13. Mau S-L, Williams EG, Atkinson A, Anderson MA, Cornish EC, Grego B, Simpson RJ, Kheyr-Pour A, Clarke AE (1986). Style proteins of a wild tomato (*Lycopersicon peruvianum*) associated with expression of self-incompatibility. Planta 169: 184-191.
14. Kamboj RK and Jackson JF (1986). Self-incompatibility alleles control a low molecular weight, basic protein in pistils of *Petunia hybrida*. Theor Appl Genet 71: 815-819.
15. Mau S-L, Raff JW, Clarke AE (1982). Isolation and partial characterization of components of *Prunus avium* L. styles, including an antigenic glycoprotein associated with a self-incompatibility genotype. Planta 156: 505-516.
16. Raff JW, Knox RB, Clarke AE (1981). Style antigens of *Prunus avium* L. Planta 153: 125-129.
17. Dickinson HG, Moriarty J, Lawson J (1982). Pollen-pistil interaction in *Lilium longiflorum*: the role of the pistil in controlling pollen tube growth following

cross-pollination and self-pollination. Proc Roy Soc Lond B 215: 45-62.
18. Nasrallah ME, Barber JT, Wallace DH (1970). Self-incompatibility proteins in plants: detection, genetics and possible mode of action. Heredity 25: 25-27.
19. Nasrallah JB, Nasrallah ME (1984). Electrophoretic heterogeneity exhibited by *S*-allele specific glycoproteins of *Brassica*. Experentia 40: 279-281.
20. Nishio T, Hinata K (1977). Analysis of *S*-specific proteins in stigmas of *Brassica oleracea* L. by isoelectric focussing. Heredity 38: 391-396.
21. Nishio T, Hinata K (1982). Comparative studies on *S*-glycoproteins purified from different *S*-genotypes in self-incompatible *Brassica* species. I. Purification and chemical properties. Genetics 100: 641-647.
22. Anderson MA, Cornish EC, Mau S-L, Williams EG, Hoggart R, Atkinson A, Bonig I, Grego B, Simpson R, Roche PJ, Haley JD, Penschow JD, Niall HD, Tregear GW, Coghlan JP, Crawford RJ, Clarke AE (1986). Cloning of cDNA for a stylar glycoprotein associated with expression of self-incompatibility in *Nicotiana alata*. Nature 321: 38-42.
23. Rae AE, Harris PJ, Bacic A, Clarke AE (1985). Composition of the cell walls of *Nicotiana alata* Link et Otto pollen tubes. Planta 166: 128-133.
24. Gell AC, Bacic A, Clarke AE (1986). Arabinogalactan-proteins of the female sexual tissue of *Nicotiana alata*. I. Changes during flower development and pollination. Plant Physiology (in press).
25. Cornish EC, Pettitt JM, Bonig I, Clarke AE (1987). Developmentally-controlled, tissue-specific expression of a gene associated with self-incompatibility in *Nicotiana alata*. Nature (in press).
26. Nasrallah JB, Kao T-H, Goldberg ML, Nasrallah ME (1985). A cDNA clone encoding an *S*-locus-specific glycoprotein from *Brassica oleracea*. Nature 318: 263-267.
27. Takayama S, Isogai A, Tsukamoto C, Ueda Y, Hinata K, Okazaki K, Suzuki A (1986). Structure of carbohydrate chains of *S*-glycoproteins in *Brassica campestris* associated with self-incompatibility. Agric Biol Chem 50(6): 1673-1676.
28. Takayama S, Isogai A, Tsukamoto C, Ueda Y, Hinata H, Okazaki K, Suzuki A (1986). Isolation and some characterization of *S*-locus-specific glycoproteins associated with self-incompatibility in *Brassica campestris*. Agric Biol Chem 50(5): 1365-1367

29. Plummer TH, Elder JH, Alexander S, Phelan AW, Tarentino AL (1984). Demonstration of peptide:N-glycosidase F activity in endo-β-N-acetylglucosaminidase F preparations. J Biol Chem 259: 10700-10704.
30. Hoggart RM, Clarke AE (1984). Arabinogalactans are common components of angiosperm styles. Phytochemistry 23: 1571-1573.
31. Gell AC (1985). Arabinogalactan-proteins of the female sexual tissue of *Nicotiana alata*. Masters Degree Thesis. The University of Melbourne (1985).
32. Sedgley M, Blesing MA, Bonig I, Anderson MA, Clarke AE (1985). Arabinogalactan-proteins are localized extracellularly in the transmitting tissue of *Nicotiana alata* Link and Otto, an ornamental tobacco. Micron and Microscopica Acta 16 (4): 247-254.
33. Harris PJ, Anderson MA, Bacic A, Clarke AE (1984). Cell-cell recognition in plants with special reference to the pollen-stigma interaction. Oxford Surveys of Plant Molecular & Cell Biology, Vol. I: 161-203.
34. Anderson MA, Harris PJ, Bonig I, Clarke AE (1986). Immunogold localization of α-L-arabinofuranosyl residues in pollen tubes of *Nicotiana alata* Link et Otto. Planta (submitted).

Plant Gene Systems and Their Biology, pages 65–74

THE URF13-T MITOCHONDRIAL GENE OF cms-T MAIZE[1]

R. E. Dewey, W. W. L. Hu and C. S. Levings III[*]

Department of Crop Science and Genetics[*]
North Carolina State University
Raleigh, North Carolina, 27695-7614, USA

ABSTRACT Evidence indicates that the cms-T type of cytoplasmic male sterility is due to a mitochondrial gene, URF13-T. The URF13-T sequence has arisen by rearrangements involving other mitochondrial genes and is unique to cms-T. The gene, which is abundantly transcribed, encodes a 13,000 Mr polypeptide located in mitochondrial membranes. Cms-T is restored to pollen fertility by two nuclear genes, Rf1 and Rf2. The Rf1 locus alters the expression of the URF13-T mitochondrial gene and reduces the abundance of the 13 kd -URF13-T polypeptide.

INTRODUCTION

The trait, cytoplasmic male sterility (CMS), is observed among higher plants where it has been reported in over 140 different species (1). CMS causes pollen abortion and is inherited in a uniparental fashion through the female parent. Normally, female fertility is unaffected by the CMS trait. In the past it has been suggested that CMS may be due to mitochondrial, chloroplast or viral factors, however, recent investigations have increasingly implicated mitochondrial genes. Nevertheless, a viral component may be responsible for CMS in a few species. Substantial evidence indicates that mutant mitochondrial genes encode the CMS trait in maize.

[1]This work was partially supported by grants from the Agrigenetics Corporation and the National Science Foundation

TABLE 1
CHARACTERISTICS OF THE MAIZE CMS CYTOPLASMS AND THEIR RESTORER GENES

Male-sterile cytoplasm	Nuclear Restorer genes	Chromosomal location
T	Rf1 and Rf2	Chr. 3, 9
C	Rf4	Chr. 2 or 8
S	Rf3	Chr. 2

Three major CMS types are recognized in maize, cms-T, cms-C and cms-S. These sterile cytoplasms are differentiated on the basis of nuclear genes (Rf) that restore pollen fertility (Table 1). More recently other differences have been discovered that distinguish the normal and the three male-sterile cytoplasms. For example, the various cytoplasms can be differentiated by restriction enzyme length polymorphism of their mtDNA, by differences in their mtRNAs with Northern blot analyses and by differences in their mitochondrial translational products (see review 2,3). In some instances, the various cytoplasms may be distinguished according to disease resistance, toxin and insecticide resistance, reversion to fertility, histological differences, and mitochondrial DNA and RNA plasmids.

Cms-T differs from the other cytoplasms in that it is susceptible to the fungal disease, *Helminthosporium maydis* race T (Southern corn leaf blight) (4). Race T produces a pathotoxin, the T toxin, that specifically affects mitochondria from cms-T plants. The toxin causes swelling of the mitochondria, leakage of NAD^+ and Ca^{++}, and uncoupling of oxidative phosphorylation (5,6). In cms-T, disease susceptibility and pollen sterility appear to be inseparable. It is not known if these traits are caused by a single locus or if they are due to closely linked genes. Other maize cytoplasms are resistant to the *H. maydis* race T pathogen and its toxin.

In the following sections we describe some recent findings regarding the cms-T type of male sterility in maize.

RESULTS AND DISCUSSION

The TURF 2H3 Sequence

Investigations of mtRNAs from the male-sterile cytoplasms of maize have led to the discovery of mutant mitochondrial genes. Northern blot analysis has identified a mtDNA sequence, termed TURF 2B, that is abundantly transcribed (steady-state levels) in cms-T mitochondria (7). Although mitochondrial transcription is detected in other fertile and sterile cytoplasms, the levels are relatively low and are due to the peculiar nature of the TURF 2B sequence. In cms-T, transcription of TURF 2B is limited to three adjacent Hind III DNA fragments, which we have called TURF 2H3. We have characterized the 3547 nucleotide sequence making up TURF 2H3.

TURF 2H3 contains two long open reading frames that are designated URF 13 and URF 25. The URF 13 and URF 25 sequences predict polypeptides of 12,961 and 24,675 Mr, respectively. An analysis of the TURF 2H3 sequence has identified nucleotide homology with portions of other mitochondrial genes (Fig. 1). TURF 2H3 contains sequences homologous with the 5' flanking region of the atp 6 gene, the 3' flanking region of the 26S rRNA gene, a portion of the coding region of the 26S rRNA gene and a chloroplast tRNAarg gene. Thus, the chimeric sequence, TURF 2H3, is the product of rearrangements involving both intra- and intermolecular recombinational events. This analysis indicates that at least seven recombinational sites occur in TURF 2H3.

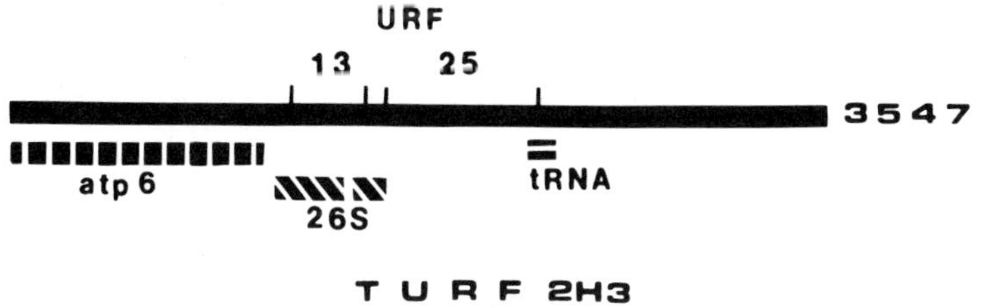

FIGURE 1. Diagram of the TURF 2H3 fragment. Segments of TURF 2H3 homologous with the 5' flanking region of the ATPase subunit 6 gene (atp6), 3' flanking and coding regions of the 26S rRNA gene (26S) and a chloroplast tRNAarg gene (tRNA) are indicated.

The URF 25 Sequence

The URF 25 sequence is common to mtDNAs of normal (fertile) and sterile maize cytoplasms (*cms-C* and *cms-S*) (Susan Stamper, personal communication). It is also found in the tobacco (*N. tabacum*) mitochondrial genome (Molly Bland, personal communication). This finding suggests that URF 25 is a mitochondrial gene common to higher plants. A comparison of the URF 25 sequence from *cms-T* and normal maize shows nucleotide differences that would cause nine changes in their predicted amino acid sequences. In spite of these amino acid variations, we do not believe that URF 25 plays a direct role in pollen abortion in *cms-T*. This view is based largely on our analysis of URF 13 rather than an understanding of URF 25.

The nucleotide sequence of URF 25 predicts that it could encode a polypeptide of about 25,000 Mr. The hydropathy profile of the predicted URF 25 polypeptide indicates the protein is hydrophilic in nature and thus is unlikely to be associated with mitochondrial membranes. Although URF 25 is transcribed, a translational product has not yet been demonstrated. Perhaps the most interesting feature of URF 25 is that it is partially composed of chloroplast DNA. In maize, the last 14 codons and the termination codon are contributed by the chloroplast-derived DNA, which consists of a chloroplast tRNAarg gene with 38bp of its 5' flanking sequence (7). Thus it appears that the integrated chloroplast DNA has become a part of a mitochondrial gene. Since the tobacco URF 25 sequence does not contain the chloroplast tRNA gene, this association of chloroplast DNA with URF 25 may be limited to mitochondrial genomes of maize and its relatives.

The URF 13 Sequence

The URF 13 reading frame has been detected only in the mitochondrial genome of *cms-T* plants (7); it is not observed in the mitochondria of other maize cytoplasms (sterile or fertile) or in other plant species. Because of the peculiar rearrangements involved in its origin, it is not surprising to find that the sequence is novel. Abundant levels of URF 13 steady-state transcript are observed in *cms-T* mitochondria. Northern blot analysis has identified major transcripts of 3900, 2000, 1800 and 1500 nucleotides. The largest transcript, 3900 nucleotides, is presumed to be the primary transcript. It is not clear which RNA species is the mature messenger. Primer extension studies have shown that the

2000 and 1800 nucleotide RNAs contain the complete coding region of URF 13 and could be URF 13 messengers. The 1500 nucleotide RNA species could not be the messenger because it lacks some of the coding sequence on the 5' end of URF 13.

The URF 13 sequence encodes a 13,000 Mr polypeptide (115 amino acids). This has been demonstrated with an antiserum prepared against a chemically synthesized oligopeptide (15 amino acids) based upon the deduced amino acid sequence of URF 13 (manuscript in preparation). This particular oligopeptide sequence occurs in a hydrophilic coding region near the C-terminus of URF 13. Western blot analysis has shown that the 13,000 Mr polypeptide is abundantly synthesized in the mitochondria of cms-T; it is not detected in maize mitochondria from normal (fertile) or the sterile cytoplasms, cms-S and cms-C. The 13,000 Mr protein has been found in every organ type of cms-T examined; this includes etiolated shoots, roots, immature ears and leaves. We have also confirmed that the 13,000 Mr polypeptide is mitochondrially encoded. This has been shown by the immunoprecipitation of a labeled 13,000 Mr protein from the translational products of an *in vitro* protein synthesis with mitochondria isolated from cms-T. Similar experiments with normal (fertile) mitochondria do not detect the 13 kd polypeptide. Leaver and co-workers, using the *in vitro* protein synthesizing system with isolated mitochondria, have reported a 13 kd protein that is produced in mitochondria from cms-T but not in normal (fertile) (8). We have concluded, based on the similarity in size and other characteristics, that their 13 kd protein is the same one encoded by the URF 13 gene.

We have localized the 13 kd polypeptide to the mitochondrial membranes by Western blot analysis with the URF 13-specific antiserum. This has been shown by comparing soluble and membrane fractions of cms-T mitochondria. The membrane fraction contains large amounts of the 13 kd protein, whereas the soluble fraction contains only a trace amount, which is probably due to cross-contamination between fractions.

Based upon the above findings, we have chosen the symbol, URF13-T, to designate the mitochondrial gene encoding the 13,000 Mr polypeptide of cms-T.

Restorer Genes

Pollen abortion in cms-T is suppressed by two nuclear restorer genes, *Rf1* and *Rf2* (9). Leaver's group has demonstrated that the 13,000 Mr mitochondrial protein, which is abundant in cms-T male-sterile plants, is drastically reduced

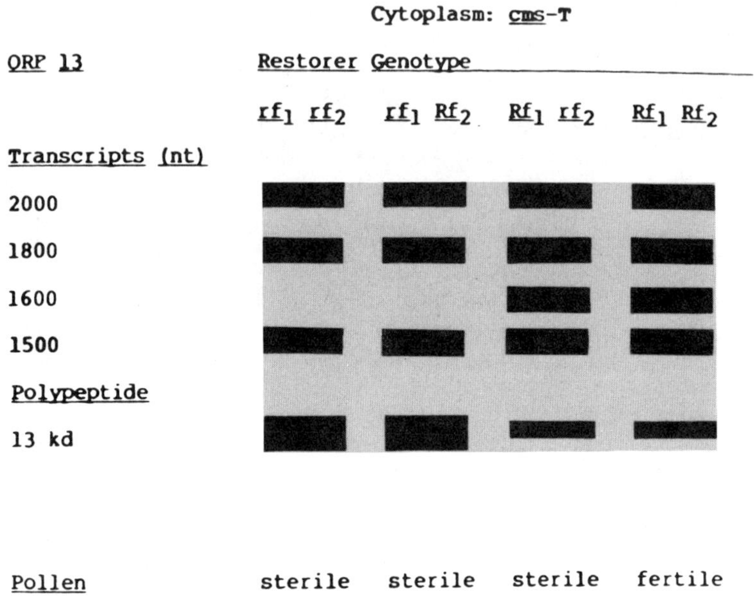

FIGURE 2. Diagrammatic summary of Northern and Western blot analyses of mitochondria from cms-T plants with different combinations of nuclear restorer genes. The 3900 nucleotide transcript and a few minor transcripts are not depicted.

in cms-T restored (pollen fertile) plants (10). We have studied the effect of the various combinations of restorer genes on the expression of the URF13-T gene (Fig. 2). These investigations indicate that the Rf1 gene modifies the expression of the URF13-T gene and its gene product. In contrast, the Rf2 locus does not appear to effect the expression of URF13-T (manuscript in preparation).

In cms-T plants with the nuclear genotypes rf1 rf1, rf2 rf2 and rf1 rf1, Rf2 ___, four major URF13-T transcripts (steady-state) are observed, 3900, 2000, 1800 and 1500 nucleotides. In cms-T plants with the genotypes, Rf1 ___, rf2 rf2 and Rf1 ___, Rf ___, five major URF13-T transcripts are seen. They differ in that a novel 1600 nucleotide transcript is observed in those genotypes carrying the dominant allele of Rf1; this RNA species is absent in the mitochondria of cms-T plants homozygous recessive for rf1. These results indicate that the Rf1 allele is able to modify the expression of URF13-T in cms-T, whereas the Rf2 allele does not effect URF13-T expression. It is not yet clear how Rf1 alters the expression of URF13-T. We have compared the DNA sequence of URF13-T in restored and nonrestored genotypes; these comparisons show no changes in the nucleotide sequences. Thus, restorer gene activity does not appear to be due to gene sequence modification. Changes in the URF13-T transcripts could be due to additional RNA processing or transcriptional initiation events. We have no evidence supporting either possibility.

By Western blot analysis we have investigated the effect of various combinations of restorer genes on the expression of the URF13-T polypeptide. The 13 kd mitochondrial protein is abundant in cms-T plants carrying the recessive alleles of Rf1, i.e. rf1 rf1, rf2 rf2 and rf1 rf1, Rf2 ___. The abundance of the protein is substantially reduced in cms-T plants with the dominant allele of Rf1, i.e. Rf1 ___, rf2 rf2 and Rf1 ___, Rf2, ___ (Fig. 2). Clearly, the dominant allele of Rf1 is responsible for the reduced levels of the 13,000 Mr protein. Even though the mechanism remains unclear, it is evident that the effect of the Rf1 gene on the expression of the URF13-T gene is related to the abundance of the 13,000 Mr mitochondrial protein.

Tissue Culture Experiments

Additional evidence from tissue culture experiments has also implicated the URF13-T gene with the pollen sterility trait of cms-T. When embryonic callus from cms-T plants are selected for resistance to the T toxin and are regenerated to whole plants, revertant types are found that are pollen fertile and resistant to the H. maydis race T pathogen and its toxin (11). Selection for toxin resistance is not absolutely required for the reversion event since passage through tissue culture and regeneration to whole plant without toxin selection has also yielded fertile and disease resistant revertants (12). Correlated with the reversion phenomena is a change in the mtDNA of revertants. Cms-T mtDNA characteristically yields a 6.6 kb fragment when it is digested with the endonuclease, Xho 1. After reversion to disease resistance and pollen fertility, the 6.6 kb Xho 1 mtDNA fragment is absent in most instances (13). This observation is relevant because the URF13-T gene is located in the 6.6 kb Xho 1 mtDNA fragment (7). Several laboratories are now studying the effect of reversion on the _URF13-T_ gene.

EPILOGUE

We have characterized the _URF13-T_ gene, its gene products, its restorer genes, and its relationship to pollen sterility in _cms-T_ (see summary in Table 2). Although there is compelling evidence that _URF13-T_ is the gene responsible for the male sterility and disease susceptibility traits, it remains unclear how the 13,000 Mr polypeptide causes these traits. Collectively, the data suggest that the lesion occurs in the inner mitochondrial membrane and perhaps interferes with oxidative phosphorylation or mitochondrial biogenesis during pollen development. We have learned that the restorer gene, _Rf1_, affects the expression of _URF13-T_, however, we do not understand the role _Rf2_ plays in suppressing pollen abortion.

TABLE 2
CHARACTERISTICS OF THE _URF13-T_ GENE, ITS GENE PRODUCT AND ITS INTERACTION WITH RESTORER GENES

Characteristics
1. Uniquely coded by the mtDNA of _cms-T_ maize
2. Arose by recombinational events
3. Abundantly and uniquely expressed in _cms-T_
4. Encodes a 13,000 Mr mitochondrial membrane polypeptide (115 amino acids)
5. _URF13-T_ transcripts are modified by the restorer gene (_Rf1_)
6. Abundance of _URF13-T_ polypeptide is reduced by the restorer gene (_Rf1_)
7. _URF13-T_ expression may not be effected by the restorer gene (_Rf2_).

ACKNOWLEDGMENTS

We wish to thank Ms. Carol Griffin and Ms. Jane Suddith for outstanding technical assistance and Ms. Amy Hamrick for manuscript preparation.

REFERENCES

1. Laser KD, Lersten NR (1972). Anatomy and cytology of microsporogenesis in cytoplasmic male sterile angiosperms. Bot Rev 38:425.
2. Laughnan JR, Gabay-Laughnan S (1983). Cytoplasmic male sterility in maize. Ann Rev Genet 17:27.
3. Levings CS III (1983). Cytoplasmic male sterility. In Kosuge T, Meredith CP, Hollaender A (eds): "Genetic Engineering of Plants," New York:Plenum, p 81.
4. Hooker AL, Smith DR, Lim SM, Beckett JB (1970). Reaction of corn seedlings with male-sterile cytoplasm to *Helminthosporium maydis*. Plant Dis Rep 54:708.
5. Payne G, Kono Y, Daly JM (1980). A comparison of purified host specific toxin from *Helminthosporium maydis*, race T, and its acetate derivative on oxidation by mitochondria from susceptible and resistant plants. Plant Physiol 65:785.
6. Holden MJ, Sze H (1984). *Helminthosporium maydis* T toxin increased membrane permeability to Ca^{2+} in susceptible corn mitochondria. Plant Physiol 75:235.
7. Dewey RE, Levings CS III, Timothy DH (1986). Novel recombinations in the maize mitochondrial genome produce a unique transcriptional unit in the Texas male-sterile cytoplasm. Cell 44:439.
8. Forde BG, Oliver RJC, Leaver CJ (1978). Variation in mitochondrial translation products associated with male-sterile cytoplasms in maize. Proc Natl Acad Sci USA 75:3841.
9. Duvick DN (1965). Cytoplasmic pollen sterility in corn. Adv Genet 13:1.
10. Forde BG, Leaver CJ (1980). Nuclear and cytoplasmic genes controlling synthesis of varient mitochondrial polypeptides in male-sterile maize. Proc Natl Acad Sci USA 77:418.
11. Gengenbach BG, Green CE, Donovan CM (1977). Inheritance of selected pathotoxin resistance in maize plants regenerated from cell cultures. Proc Natl Acad Sci USA 74:5113.

12. Brettell RIS, Goddard BVD, Ingram DS (1979). Selection of Tms-cytoplasm maize tissue cultures resistant to _Drechslera maydis_ T toxin. Maydica 24:203.

13. Umbeck PF, Gengenbach BG (1983). Reversion of male-sterile T-cytoplasm to male fertility in tissue culture. Crop Sci 23:584.

Plant Gene Systems and Their Biology, pages 75–83

APOMIXIS IN PLANT IMPROVEMENT[1]

Wayne W. Hanna[2]

USDA-ARS, Coastal Plain Experiment Station,
Tifton, GA 31793

ABSTRACT Apomixis is a genetically controlled method of reproduction in plants where an embryo is formed without union of an egg and a sperm. There are three basic types of apomictic reproduction: 1) apospory - embryo develops from a chromosomally unreduced egg in an embryo sac derived from the nucellus, 2) diplospory - embryo develops from an unreduced egg in an embryo sac derived from the megaspore mother cell, and 3) adventitious embryony - embryo develops directly from a somatic cell. In most forms of apomixis, pseudogamy or fertilization of the polar nuclei to produce endosperm is necessary for seed viability. These types of apomixis have economic potential because they can cause any genotype regardless of how heterozygous to breed true. In addition to fixing hybrid vigor, apomixis can make possible commercial hybrid production in crops where efficient male sterility or fertility restoration systems for producing hybrids are not known or developed. Apomicts are usually highly heterozygous because of their asexual reproduction and release large amounts of genetic variability when crossed as males with sexual plants. The apomictic mechanism can be: 1) obligate - reproduce only by apomixis, or 2) facultative - a plant can reproduce by apomixis and/or sexually. Obligate apomixis is the most desirable in commercially economical crops especially if cross-compatible sexual

[1]This work was supported in part by the U.S. Dept. of Energy Contract no. DE-AS09-76-EV00637 and Competitive Research Grant no. 81-CRCR-1-1002.

[2]Present address: Department of Agronomy, Coastal Plain Experiment Station, Tifton, GA 31793.

genotypes, needed to produce hybrids, are available in the species. Although facultative apomixis has been reported in cereals such as sorghum and pearl millet, genes controlling obligate apomixis are mainly found in distantly related wild species. Therefore, novel and unique procedures are needed to transfer gene(s) controlling obligate apomixis to cultivated species. Apomixis has been reported to be mainly controlled by one or a few dominant or recessive gene(s) in most species which improves the possibility of gene transfer from the wild to cultivated species. More success can be expected when working with apomixis if the species or genotype with the greatest number of ideal characteristics is selected. These include: simple, dominant inheritance; strong expression; environmentally stable; effects good seed set; obligate reproduction and some male fertility. The transfer, development, and use of gene(s) controlling apomixis in important world food crops provides a unique opportunity to researchers for producing superior plants in the future. The challenge can be met by combining plant breeding, cytogenetic, cell culture, and molecular techniques.

INTRODUCTION

Apomixis is a unique reproductive behavior through the seed that allows a chromosomally unreduced egg cell to develop into a zygote without fertilization by a sperm. Fertilization of the polar nuclei (pseudogamy) is needed in many species for endosperm development in the seed. This method allows a plant to reproduce identical clones of itself through the seed without union of male and female gametes. It is a method of reproduction found mainly in the wild relatives of our cultivated crops. It could be used to cause any genotype, regardless of heterozygosity, to breed true through the seed.

APOMICTIC MECHANISMS

In sexual reproduction of most grasses, a megaspore mother cell arising from the hypodermal layer of the ovule enlarges and goes through meiosis and two cell divisions to form a linear tetrad of megaspores each with a haploid chromosome number. The three micropylar spores degenerate while the functional chalazal spore enlarges to form an embryo sac with an egg, two polar nuclei, two synergids, and three antipodals.

The three major apomictic mechanisms of potential economic importance, apospory, diplospory, and adventitious embryony deviate from the sexual process. Classification of each mechanism is based on the origin and development of the cell from which the embryo develops.

In apospory, a megaspore mother cell may begin enlarging and even produce reduced megaspores but this sexual tissue usually degenerates before embryo sac development. Instead, somatic cells of the nucellus enlarge and the nuclei of these cells go through mitotic divisions to form one to many embryo sacs per ovule each with one to eight chromosomally unreduced nuclei.

The main difference in diplospory compared to sexual development is that only two megaspores are produced and the functional megaspore has the somatic chromosome number which results in an embryo sac similar in appearance to a sexual embryo sac but with an egg containing the somatic chromosome number.

In adventitious embryony, embryos develop directly from somatic cells of the ovule without formation of embryo sacs. Sexual sacs which are needed for endosperm formation may also form in the same ovule.

Although apomixis can be detected by various test-crosses and progeny tests, cytological analyses are necessary for identifying the apomictic mechanism. More detailed descriptions of the apomictic mechanism have been previously presented (1, 2, 3).

ADVANTAGES OF APOMIXIS

The most obvious advantage of obligate apomixis is that it can be used to produce true-breeding hybrids which would simplify the production and increase of commercial F_1 seed. A single superior obligate apomictic F_1 plant produced by hand emasculation or any other method could be rapidly increased for commercial production through open-pollination. It would eliminate the need: 1) for isolation to produce the hybrids and increase the female lines, 2) for increasing and maintaining large quantities of the male sterile, maintainer, and restorer lines, and 3) for planting large acreages of both male and female parents of a hybrid to produce the commercial hybrid when hybrid seed is only harvested from the female parent.

Apomixis would increase the opportunity for producing superior gene combinations. Any obligate apomictic genotype, regardless of heterozygosity, would breed true. New true-breeding genotypes could be rapidly developed by crossing sexual genotypes (as female) with apomictic genotypes. Unique genetic characters could be readily fixed in apomictic genotypes. Apomixis would broaden the gene pool and lessen the genetic vulnerability of commercial hybrids because one would not be limited to females with the male sterility inducing system to produce commercial hybrids. New genotypes could be rapidly produced for testing. Progeny testing for genotype stability would be eliminated. It would make possible efficient hybrid production in crops where good cytoplasmic male sterile and/or restorer systems are not available.

Apomixis was once considered an evolutionary dead end because sexual recombination on the female side is eliminated. However, it is now becoming recognized as a powerful genetic mechanism as more information is known about its genetics and manipulation. It can have an important role in genome preservation, polyploid buildup, evolution, and speciation (4).

Apomixis is effectively used in *Citrus* to produce uniform and disease- and virus-free rootstock (5) and in buffelgrass (6) and *Poa* (7) to produce improved cultivars.

HYBRID SEED PRODUCTION

Most cultivated crops reproduce sexually where chromosomally reduced (n) female (egg) and male (sperm from pollen) gametes unite to form a zygote (2n). In some sexual plants, fertilization of the polar nuclei (double fertilization) by another sperm nucleus is needed for endosperm formation and seed viability.

Commercial F_1 hybrid seed for most field crops is produced by alternating rows of a sexual cytoplasmic-genic male sterile (cms) with rows of a sexual male fertile (pollen shedding) pollinator. All seed produced on the cms is F_1 hybrid seed that is sold to the consumer. This method, repeated each year, is used to produce commercial hybrids for such crops as maize, sorghum, pearl millet, wheat, rice and onion. In some crops such as vegetables, genetic male steriles, self-incompatibility, and gametocides are used instead of cms in the female line to facilitate hybrid production (8). These methods of hybrid production require isolation from other fields of the same species, large acreages, and maintenance of large seed quantities of male and female lines to remake the hybrid each year. Hand emasculation of the female parents in high-value vegetable, ornamental, and fruit crops is also used to produce hybrids. This latter method becomes less expensive if the crop is perennial and/or can be vegetatively propagated through leaf or stem cuttings, budding, grafting or other vegetative propagation methods. Apomixis would greatly simplify these methods.

Hybrid vigor is maximized in the F_1 hybrid generation. Plants established from seed produced on sexual F_1 plants (F_2 generation) usually lose a third or more of the vigor or productivity of the F_1 hybrid. The vigor continues to decrease in succeeding generations.

An obligately apomictic F_1 hybrid would breed true through the seed indefinitely. It could be considered a vegetative or cloning method of reproduction through the seed.

GENETICS OF APOMIXIS

Inheritance studies on apomixis is limited and linkage information is non-existent. One reason for this is that genes controlling apomixis are usually found in the wild species where little genetic research has been conducted. Another reason is that precise genetic studies are difficult to conduct with apomictic species, especially if apomixis is controlled by dominant gene(s). A plant heterozygous for obligate apomictic method of reproduction controlled by dominant gene action will always breed true.

Apomixis has been shown to be controlled by different mechanisms both within and between species and genera. It has usually been shown to be controlled by one or a few genes that can be either dominant or recessive (1, 2, 9, 10, 11). The inheritance may be more complex in species where expression of apomixis is also controlled by modifying genes and affected by genotype interactions.

High levels of apomictic reproduction (>99%) is the most desirable but facultative types with varying degrees of apomixis occurs both within and between species and genotypes.

SOURCES OF APOMIXIS

Apomixis has been found in over 300 species of plants in more than 30 families (1, 12, 13, 14). It can occur in the cultivated or wild species. In the major grain crop genera, apomictic types are usually found in the wild species. Examples are: a) *Tripsacum dactyloides*, a relative of maize (15); *Elymus scabrus*, a relative of wheat (16); and many wild *Pennisetum* species related to pearl millet (17, 18). Apomixis exists in many of the tropical forage grass genera such as *Eragrostis* (19), *Panicum* (20), *Paspalum* (21), and *Cenchrus* (1) and in the wild relatives of beet and strawberry (2). Facultative apomictic types have been found in sorghum (22, 23) and pearl millet (24). Many fruit and nut plants also reproduce by apomixis.

GENE TRANSFER FROM WILD TO CULTIVATED SPECIES

It would be ideal to find genes controlling obligate or a high level apomixis in the cultivated species and be able to readily hybridize cross-compatible sexual x apomictic genotypes to produce true-breeding F_1 hybrids. In reality, most desirable genes controlling apomixis are found in the wild species which are distantly related to the cultivated species. Although interspecific crosses may be possible between the cultivated and wild species, chromosome pairing between genomes is usually low or non-existent. Therefore, methods other than traditional gene recombination are needed to transfer genes for apomixis from chromosomes of the wild species to those of the cultivated species.

In maize x *Tripsacum dactyloides* crosses and backcrosses, most progenies were either sexual or facultative apomicts with a low level of apomixis (25). Similar results were obtained from pearl millet - *P. orientale* crosses and back-crosses (18). However, crosses between pearl millet and apomictic *P. squamulatum* produced some F_1 hybrids and BC_1 progenies that reproduced by obligate apomixis (26, 27) indicating that different genes or genetic systems may control apomictic reproduction in the same genus. Male fertility which is necessary for transferring a dominant gene controlling apomixis at each generation was low in most of the BC_1 progenies. Apomixis in *P. squamulatum* appears to be controlled by one or a few dominant genes.

Improved male fertility along with obligate apomixis was obtained by producing trispecific pearl millet - *P. squamulatum* x pearl millet - *P. purpureum* hybrids (10). The male fertile apomictic trispecific hybrids backcrossed again to tetraploid (2n = 4x = 28) pearl millet resulted in some male fertile obligate apomictic BC_1 plants. Tetraploid pearl millet pollinated with BC_1 apomictic plants produced 30 - 32 chromosome sexual and obligate apomictic plants. Some BC_2 plants were partially male fertile. These BC_2 plants combine about 28 pearl millet chromosomes and two - four *P. squamulatum* chromosomes.

At this stage of backcrossing, transfer of genes controlling apomixis from *P. squamulatum* to pearl millet may be accomplished by: A) gene recombination due to residual homology, B) induced or natural translocations, C) Chromosomal substitution - with tetraploid pearl millet but probably not at the diploid level, or D) mobile genetic factors. In cooperation with Dr. Peggy Ozias-Akins at Tifton, Georgia, we are using cell culture techniques to enhance the gene transfer process. Even if the above procedures are successfully used to transfer genes controlling apomixis to pearl millet, molecular methods would probably be needed to transfer the gene(s) controlling apomixis from *P. squamulatum* to other genera and possibly families of plants. Molecular studies are being initiated in 1987 on this problem.

REFERENCES

1. Bashaw EC (1980). Apomixis and its application in crop improvement. In Fehr WR, Hadley HH (eds): "Hybridization of Crop Plants", American Society of Agronomy and Crop Science of America, Madison, WI, pp 45-63.
2. Nogler GA (1984). Gametophytic apomixis. In Johri BM (ed): "Embryology of Angiosperms", Springer-Verlag, New York, pp 475-518.
3. Gustafsson A (1946). "Apomixis in Higher Plants. Part I. The Mechanism of Apomixis". Lunds Universitets Arsskrift, N.F. Avd. 2, 42:1-66.
4. Bashaw EC (1975). Problems and possibilities of apomixis in the improvement of tropical forage grasses. In Doll EC, Mott GO (eds): "Tropical Forages in Livestock Production Systems", American Society of Agronomy Publication No. 24, pp 23-30.

5. Parlevliet JE, Cameron JW (1959). Evidence on the inheritance of nucellar embryony in Citrus. Proc. Am Soc Hort Sci 74:252-260.
6. Bashaw EC (1980). Registration of Nueces and Llano buffelgrass. Crop Sci 20:112.
7. Pepin GW, Funk CR (1971). Interspecific hybridization as a method of breeding Kentucky bluegrass (*Poa pratensis* L.) for turf. Crop Sci 11:445-448.
8. Driscoll CJ (1986). Nuclear male sterility systems in seed production of hybrid varieties. CRC Critical Review in Plant Sci 3:227-256.
9. Asker S (1979). Progress in apomixis research. Hereditas 91:231-240.
10. Dujardin M, Hanna W (1984). Cytogenetics of double cross hybrids between *Pennisetum americanum* - *P. purpureum* amphiploids and *P. americanum* x *Pennisetum squamulatum* interspecific hybrids. Theor and Appl Gen 69:97-100.
11. Hanna WW, Powell JB, Millot JC, Burton GW (1973). Cytology of obligate sexual plants in *Panicum maximum* Jacq. and their use in controlled hybrids. Crop Sci 13:695-697.
12. Marshall DR, Brown AHD (1981). The evolution of apomixis. Heredity 47:1-15.
13. Nygren A (1954). Apomixis in the Angiosperms. Bot Rev 20:577-649.
14. Davis GL (1966). "Systematic embryology of the Angiosperms." John Wiley and Sons, New York, 529 pp.
15. Farguharson LI (1955). Apomixis and polyembryony in *Tripsacum dactyloides*. Am Jour Bot 42:737-743.
16. Hair JB (1956). Subsexual reproduction in *Agropyron*. Heredity 10:143-179.
17. Dujardin M, Hanna W (1984). Microsporogenesis, reproductive behavior, and fertility in five *Pennisetum* species. Theor and Appl Gen 67:197-201.
18. Hanna WW, Dujardin M (1986). Interspecific transfer of apomixis in *Pennisetum*. Proc XV Intern'l Grassl Congr, pp 249-250.
19. Voigt PW, Bashaw EC (1976). Facultative apomixis in *Eragrostis curvula*. Crop Sci 16:803-806.
20. Warmke HE (1954). Apomixis in *Panicum maximum*. Am Jour Bot 41:5-11.
21. Burson BL (1975). Cytology of some apomictic *Paspalum* species. Crop Sci 15:229-232.
22. Hanna WW, Schertz KF, Bashaw EC (1970). Apospory in *Sorghum bicolor* (L.) Moench. Sci 170:338-339.

23. Tang CY, Schertz KF, Bashaw EC (1980). Apomixis in *Sorghum* lines and their F_1 progenies. Bot Gazette 141:294-299.
24. Hanna WW, Powell JB (1973). Stubby head, an induced facultative apomict in pearl millet. Crop Sci 13:726-728.
25. DeWet JMJ, Harlan JR, Engle LM, Grant CA (1973). Breeding behavior of Maize - *Tripsacum* hybrids. Crop Sci 13:254-256.
26. Dujardin M, Hanna WW (1983). Apomictic and sexual pearl millet x *Pennisetum squamulatum* hybrids. Jour Hered 74:277-279.
27. Dujardin M, Hanna WW (1985). Cytology and reproduction of reciprocal backcrosses between pearl millet and sexual and apomictic hybrids of pearl millet x *Pennisetum squamulatum*. Crop Sci 25:55-62.

III. DEVELOPMENTAL AND STRESS REGULATED GENE EXPRESSION

Plant Gene Systems and Their Biology, pages 87–97

HEAT SHOCK GENE FAMILIES OF SOYBEAN AND THE REGULATION OF THEIR EXPRESSION [1]

Joe L. Key, Janice Kimpel* and Ronald T. Nagao

Department of Botany, University of Georgia
Athens, Georgia 30602; *Agrigenetics Advanced Science Co.,
Madison, WI 53716

ABSTRACT When an organism is shifted from its normal growing temperature to some higher temperature (usually an increase of about 10 ± 4°C), most cells/tissues of the organism undergo a stress response referred to as heat shock (HS). The physiological and biochemical responses elicited by HS are numerous. Fundamental to this response, however, are changes in gene expression; the transcription of most genes expressed at the normal growing temperature and the translation of most of the related mRNAs slow dramatically or cease. The expression of several families of genes (HS genes) is rapidly turned on which were either not expressed at a detectable level before HS or which were enhanced several-fold in their expression. These HS mRNAs accumulate to high levels and are preferentially translated. With continuous HS (for example, at 40°C with soybean) transcription of the HS genes slows or ceases after 1 to 2 hr and HS protein accumulation slows after several hours. The slowing or shut-down of synthesis of HS mRNAs and proteins occurs as a result of mechanisms which may be referred to as autoregulation, apparently as a result of the accumulation of functional HS proteins. The expression of HS genes and the production of HS proteins at a permissive HS temperature correlates with the acquisition of thermotolerance, i.e., tolerance to an otherwise lethal temperature. The HS response may be elicited 1) by a rapid shift from normal to the optimum permissive HS temperature with maintenance at that temperature, 2) by a gradual increase in temperature (e.g. 3° per hr) and 3) by a brief (3 to 10 min) exposure to a lethal (if given for 1 to 2 hr) HS temperature.

---

[1]This work was supported by U.S. Department of Energy Contract DE-FG09-86ER13602.

INTRODUCTION

The heat shock (HS) response is one of the most conserved phenomena known in biology, occurring in a diverse and inclusive range of prokaryotic and eukaryotic organisms. From research that has been done on many different organisms, the HS response can be characterized by several parameters: 1) a fast switch from transcription of normal mRNAs to HS mRNAs, 2) a rapid switch from the translation of normal mRNAs to the translation of HS mRNAs, 3) the accumulation of HS mRNAs for 1 to 2 hours, in some cases to over 20,000 copies per cell for the more abundant HS mRNAs, 4) the accumulation of HS proteins to very high levels over a few hours, the majority of which are stable for many hours at both HS and non-HS temperatures, 5) the acquisition of thermotolerance to otherwise lethal temperatures, which is strongly correlated with the accumulation of HS proteins, 6) the selective localization of HS proteins only at HS temperatures, which is reversed when the temperature returns to normal, and 7) a gradual return to normal protein synthesis, with a concomitant decline in HS protein synthesis, during prolonged heat treatments.

In the several plant species that have been investigated in this laboratory and others (see Mansfield and Key, 1987), a diverse range of low molecular weight (15 to 25 kD) HS proteins are synthesized (Table 1). A few of these HS proteins

TABLE 1
LOW MOLECULAR WEIGHT HSP
SYNTHESIZED BY VARIOUS PLANT SPECIES [a]

Species	Number of Induced Polypeptides	Number of Enhanced Polypeptides	Total Number of HSP
Soybean	21	6	27
Pea	13	9	22
Sunflower	19	4	23
Wheat	11	1	12
Rice	13	2	15
Maize	23	1	24
Millet	16	1	17
P. miliaceum	21	2	23

[a]Values represent the number of polypeptides characterized as low molecular weight hsp of M_r 15 to 25 kD. Induced polypeptides are detectable only after hs. Enhanced polypeptides are detectable in control tissue but exhibit increased stainability and radiolabel incorporation after hs.

are detectable by silver staining at normal temperatures in 2 day old soybean seedlings, and they are induced to accumulate to higher levels by HS (Figure 1).

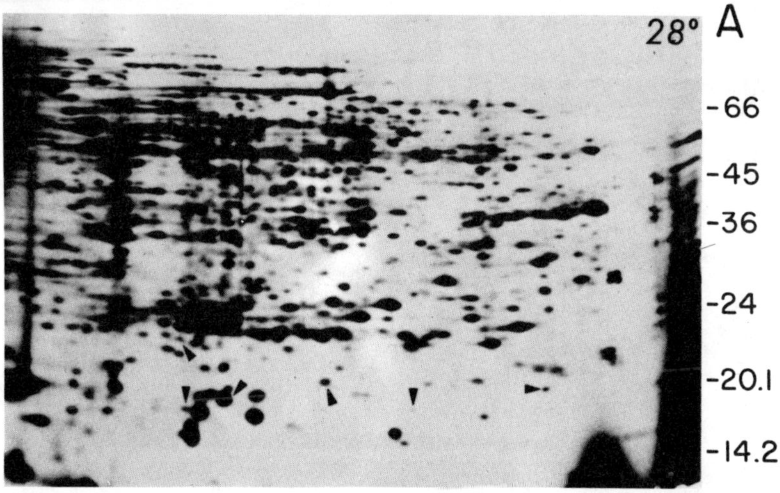

Figure 1A. Two-dimensional analysis of newly synthesized proteins in control and heat-shocked soybean seedlings. Two day-old soybean seedlings (without cotyledons) were labeled with ^{3}H-leucine for 3 hr in shaker culture at 28°C. Total seedling protein was extracted as described in Key et al. (1981) and analyzed by electrophoresis in two dimensions according to O'Farrell (1975). Protein was visualized by silver staining (1A and 1B). Molecular weights (kD) of protein standards are indicated.

The majority of the HS proteins are not detectable prior to a HS, but many of these become major stainable proteins after 2 to 6 hours of HS. The synthesis of the abundant, low molecular weight group of HS proteins observed in plants contrast the profile of HS proteins in most other organisms where only a small number (e.g. four major ones in Drosophila) of low molecular weight HS proteins accumulate. On the other hand, the synthesis and two-dimensional electropherogram patterns of high molecular weight (68-110 kD) HS proteins is much more conserved between plants species and between plants and other groups of organisms; this conservation has allowed the use of DNA probes for the Drosophila HS proteins of 70 and 83 kD in isolating the analogous sequences in plant tissues (Vierling, Roberts and Key, unpublished). In Drosophila, yeast and mammalian tissues, the synthesis of the hsp70 protein(s) represents the large majority of total HS protein synthesis. However, the accumulation of these high molecular weight proteins is quite low in plants relative to the abundant accumulation in most other organisms. Rather, the major HS proteins that accumulate in plants are the 15 to 25 kD group. Thus, while many aspects of the HS response are highly conserved across a wide range of organisms, some

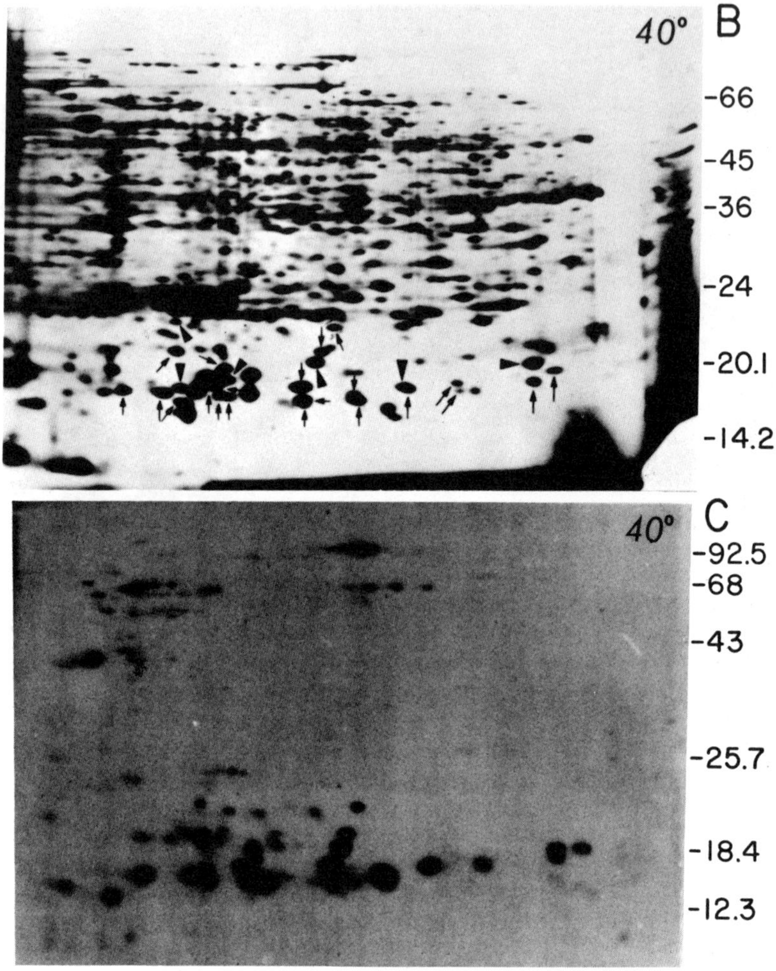

Figures 1B and 1C. Experimental conditions and procedures were the same as (A) except that the tissue was incubated for 3 hr at 40°C. Labeled proteins were visualized by fluorography (C). Arrows in B indicate low molecular weight HS proteins induced by HS. Arrowheads in A and B indicate low molecular weight HS proteins which are present in low amounts in control tissue.

aspects, such as the relative levels of subsets of HS proteins that accumulate, seem relatively unique to plants. Still, the plant HS proteins and their respective genes do have many properties in common with HS proteins and genes of other organisms.

Research results are accumulating on the transcription of HS genes in relation to both cis- and trans-acting factors in Drosophila and Xenopus (see Bienz and Pelham, 1986; Riddihough and Pelham, 1986; Pelham, 1985; Parker and Topol, 1984; Topol et al., 1985; Shuey and Parker, 1986; Wiederrecht et al., 1987). Important regulatory regions in the HS genes of plants are also being defined (Schöffl et al., 1984; Nagao et al., 1985; Czarnecka et al., 1985; Gurley et al., 1986; Gurley et al., this volume). For extensive review regarding these introductory remarks, see the reviews by Ashburner and Bonner, 1979; Bienz, 1985; Craig, 1985; Key et al., 1985; Kimpel and Key, 1985; Lindquist, 1986; Neidhardt et al., 1984; Nagao et al., 1986; Nover and Scharf, 1984; Pelham, 1985; and Schlesinger, 1986. The research presented in this paper is based on studies in higher plant species, with the bulk of the results derived from studies on soybean (Glycine max) seedlings.

PHYSIOLOGY OF THE HS RESPONSE IN SOYBEAN

In addition to research on the molecular biology of the HS genes in soybean, this laboratory has been studying the physiologically relevant parameters of the HS response in plants. In general, each plant species has a characteristic optimum or "inductive breakpoint" temperature at which HS protein synthesis becomes maximal. This temperature is about 5 to 10°C above the optimal growing temperature of the plants, tissues, or cultured cells. For 2 day old soybean seedlings, the breakpoint temperature is 40°C. If seedlings are incubated continuously at this temperature, they accumulate HS mRNAs for the first one to two hours. After two to three hours at 40°C, the steady-state levels of HS mRNAs (determined by Northern analysis, Figure 2) begin to gradually decline (Schöffl and Key, 1982; Nagao et al., 1986). These steady-state RNA levels reflect the rapid, transient activation of HS gene transcription which is turned off after 2 to 4 hours at 40°C (Kimpel and Key, unpublished). The accumulated HS mRNAs gradually decay, but they are still detectable after continuous treatment for 12 hours at 40°C. However, if these seedlings are transferred back to 30°C after 2 to 3 hours at 40°C, the HS mRNAs rapidly deplete and are essentially non-detectable after 3 to 4 hours. Two points may be inferred from these observations: 1) HS mRNAs are several-fold more stable at HS temperatures than at the normal growing temperature, and 2) there is additional regulation of these genes, beyond the direct response to ambient temperature that is responsible for limiting HS gene transcription during continued heat treatment. It has been suggested from other researchers that HS proteins themselves may be involved in this regulation. In E. coli, the inductive synthesis of HS mRNAs is controlled by a 32 kD sigma factor (the product of the htpR locus), the synthesis of which is increased during HS. The 32 kD sigma factor allows RNA polymerase to transcribe the HS genes, making it a positive regulator of the HS response (Neidhardt et al., 1984). In contrast, the dnaK protein of E. coli, a HS protein that is considered the bacterial counterpart of the hsp70 protein of Drosophila, yeast, etc., is a negative regulator of the HS response in bacteria. As the dnaK protein accumulates, it represses the HS

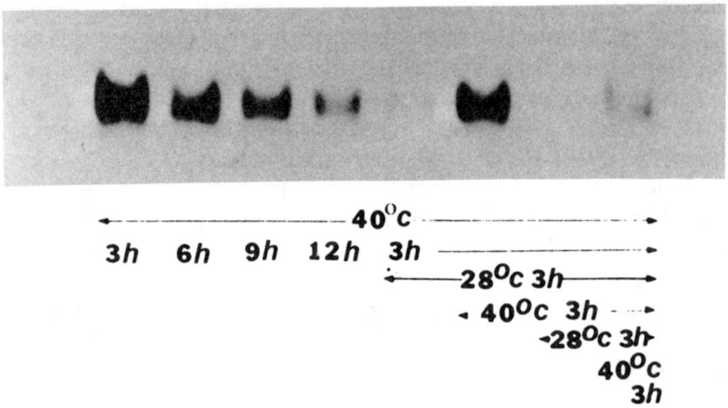

Figure 2. Northern hybridization analysis of HS mRNA after varying lengths of heat treatment. Poly(A)RNA was isolated from soybean seedlings incubated at 40°C for 3, 6, 9, 12 hr, 3 hr at 40°C/3 hr at 28°C, 3 hr at 40°C/3 hr at 28°C/3 hr at 40°C, 3 hr at 40°C/3 hr at 28°C/3 hr at 40°C/3 hr at 28°C, 3 hr at 40°C/3 hr at 28°C/3 hr at 40°C/3 hr at 28°C/3 hr at 40°C. One µg of poly(A)RNA was analyzed by Northern blotting and probed with ^{32}P-labeled HS cDNA clone pFS2005.

response, perhaps by mediating the degradation of the 32 kD sigma factor (Tilly et al., 1983). In Drosophila, the accumulation of high levels of the hsp70 protein also correlates with the repression of HS gene transcription and the decline and subsequent cessation of HS protein synthesis (DiDomenico, et al., 1982; Lindquist and DiDomenico, 1985). The incorporation of amino acid analogues (canavanine and azetidine-2-carboxylic acid) into HS proteins during HS is sufficient to prevent proper regulation of the response, suggesting that functional HS proteins are required (DiDomenico et al., 1982). Studies using amino acid analogues in plants to date have not been informative in this respect.

In addition to a "breakpoint" temperature for the induction of HS gene expression, all species have a characteristic temperature above which they simply cannot survive. This threshold is usually defined in a dosage-dependent way. For example, a 1 to 2 hour exposure of soybean seedlings to 45°C is lethal. That is, while HS gene transcription is induced, only very small amounts of HS proteins actually accumulate, presumably because this heat stress is too severe to allow sufficient protein synthesis to acquire thermotolerance (Lin et al., 1984). However, if the seedlings are exposed to a brief 5 to 10 minute exposure at 45°C and then transferred back to 30°C, they develop a complete HS response (Figure 3). That is, during the subsequent incubation at 30°C, HS mRNAs and HS proteins are synthesized and accumulate to allow this tissue to become thermotolerant. In contrast, a 5 to 10 minute treatment at 40°C with subsequent incubation at 30 causes only a limited and brief HS response. While the inductive temperature was perceived and HS mRNAs and HS protein synthesis

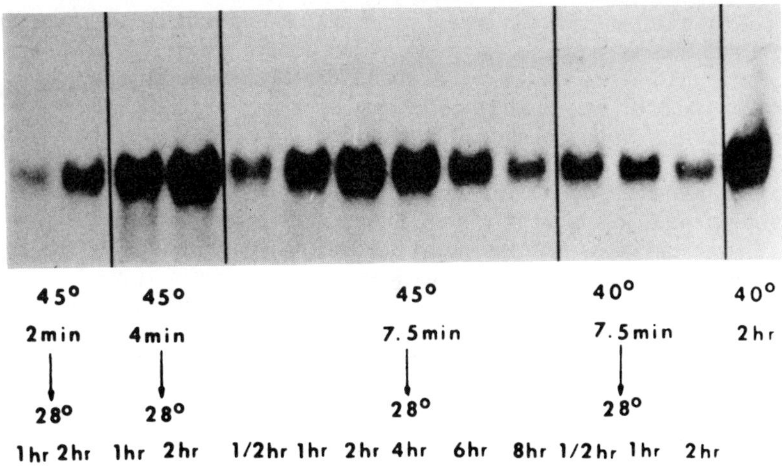

Figure 3. Northern blot analysis of HS mRNA induction after brief heat treatments. Soybean seedlings were incubated at 45°C for 2, 4, 7.5 min or 40°C for 7.5 min and then incubated at 28°C for the indicated times. Poly(A)RNA was isolated, and 1 μg was analyzed by Northern blot hybridization using ^{32}P-labeled pFS2033 as probe. For comparison 1 μg poly(A)RNA from seedlings incubated at 40°C for 2 hr was included (far right lane).

initiated at 40°C, when the tissue is transferred to 30°C, HS mRNAs rapidly deplete and significant levels of HS protein do not accumulate. This tissue does not become thermotolerant. As discussed below, these results suggest that the HS response is quantitatively regulated; that is, the level of induction is related to the severity of the stress. Additionally, these responses to a short burst of high heat that is followed by incubation at normal temperatures indicate the involvement of other regulatory control signals for the HS genes in addition to the sensing of the ambient temperature. While temperature is clearly a trigger for the induction of the HS genes, their continued expression and eventual repression must be under additional, regulatory controls. It is also noteworthy that HS mRNAs induced by a brief 45°C HS apparently are stable and accumulate at 30°C in contrast to the observations on HS mRNA stability under continuous 40°C HS versus return to 30°C after a 2 to 3 hr 40°C HS as noted above.

Soybean seedlings respond to a sequential series of HS treatments (cycles of 2 to 3 hours at 40°C followed by 4 hours at 30°C) by repetitive induction of the HS gene transcription (Nagao et al., 1986). The level of HS mRNAs that accumulate diminishes somewhat with each successive round of HS. Interestingly, the HS proteins themselves are generally stable for many hours following either a single HS or a series of HS treatments (Lin et al., 1984); analogous results have been reported for Chinese hamster HA-1 cells by Li and Laszlo (1985). Apparently, the presence of previously synthesized HS proteins in soybean tissue is not in itself sufficient to prevent the typical HS response. This may be reconciled with the previous discussion on the putative role of HS proteins in regulating the HS response by several possibilities: 1) the loss in autoregulatory function could be due to a change in state of one or more HS proteins that occurs during the intervening 30°C treatment, 2) a specific, regulatory HS protein(s) is much more short-lived than the bulk of the HS proteins, or 3) HS proteins themselves do not directly determine the regulation of the response. This latter possibility is unlikely, given the recent evidence from genetic mutants in *E. coli* and yeast that are altered in their regulatory controls of the HS response. In *E. coli*, mutants at the *dnaK* locus show prolonged synthesis of the HS proteins, indicating a loss of the negative (or feedback) control (Tilly et al., 1983). In yeast, deletion of two hsp70 genes results in the constitutive synthesis of other HS proteins (Craig et al., 1985). Attempts to identify HS regulatory mutants in *Arabidopsis* are underway in this laboratory.

As noted above, the steady-state level of HS mRNAs is highest on the first cycle of HS and diminishes with subsequent cycles. This has also been demonstrated with different types of HS treatments. An initial HS for 5 to 10 minutes at 45°C triggers a strong response (Figure 3); however, if such a short HS at 45°C occurs after a prior cycle of a 2 hour HS at 40°C followed by an intervening 4 hours at 30°C, the response is substantially reduced (Figure 4). If

1. 40°(2h)
2. 40°(2h) → 28°(4h)
3. 40°(2h) → 28°(4h) → 40°(2h)
4. 40°(2h) → 28°(4h) → 45°(2h)
5. 40°(2h) → 28°(4h) → 45°(7') → 28°(113')
6. 40°(2h) → 28°(4h) → 50°(7') → 28°(113')

Figure 4. Dot blot hybridization analysis of HS mRNA following heat shock, recovery, and re-heat shock at various temperatures and lengths of time. Poly(A)RNA was isolated from soybean seedlings incubated under the indicated regimes, blotted and hybridized with ^{32}P-labeled HS cDNA clone pCE53.

the short HS is done at 50°C instead of 45°C in the second cycle, the HS mRNAs accumulate to substantial levels over the next 2 hr. These results, coupled with several other permutations that have been investigated in our laboratory, suggest that the plant is able to measure the extent of the heat treatment both in terms of the duration and the severity, and the level of the response reflects these parameters. These results point out the complexity of the regulatory mechanisms which operate in the elicitation of a HS response. An understanding of these differential responses to sequential HS treatments and the intensity/severity of those treatments will contribute significantly to an explanation of the regulatory aspects of the HS response. Clearly, high temperature will trigger the response, but the regulation involves more than a simple sensing of a critical temperature by the cells or tissues.

It may be that the fundamental difference in response to a given temperature during a second cycle of HS relates to the induced state of thermotolerance. Thermotolerance is the widely predicted biological manifestation of the HS response. It is defined as the acquired, transient resistance of an organism to normally lethal temperatures by prior exposure to a moderate HS. There is now a substantial amount of circumstantial evidence that supports the hypothesis that HS proteins confer thermotolerance, although absolute proof is still lacking, and there are some conflicting reports in the literature (reviewed by Lindquist, 1986). Accepting the hypothesis that HS proteins are involved in thermotolerance, a reasonable corollary is that the level and/or rate of synthesis of these proteins would determine the degree of thermotolerance conferred. Furthermore, it is reasonable to expect that the degree of thermotolerance needed would be a function of the severity of the heat stress. The inference is that the level and/or rate of HS protein synthesis will determine the extent of development of thermotolerance. The observations described above on soybean seedlings are consistent with these ideas; strengthening of this hypothesis will come as methods for quantifying HS proteins and the level of thermotolerance are developed.

DISCUSSION

While much evidence has accumulated regarding a putative biological role for HS proteins in the acquisition of thermal tolerance, the molecular function of eukaryotic HS proteins remains unresolved. Pelham (1986) has recently suggested that the 70 kD HS protein may be involved in protein folding and/or assembly. During a heat shock (or other stress known to induce HS proteins), the 70 kD HS protein may participate in reassembly or protection of damaged, disrupted cellular machinery, a process presumed to be ATP-dependent (the families of eukaryotic 70 kD HS proteins bind ATP tightly, and at least some of them, in analogy to the dnaK protein of E. coli, mediate its hydrolysis). The high abundance of the 70 kD HS protein in many organisms is consistent with this hypothesis, since it might be expected that many cellular proteins may be damaged during heat stress. In plants, the 70 kD protein is not as abundant, but it may be reasonable to hypothesize that the 15 to 18 kD HS proteins, which are very abundant, contribute this function; the relatively low level of the 70 kD HS protein of plants might be sufficient to fulfill this putative role, however. Fractionation studies in plants indicate that these 15 to 18 kD HS proteins do 'aggregate' with several cellular structures (Nover and Scharf, 1984; Lin et al., 1984; Mansfield and Key, unpublished). Cytochemical localization of these proteins in plant tissues and biochemical studies on possible binding or enzymatic activities possessed by these proteins should yield valuable information regarding the function of these proteins in plants.

REFERENCES

1. Ashburner M, Bonner JJ (1979). The induction of gene activity in Drosophila by heat shock. Cell 17:241.
2. Bienz M (1985). Transient and develomental activation of heat-shock genes. Trends in Biochem Sci 10:157.
3. Bienz M, Pelham HRB (1986). Heat shock regulatory elements function as an inducible enhancer in the Xenopus hsp 70 gene and when linked to a heterologous promoter.
4. Craig EA (1985). The heat shock response. CRC Critical Reviews in Biochem 18:239.
5. Czarnecka E, Gurley WB, Nagao RT, Mosquera L, Key JL (1985). DNA sequence and transcript mapping of a soybean gene encoding a small heat shock protein. Proc Natl Acad Sci USA 82:3726.
6. DiDomenico BJ, Bugaisky GE, Lindquist S (1982). The heat shock response is self-regulated at both the transcriptional and posttranscriptional levels. Cell Cell 31:593.
7. Gurley WB, Czarnecka E, Nagao RT, Key JL (1986). Upstream sequences required for efficient expression of a soybean heat shock gene. Mol Cell Biol 6:559.
8. Key JL, Kimpel J, Vierling E, Lin CY, Nagao RT, Czarnecka E, Schöffl F (1985). Physiological and molecular analyses of the heat shock response in plants. In Atkinson BG, Walden DB (eds): "Changes in Eukaryotic Gene expression in Response to Environmental Stress", New York: Academic Press Inc, p 327.
9. Kimpel JA, Key JL (1985). Heat shock in plants. Trends in Biochem Sci 10:353.
10. Li GC, Laszlo A (1985). Thermotolerance in mammalian cells: A possible role for heat shock proteins. In Atkinson BG, Walden DB (eds): "Changes in Eukaryotic Gene Expression in Response to Environmental Stress" New York: Academic Press, p 227.
11. Lin CY, Roberts JK, Key JL (1984). Acquisition of thermotolerance in soybean seedlings: Synthesis and accumulation of heat shock proteins and their cellular localization. Plant Physiol 74:152.
12. Lindquist S (1986). The heat-shock response. Ann Rev Biochem 55:1151.
13. Lindquist S, DiDomenico BJ (1985). Coordinate and noncoordinate gene expression during heat shock: A model for regulation. In Atkinson BG, Walden DB (eds): "Changes in Eukaryotic Gene Expression in Response to Environmental Stress" New York: Academic Press, p 72.
14. Mansfield MA, Key JL (1987). Synthesis of the low molecular weight heat shock proteins in plants. (Submitted to Plant Phys).

15. Nagao RT, Czarnecka E, Gurley WB, Schöffl F, Key JL (1985). Genes for low molecular weight heat shock proteins of soybean: sequence analysis of a multigene family. Mol Cell Biol 5:3417.
16. Nagao RT, Kimpel JA, Vierling E, Key JL (1986). The heat shock response: a comparative analysis. In Miflin BJ (ed): "Oxford Surveys of Plant Molecular and Cell Biology," Oxford: Oxford Univ Press, p 384.
17. Neidhardt FC, VanBogelen RA, Vaughn V (1984). The genetics and regulation of heat-shock proteins. Ann Rev Genet 18:295.
18. Nover L, Scharf KD (1984). Synthesis, modification and structural binding of heat-shock proteins in tomato cell cultures. Eur J Biochem 139:303.
19. O'Farrell PH (1975). High resolution two-dimensional electrophoresis of proteins. J Biol Chem 250:4007.
20. Parker CS, Topol J (1984). A Drosophila RNA polymerase II transcription factor binds to the regulatory site of an hsp70 gene. Cell 37:273.
21. Pelham HRB (1985). Activation of heat-shock genes in eukaryotes. Trends in Gen Sci 1:31.
22. Pelham HRB (1986). Speculations on the functions of the major heat shock and glucose-regulated proteins. Cell 46:959.
23. Riddihough G, Pelham HRB (1986). Activation of the Drosophila hsp27 promoter by heat shock and by ecdysone involves independent and remote regulatory sequences. Embo J 5:1653.
24. Schlesinger MJ (1986). Heat shock proteins: the search for functions. J Cell Biol 103:321.
25. Schöffl F, Key JL (1982). An analysis of mRNAs for a group of heat shock proteins of soybean using cloned cDNAs. J Mol Appl Genet 1:301.
26. Schöffl F, Raschke E, Nagao RT (1984). The DNA sequence analysis of soybean heat-shock genes and identification of possible regulatory promoter elements. Embo J 3:2491.
27. Shuey DJ, Parker CS (1986). Binding of Drosophila heat-shock gene transcription factor to the hsp 70 promoter. J Biol Chem 261:7934.
28. Tilly KN, McKittrick N, Zylicz M, Georgopoulos C (1983). The dnaK protein modulates the heat shock response in Escherichia coli. Cell 34:641.
29. Topol J, Ruden DM, Parker CS (1985). Sequences required for in vitro transcriptional activation of a Drosophila hsp70 gene. Cell 42:527.
30. Wiederrecht G, Shuey DJ, Kibbe WA, Parker CS (1987). The Saccharomyces and Drosophila heat shock transcription factors are identical in size and DNA binding properties. Cell 48:507.

Plant Gene Systems and Their Biology, pages 99–108

CHARACTERIZATION OF CHLOROPLAST-LOCALIZED HEAT SHOCK PROTEINS

Elizabeth Vierling

Department of Biochemistry, University of Arizona, Tucson, Arizona, 85721.

ABSTRACT

During high temperature stress, or heat shock, specific nuclear-encoded heat shock proteins are transported into chloroplasts. The function of these proteins is unknown. We have isolated a cDNA encoding chloroplast HSP22 from soybean. DNA sequence analysis indicates there is amino acid homology in the carboxy terminal region between the soybean chloroplast HSP and low molecular weight cytoplasmic HSPs of soybean. These data suggest possible functional homologies as well. We have now used the soybean HSP22 cDNA to isolate homologous cDNAs from pea and petunia, and are using these clones to investigate further the structure and expression of these proteins. We hypothesize that chloroplast HSPs function in all types of plastids and may play a role similar to that of small HSPs found in other parts of the cell.

INTRODUCTION

The response of plants to high temperature stress, or heat shock (HS), exhibits many of the characteristics of the HS response in other organisms (1). When plant cells or tissues are shifted either rapidly or gradually to temperatures five or more degrees above normal growth temperatures, they respond by synthesizing a new set of proteins, the heat shock proteins (HSPs). There are three major classes of HSPs found in all eukaryotes studied to date: two classes of high molecular weight HSPs, HSP83 and HSP70, and a class of low molecular weight (LMW) HSPs between 15 to 30 kDa (2). The HSP83 and HSP70 proteins are highly conserved between plant and animal species having 70%

or greater homology at the nucleic acid level. The LMW HSPs have apparent structural similarities as determined by limited amino acid homology and hydropathy characteristics (2,3), but show considerable heterogeneity between species, having no significant nucleic acid homology across phyla. The specific function of HSPs remains unclear, but it is currently hypothesized that they protect cells from high temperature damage.

A unique characteristic of the HS response in plants is the complexity and abundance of the LMW HSPs (1,3). In soybean there are at least five multigene families of LMW HSPs, each with two or more members (3). Although the majority of LMW HSPs are found in the plant cytoplasm, we now have evidence that one of the LMW HSPs localizes to the chloroplast (4,5,6). My laboratory is investigating the structure and expression of chloroplast HSPs in several plant species with the ultimate goal of understanding the function of these proteins. Our work identifying chloroplast HSPs, the isolation and sequence analysis of cDNAs encoding these proteins, and initial studies of chloroplast HSP expression are descibed in this report.

METHODS

Methods for RNA isolation, *in vitro* translation, and *in vitro* import of proteins into chloroplasts have been described (7). cDNA synthesis, DNA sequencing and sequence analysis were performed using standard, well established techniques.

RESULTS

Identification of Chloroplast HSPs

Nuclear-encoded HSPs which localize to chloroplasts were first identified using an *in vitro* system for import of proteins into chloroplasts (4,5). Poly(A)RNA was isolated from control or heat stressed plants (39-45 C, 2 hr) and translated *in vitro* to produce ^{35}S-methionine labeled polypeptides. These polypeptides were incubated with isolated, intact pea chloroplasts for one hour in the light. Following incubation, chloroplasts were protease treated, re-isolated and fractionated into soluble and membrane components. Figure 1 shows the results of this experiment performed with soybean and pea. In the HS import samples (lanes 4 and 8) there are one or two abundant polypeptides which are not present in the control samples. These results demonstrate that HS induces the accumulation of mRNAs specifying proteins which are post-translationally transported into chloroplasts. Because poly(A)RNA was used in these experiments it could be assumed these proteins are encoded by the nuclear genome.

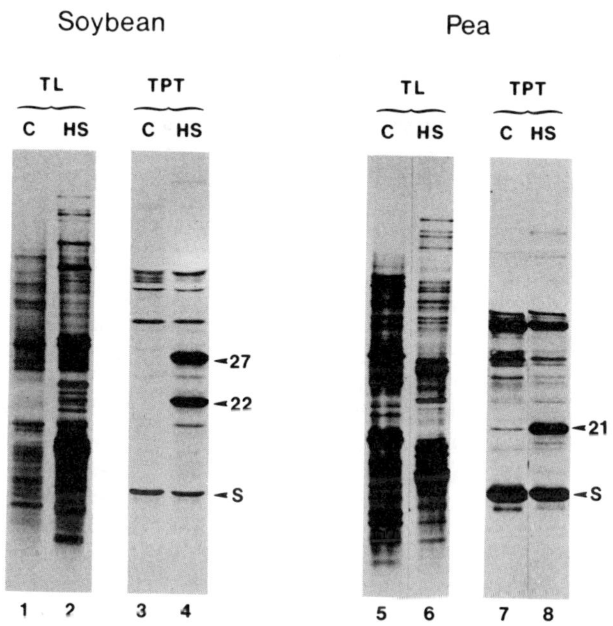

Figure 1. HSPs From Soybean and Pea Imported into Chloroplasts In Vitro. Lanes 1,2,5 and 6 compare total *in vitro* translation products of poly(A)RNA (TL) and lanes 3,4,7 and 8 compare proteins imported into chloroplasts (TPT) from control (C) or heat shock (HS) samples. Only the soluble fractions from the chloroplasts are shown. Positions of major HSPs and the small subunit of RuBPCase (S) are marked.

We have observed chloroplast HSPs in five plant species, soybean, pea, corn, petunia and tobacco using the *in vitro* import technique. All species have one major polypeptide between 21 and 24 kDa, except soybean which has two polypeptides of 22 and 27 kDa. Kloppstech, et al. have also identified chloroplast HSPs in pea and *Chlamydomonas* (5), and Suss has reported their appearance in *Phaseolus* (6).

To establish that HSPs localize to chloroplasts *in vivo* and that the HSPs observed *in vitro* were correctly processed, we performed *in vivo* labeling experiments (see Ref. 4). Chloroplasts were isolated from control or heat shocked peas which had been labeled with ^{35}S-methionine. Two dimensional gel analysis demonstrated that the soluble protein fraction from stressed chloroplasts contained an abundant, newly synthesized 21 kDa polypeptide. The *in vitro* imported HSP and the *in vivo* synthesized HSP had the same pI and molecular weight.

cDNA Cloning

To enable further characterization of chloroplast HSPs, a cDNA clone encoding soybean HSP22 was isolated (7, Vierling, in prep). HS poly(A)RNA was size-fractionated to enrich for chloroplast HSP mRNA. Each RNA fraction was translated *in vitro* using wheat germ extracts and analyzed by import into isolated chloroplasts. Figure 2 shows the soluble proteins recovered from chloroplasts incubated with polypeptides synthesized using individual RNA fractions obtained by sucrose density gradient centrifugation. Samples 3 and 4 from the HS experiment show good enrichment for the 22 and 27 kDa HSPs. HS RNA from fractions 3 and 4 was used to prepare cDNA which was cloned into pUC9.

Clones were screened by differential hybridization with ^{32}P-labeled single-stranded control or HS cDNA. HS positive clones were isolated and used for hybrid-selection translation experiments to identify cDNAs which encoded polypeptides large enough to be the chloroplast HSP precursor protein. These polypeptides were assayed for their ability to be imported into isolated chloroplasts. A cDNA was obtained which encoded a 27 kDa polypeptide that is processed to 22 kDa during import into chloroplasts (7, Vierling, in prep). We were not successful in isolating a cDNA encoding the other chloroplast HSP in soybean.

We have now used the soybean HSP22 cDNA to isolate homologous cDNAs from pea and petunia. The pea and petunia cDNAs hybrid-select mRNAs which translate to polypeptides of 27.5 and 26 kDa (pea) or 30 kDa (petunia). Chloroplast import experiments with the pea polypeptides yield a single 21 kDa protein, suggesting molecular weight heterogeneity in the transit peptide, or that only one of the polypeptides observed in the hybrid-selection experiments localizes to chloroplasts. We have not yet tested the petunia protein by import experiments, but based on observations from import of total translation products, we would predict the 30 kDa precursor is processed to 23.5 kDa during import.

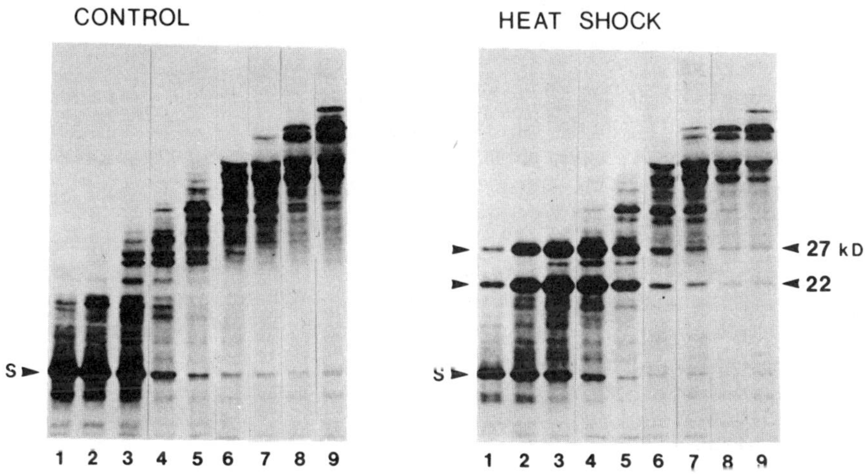

Figure 2. Proteins Imported into Chloroplasts from Translation Products of Size-fractionated Poly(A)RNA. Poly(A)RNA from control or HS soybean tissue was fractionated by size on 5-20% sucrose gradients. An aliquot of RNA from each fraction was translated _in vitro_ and resultant polypeptides imported into chloroplasts. The soluble fraction of the chloroplasts was analyzed by SDS-PAGE and fluorography.

DNA Sequence Analysis

The cDNA encoding soybean chloroplast HSP22 was sequenced and shown to contain the coding region for approximately 90% of the mature protein (181 amino acids) and 155 bp of 3′ non-coding region. The amino terminal portion of the protein, including the transit peptide is not present in this cDNA. Attempts to screen for longer cDNAs were unsuccessful. The amino acid sequence of the open reading frame was compared to the complete amino acid sequence for a soybean cytoplasmic HSP of 17.5 kDa (3) using the computer program FASTP (8). A region of high homology between the two proteins was identified between aa 77 and 173 of HSP22 (aa 48 to 143 of the cytoplasmic HSP). In this region 78% of the residues are identical or conservative replacements. The z value, a measure of the significance of the homology is 25 (>10 = significant). There is no detectable homology in the amino terminal 77 amino acids. The overall nucleic acid homology of these two proteins is less than 52%.

Comparisons of HSP22 to the total NBRF PIR data bank identified weak homology to *Drosophila* HSP22 and to alpha crystallins from several mammals (8). These homologies were restricted to the same region of the protein that is homologous to the soybean cytoplasmic HPSs. These results were not surprising in that the low molecular weight *Drosophila* HSPs were already known to be homologous to the alpha cystallins, and similarities of soybean and *Drosophila* HSPs have been noted (3). No homologies to any other proteins were identified.

Chloroplast HSP Expression

The HSP22 cDNA from soybean and the HSP22 cDNA from pea have been used to study the level of chloroplast HSP mRNA in several plant organs under both control and HS conditions. The HSP22 cDNA from soybean recognizes an mRNA that is only detectable at high levels in cells which have been heat shocked (Figure 3). Low or undetectable mRNA levels at control temperatures and rapid accumulation of high mRNA levels during stress are characteristic of HSPs. At high stringency, the soybean cDNA also hybridizes with a heat induced mRNA from pea, but not from corn. This result indicates the pea and soybean cDNAs are greater than 90% homologous, while the soybean and corn sequences are less homologous. This not surprising given the evolutionary relationship of these species.

Toward understanding the function of chloroplast HSPs we have begun studies of their expression in different parts of the plant. In soybean, chloroplast HSP mRNAs can be detected in the non-photosynthetic etiolated hypocotyl (4). HSP23 mRNA in pea appears to be equally abundant in heat stressed root tissues as it is in stressed leaf tissues. We do not yet know if the protein is made and localized to plastids in these tissues.

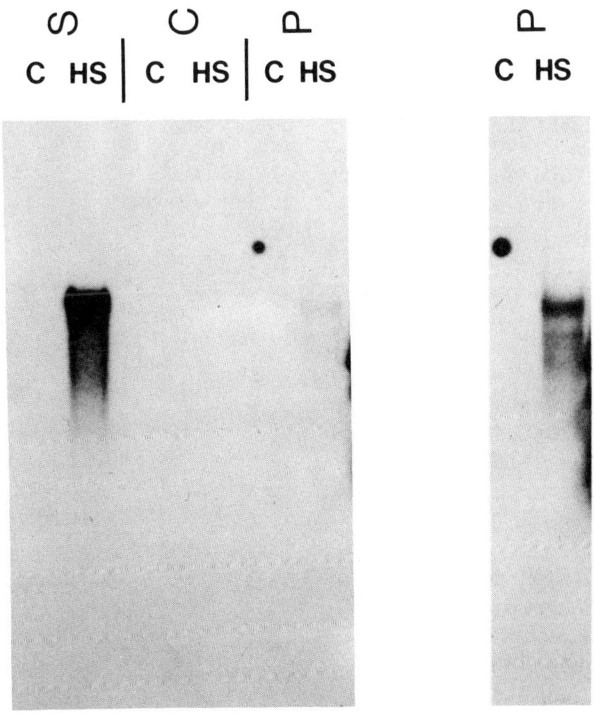

Figure 3. Northern Analysis of Chloroplast HSP mRNA Levels. Northern blots of poly(A)RNA from control (C) or heat shocked (HS) tissues of soybean (S), corn (C) or pea (P) were prepared. Heat treatment was for two hours at maximum HSP induction temperatures. The blots were hybridized with ^{32}P-labeled soybean HSP22 as a probe at a stringency requiring nucleic acid homology of 90% or greater. Panel on the left was exposed for 12 hr; panel on the right was exposed for 48 hr.

DISCUSSION

The synthesis of HSPs in response to elevated temperature has been described in all organisms studied to date (2). Data indicate HSPs are essential for survival during, or recovery from, high temperature stress. We and others have shown that in plants, specific HSPs localize to chloroplasts (4,5,6). These HSPs may be important for maintaining the function and integrity of chloroplasts under stress conditions.

We have isolated cDNAs which encode chloroplast HSPs from soybean, pea and petunia and have begun molecular characterization of their structure and expression. These cDNAs have been identified based on the criteria that they hybrid-select mRNAs which translated to precursor polypeptides of chloroplast proteins. An interesting clue to the function of chlorolast HSPs is the fact that the soybean protein has homology to the carboxy-terminal half of LMW cytoplasmic HSPs. We are currently completing sequence analysis of chloroplast and cytoplasmic LMW HSPs from pea and petunia. Comparisons of these proteins will define further the homologous region. This potential common structural feature suggests chloroplast and cytoplasmic HSPs may function in similar ways during HS, but in different cellular compartments.

Hybrid-selection translation experiments indicate the chloroplast HSPs in soybean, pea and petunia are synthesized as precursor proteins which are processed to their mature molecular weight upon import into chloroplasts. The predicted size of the transit peptides are 5 kDa for soybean, 6.5 and 5 kDa for pea and 6.5 kDa for petunia. In both soybean and pea the precursor proteins are more basic than the mature polypeptides, which is consistent with the basic character of transit peptides for other chloroplast and mitochondrial poypeptides. The pea and petunia chloroplast HSP cDNAs are long enough to encode the entire precursor protein, and should provide the transit peptide sequences.

Preliminary studies of chloroplast HSP expression show high levels of mRNA during heat shock in two non-photosynthetic organs, roots and etiolated hypocotyls. This result suggests these HSPs are "plastid" proteins whose function is not restricted to chloroplasts. This is also consistent with homology to cytoplasmic proteins.

We are also interested in determining if chloroplast HSPs are expressed during plant development in the absence of temperature stress. LMW HSPs in *Drosophila* are known to be regulated by developmental signals implying their function is not restricted to periods of stress (2). We have begun to examine HSP mRNA levels during embryogenesis and early stages of seed germination.

Determining where chloroplast HSPs are localized both within the plant and within the chloroplast will be important to understanding possible functions. In our studies, both *in vivo* and *in vitro*, the majority of protein is recovered with the soluble protein fraction of chloroplasts. Suss et al report similar observations (6). In contrast, Ohad and colleagues observe a significant fraction associated with thylakoid membranes (5,9). The discrepancy may arise from differences in stress conditions or chloroplast fractionation procedures. Schuster et al. suggest that the chloroplast HSP in *Chlamydomonas* somehow protects the Q_B site during HS, thereby protecting photosystem II activity (9). If chloroplast HSPs do have this function, we propose the mechanism of protection is a general one, such as protein stabilization or protection from proteolysis, and that this mechanism of action should be similar for other LMW HSPs. Furthermore, the protein may also act at other sites within the chloroplast and within non-photosynthetic plastids as well.

We plan to generate antibodies against chloroplast HSPs from antigen produced as a fusion protein in *E. coli*. The antibodies will be used to examine chloroplast HSP localization by immunocytochemical techniques and biochemical fractionation procedures. We would like to determine not only with what structures in the chlorolast this protein is associated, but also whether it accumulates differentially in different cell or plastid types. In the long range these studies should lead to an understanding of the importance of chloroplast HSPs to the plant cell.

ACKNOWLEDGEMENTS

I thank Dr. J.L. Key in whose laboratory this work was initiated, Dr. M. Mishkind with whom the original observation of chloroplast HSPs was made, Dr. G. Schmidt for participation in the early studies and Dr. R. Nagao for DNA sequence data. This work has been supported by NSF DCB-8517576 (E.V.) and University of Arizona Hatch Funds.

REFERENCES

(1) Kimpel JA & Key JL (1985). Heat shock in plants. Trends Biochem Sci 10:353.

(2) Lindquist SL (1986) The heat shock response. Ann Rev Biochem 55:1151.

(3) Nagao RN, Czarnecka E, Gurley WB, Schoffl F, Key JL (1985). Genes for low-molecular-weight heat shock proteins of soybeans. Mol Cell Biol 5:3417.

(4) Vierling E, Mishkind ML, Schmidt GW, Key JL (1986). Specific heat shock proteins are transported into chloroplasts. Proc Natl Acad Sci 83:361.

(5) Kloppstech K, Meyer G, Schuster G, Ohad I (1985). Synthesis, transport and localization of a nuclear coded 22-kd heat shock protein in the chloroplast membranes of peas and Chlamydomonas. EMBO J 4:1901.

(6) Suss K-H, Yordanov IT (1986). Biosynthetic cause of in vivo acquired thermotolerance of photosynthetic light reactions and metabolic responses of chloroplasts to heat stress. Plant Physiol 81:192.

(7) Vierling E, Mishkind M, Roberts JK, Schmidt GW, Key JL (1985). Chloroplast-localized heat shock proteins. In: Abstracts, First Int Cong Plant Mol Biol, Savannah, GA, p 34.

(8) Vierling E, Roberts JK, Nagao RT, Key JL (1987) A chloroplast heat shock protein has homology to cytoplasmic heat shock proteins. In Biggins J: Proc. VIIth Int Cong Photosyn, in press.

(9) Schuster G, Even D, Kloppstech K, Ohad I (1987) Heat shock proteins partially protect against photoinhibition of Chlamydomonas reinhardtii during heat shock. In Biggins J: VIIth Int Cong Photosyn, in press.

Plant Gene Systems and Their Biology, pages 109–120

GENETIC DIVERSITY OF HEAT SHOCK PROTEIN SYNTHESIS AND ITS RELATION TO THERMAL TOLERANCE IN WHEAT LEAVES[1]

M. Krishnan, Henry T. Nguyen, and John J. Burke

Department of Plant and Soil Science, Texas Tech University, Lubbock, Texas 79409 and USDA-ARS, Plant Stress and Water Conservation Unit, Lubbock, Texas 79401

ABSTRACT Numerous studies have identified species differences in the pattern of heat shock protein synthesis during incubation at temperatures 10 to 15°C above their normal growth temperatures. The present study was undertaken (1) to determine if genetic differences in acquired thermal tolerance existed within a species, and (2) to determine if the differences in thermal tolerance were correlated with differences in the level of heat shock protein synthesis. Genetic differences in the level of acquired thermal tolerance in wheat were identified using the 2,3,5-triphenyltetrazolium chloride cell viability test following a 37 to 42°C heat shock treatment. The winter wheat cultivar Mustang exhibited a higher level of heat tolerance than the cultivar Sturdy. Analyses of heat shock protein synthesis in the green seedling leaves showed that quantitative differences existed between these cultivars with Mustang, the heat tolerant cultivar, exhibiting a greater degree of heat shock protein synthesis. These results support the concept that heat shock proteins are intimately involved in the genetic control of thermal tolerance.

[1]This work was supported by a grant from the Advanced Technology Research Program, State of Texas, and the Plant Stress and Water Conservation Program, USDA-ARS, Lubbock, Texas. TTU# T-4-205

INTRODUCTION

Wheat, (*Triticum aestivum* L.) a cool-season C_3 species, has optimal productivity under the growing conditions that provide leaf temperatures of 20 to 25°C. This temperature range has been shown to provide optimal photosynthetic gas exchange, carbon fixation, and photosynthetic electron transport activity (1,2). This temperature range has also been shown to be required for optimal grain development, and grain yields decline with elevated temperatures (3). Ambient temperatures of 30 to 35°C are not unusual in many wheat-producing areas during seedling establishment and/or grain filling. In semi-arid and arid environments foliage temperatures commonly increase above air temperatures as a result of stomatal closure and reduced transpiration (4). These elevated foliage temperatures may limit dry matter accumulation because of increased respiration, reduced photosynthesis, and other structural and/or functional cellular damage (5). Plants can resist high temperature stress either by avoidance or tolerance mechanisms. Potential mechanisms of heat tolerance include the synthesis of protectants and increased thermostability of enzymes (6).

Although high temperature stress is a major cause of yield loss in several crops, few germplasm selections have been conducted by plant breeders to specifically improve high temperature tolerance. This reflects our limited understanding of the genetic and physiological basis of heat tolerance in plants (7). Chen et al. (8) reported that genotypic differences exist in the induction of an acquired thermal tolerance in common bean, potato, soybean, and tomato. They stated that these differences, identified by the 2,3,5-triphenyltetrazolium chloride (TTC) viability test, were related to plant performance under high temperature conditions in the field. There is, however, little available information on the genetic control of the thermal tolerance or the gene products responsible for the observed genetic differences.

The discovery of specific proteins which are expressed under high temperature stress conditions, the heat shock proteins (HSPs), has sparked considerable interest in their role in an acquired thermal tolerance mechanism (9,10). The heat shock response was initially observed in *Drosophila*, and has subsequently been shown to be a ubiquitous response in both prokaryotic and eukaryotic organisms (9 - 14). Early reports of heat shock response in higher plants demonstrated the synthesis of HSPs in tobacco and soybean cells grown in suspension culture (15), and in soybean seedling tissue (16). The electrophoretic profile of HSPs in soybean seedlings consisted of 10 new bands on one-dimensional SDS gels, and a more complex pattern on two-dimensional gels (16). When the seedlings were returned to a control temperature of 28°C after 4 h of heat shock at 40°C, there was a progressive decline in the synthesis of HSPs and a reappearance of the normal pattern of protein synthesis within 3 to 4 h. To date, the incorporation of radioactive precursors into heat shock proteins has been identified in peas (17), tomato (18,19), carrot (20), tobacco (21), mung bean (22), barley (23), *Tradescantia* (24), *Gladiolus* cormels (25), *Lilium longiflorum* (26), soybean (16), corn (27 - 30), cotton (31), wheat (32), millet (32), sunflower (33), and sorghum (34). The optimal induction temperature for maximal HSP synthesis varies between plant species, but is generally 10 to 15°C above the temperature empirically determined for optimal plant growth.

Physiological studies involving higher plants, *Drosophila*, and yeast have correlated HSP synthesis with the development of thermal tolerance or thermal protection (10). Here, the thermal tolerance is defined as the ability of organisms to tolerate normally lethal high temperatures after an initial exposure to a sub-lethal, but elevated HSP-induction temperature. The development of thermal tolerance in a crop cultivar was best demonstrated in "Wayne" soybean seedlings by Lin, Roberts, and Key (35). This report shows that the growth of seedlings was protected during a subsequent incubation at 45°C for 2 h (normally a lethal treatment) by a heat shock pretreatment of 40°C for 1-2 h.

Although HSPs have been implicated in the ability of organisms to grow at high temperatures through the development of thermal tolerance, there have been no reports of differences in the quality or quantity of HSPs synthesized in relation to genotypic differences in heat tolerance in plants. In this paper, we report results on the genetic diversity in the synthesis of heat shock proteins and their relationship to the genetic differences in thermal tolerance of winter wheat.

MATERIALS AND METHODS

Research in our laboratory has identified significant genetic variability in acquired thermal tolerance among ten winter wheat (*Triticum aestivum* L., AABBDD genome) cultivars at both the seedling and flowering stages using the chloride (TTC) viability test (8). The cultivars (inbred lines) Mustang and Sturdy were chosen for the heat shock protein synthesis study because of their differences in thermal tolerance.

Seeds were planted in vermiculite in plastic dish pans, and germinated at 22°C (control temperature) under fluorescent and incandescent lighting. When the seedlings reached the two-leaf stage, they were heat shocked at 37°C for 24 h in an incubator, returned to 22°C for 6 h, and then exposed to 42°C for an additional 8 h. The heat shock treatments were conducted in the dark to ensure that leaf temperature would be equal to the chamber temperature, thereby avoiding the cooling effects of transpiration experienced in the light.

TTC Cell Viability Test.

Every 4 h during both the 37 and 42°C heat shock, 50 mg of leaf tissue was excised, cut into small sections, and transferred to a test tube containing 3 ml of 0.03% TTC in 0.05 M phosphate buffer, pH 7.4. The leaf segments were vacuum infiltrated for 5 min to ensure solution penetration and incubated for 18 to 20 h in the dark. The leaf segments were washed extensively following the TTC incubation, and 3 ml of 95% ethanol added to the test tube. The ethanol was boiled off in a dry block, and an additional 3 ml of 95% ethanol added to cooled tubes to extract the TTC reduced by the leaf segments. The reduction of TTC was monitored spectrophotometrically at 485 nm.

In vivo Labelling of Proteins, Gel Electrophoresis, and Fluorography.

Leaf segments (100 mg) were taken from the 37°C heat shock treatment at 4 and 16 h. The tissue was cut into small segments and incubated for an additional hour in 3.0 ml of 0.05 M Tris-HCl, pH 7.5, 1% sucrose, 50 μg ml^{-1} chloramphenicol and 25 μCi of ^{35}S-methionine (New England Nuclear, >800 Ci mmol^{-1}). Labelled tissue was rinsed thoroughly and blotted dry. The samples were crushed in a liquid nitrogen cooled mortar and the powder was resuspended in a solution of 10% trichloroacetic acid (TCA) in acetone with 0.07% 2-mercaptoethanol (36). Proteins were precipitated for 45 min at -18°C, centrifuged at 12000 g for 5 min in an Eppendorf centrifuge, and the pellet rinsed twice with acetone for 1 h at -18°C. The supernatant was discarded and the pellet dried under vacuum. One-dimensional SDS-polyacrylamide gel electrophoresis was performed according to the procedure of Laemmli (37) using a 7.5 to 15% polyacrylamide gradient. A ^{14}C methylated protein mixture containing lysozyme (14,300 kDa), carbonic anhydrase (30,000 kDa), ovalbumin (46,000 kDa), bovine serum albumin (69,000 kDa), phosphorylase b (92,000 kDa), and myosin (200,000 kDa) was used as molecular weight markers. Fluorography of the gels was accomplished by incorporating the fluor ENHANCE (New England Nuclear) into the gel, drying the gel under vacuum, and exposing Kodak XAR-5 film at -90°C.

RESULTS

Genotypic differences in cell viability were observed between the cultivars Mustang and Sturdy following the heat shock treatment at 37 and 42°C (Fig. 1). The genotypic

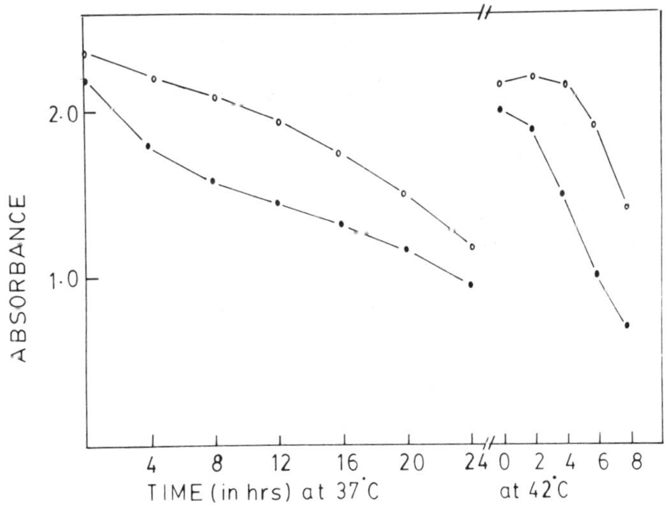

Figure 1. Comparison of the timecourse of 2,3,5-triphenyl tetrazolium chloride reduction in the wheat cultivar Mustang (o—o—o) and Sturdy (•—•—•) following seedling incubations at 37°C, 22°C, and 42°C.

differences became more apparent as the high temperature (37°C) stress progressed up to 8 h. The differences in heat tolerance between Mustang and Sturdy remained between 22 to 27% of TTC absorbance for the remainder of the 24 h incubation. The seedlings fully recovered within 6 h following the removal of the high temperature stress, as shown by the increase in the TTC reduction to levels comparable to the time zero controls. A second heat shock treatment of 42°C following the tissue recovery showed even greater genotypic differences with 2-fold higher levels of TTC reduction occurring in the cultivar Mustang.

To determine if the differences in cell viability between the cultivars Mustang and Sturdy were related to the differences in heat shock protein synthesis, the leaf proteins were electrophoretically separated, and protein synthesis determined by autoradiography. Figure 2 compares the pattern of proteins synthesized by Sturdy and Mustang leaves at 22°C (lanes 1 and 2), 37°C for 4 h (lanes 3 and 4), and 37°C for 16 h (lanes 5 and 6). No genotypic differences were detected in the pattern of the proteins synthesized by the control (non-stressed) leaf tissue (lanes 1 and 2). A 4 h heat shock treatment at 37°C resulted in the suppression of normal protein synthesis in both the thermal tolerant (Mustang) and susceptible (Sturdy) cultivars. Newly synthesized heat shock proteins had apparent molecular size of 220, 94, 85, 70, 43, 42, 40, 22, 17, and 16 kDa (Fig. 2). Although no qualitative differences in the heat shock protein pattern were observed on the one-dimensional gels, quantitative differences in several polypeptides were detected. Higher levels of incorporation were detected in the 16, 85, 94, and 220 kDa polypeptides of the thermal tolerant cultivar Mustang. The prolonged heat shock (16 h at 37°C) caused a reduction in the synthesis of the HSPs, especially in the low molecular weight range. The prolonged heat shock also caused a return to the synthesis of normal proteins observed at the control temperature (22°C).

DISCUSSION

Cellular injury in response to temperature extremes is commonly associated with reduced enzyme activity and membrane damage (38). Cell viability tests have been developed which focus on one or the other of these sites of injury. The cell viability test associated with enzyme

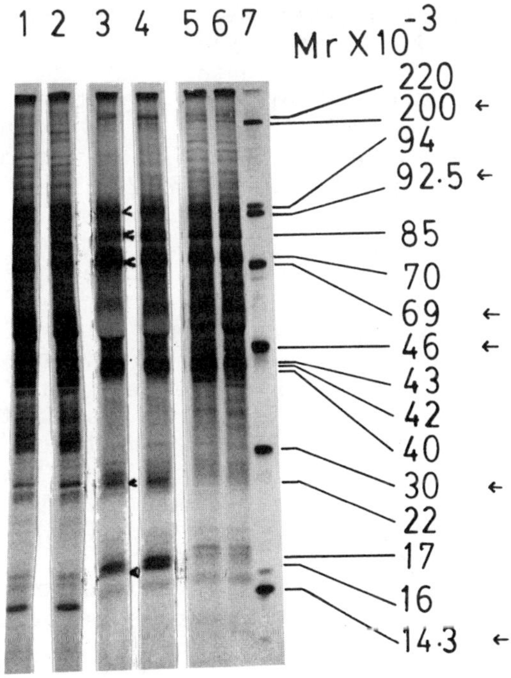

Figure 2. Analysis of the effect of elevated temperature on the proteins synthesized by the wheat cultivars Sturdy and Mustang. The polypeptide samples analyzed were from control (22°C) seedlings of Sturdy [1] and Mustang [2]; from Sturdy [3] and Mustang [4] seedlings exposed to 37°C for 4 h; and from Sturdy [5] and Mustang [6] seedlings exposed to 37°C for 16 h. Molecular weight standards are shown in lane 7.

inhibition analyzes the reduction of tetrazolium salts via the action of the mitochondrial electron transport chain (39). The ability of cells to reduce the tetrazolium salts has made this procedure a simple and accurate measurement of cell viability for large sample numbers (40). The cell viability test based upon membrane damage monitors ion leakage from cells with a conductivity meter (41). Chen et al. (8) have compared these two methods in the analysis of thermal tolerance in bean and tomato cultivars. Their results showed that both methods of determining cell viability could be used in the analysis of genotypic differences in thermal tolerance.

The results of our cell viability studies indicate that genetic differences in the ability to induce a cellular acquired thermal tolerance mechanism exist between the wheat cultivars Mustang and Sturdy. The 2-fold difference in TTC reduction observed in our study is similar to the differences previously reported by Chen et al. (8) for genetic diversity in the thermal tolerance of beans and tomatoes.

The pattern of proteins synthesized at 42°C following a 37°C heat shock period was similar to the 37°C pattern observed in the 4 h period, except that there was a quantitative decline in the labelled amino acid incorporation in the HSP 22 and the distinct appearance of HSP 34 which was not observed at 37°C (data not shown). These results suggest that different proteins may be involved in protecting cells from different levels of heat shock.

Incubation at 37°C for 16 h caused reduced levels of heat shock protein synthesis, especially in the low molecular weight region from 16 to 22 k. Our results agree with the observation that the kinetics of thermotolerance induction under many different conditions are tightly correlated with the kinetics of heat shock protein synthesis and reach maximum when the accumulation of HSPs plateaus (42).

In this paper, we report that quantitative differences exist in the synthesis of HSPs within a species exhibiting differences in thermal tolerance. Our findings support the concept that HSP synthesis and thermal tolerance are related. The relationship of HSP synthesis and thermal tolerance is supported by the finding that at the stages in development in which HSPs cannot be induced, organisms are extremely sensitive to heat and thermal tolerance cannot be induced (10).

The analysis of the response of cellular protein synthesis in the cultivars Mustang and Sturdy to heat shock temperatures was an initial attempt to identify specific protein changes that might be associated with the thermal tolerance mechanism. Our results demonstrate that quantitative differences in heat shock protein synthesis occur in these cultivars. The cultivar Mustang, not only exhibited greater cell viability following exposure to a previously lethal temperature, but it also had higher level of incorporation of radioactive methionine in the 16, 85, 94, and 200 kDa heat shock proteins. Whether the quantitative difference in the labelling patterns between the two cultivars reflects a reduction of all of the heat shock proteins detected in the two-dimensional gel electrophoresis migrating at the 16 kDa range or if it reflects the selective loss of specific heat shock proteins cannot be determined from the one-dimensional SDS-PAGE. Analysis of these cultivars using two-dimensional electrophoresis and specific clones to the low molecular weight heat shock proteins is currently being conducted to further clarify the genetic differences in the heat shock response of these cultivars. If further experiments can clearly identify the specific heat shock protein(s) that relate to the genetic differences in thermal tolerance, then they can be used as a biochemical marker to identify the genes which are involved in heat tolerance. Knowledge of these genes is essential in the development of new genetic techniques and approaches to breeding for wheat cultivars that are more heat tolerant under arid and semi-arid environments.

ACKNOWLEDGEMENTS

The authors thank David R. Porter for assistance in the initial screening for heat tolerance in winter wheat germplasm and David E. Koeppe for critical review of the manuscript.

REFERENCES

1. Planchon C (1976). Heterosis, combining ability and photosynthesis in bread wheat (Triticum aestivum L.) In Janossy A, Lupton FGH (eds): "Heterosis in Plant Breeding," Amsterdam: Elsevier.
2. Al-Khatib K, Paulsen GM (1984). Mode of high temperature injury to wheat during grain development Physiol Plant 61:363-368.
3. Evans LT, Wardlaw IF, Fischer RA (1975). Wheat. In Evans LT (ed): "Crop Physiology, Some Case Histories," London: Cambridge Univ Press, p 101.
4. Hatfield JL (1979). Canopy temperatures. The usefulness and reliability of remote measurements. Agron J 71:889-892.
5. Berry J, Bjorkman O (1980). Photosynthetic response and adaptation to temperature in higher plants. Ann Rev Plant Physiol 31:491-543.
6. Levitt E (1980). "Responses of Plants to Environmental Stresses (Vol 1)," New York: Academic Press.
7. Marshall HG (1982). Breeding for tolerance to heat and cold. In Christiansen MN, Lewis (eds):" Breeding Plants for Less Favorable Environments," New York: John Wiley and Sons, p 47.
8. Chen HH, Shen ZY, Li PH (1982). Adaptability of crop plants to high temperature stress. Crop Sci 22:719-725.
9. Sachs MM, Ho THD (1986). Alteration of gene expression during environmental stress in plants. Ann Rev Plant Physiol 37:363-376.
10. Lindquist S (1986). The heat-shock response. Ann Rev Biochem 55:1151-1191.
11. Schlesinger MJ, Ashburner M, Tissieres A (1982). "Heat shock from bacteria to man," New York: Cold Spring Harbor Lab.
12. Nover L, Hellmund D, Newmann D, Scharf KD, Serfling E (1984). The heat shock response of eukaryotic cells. Biol Zentralbl 103:357-435.
13. Atkinson BG, Walden DB (1985). "Changes in Eukaryotic Gene Expression in Response to Environmental Stress," New York: Academic Press.
14. Craig EA (1985). The heat shock response. CRC Crit Rev Biochem 18:239-280.
15. Barnett T, Altschuler M, McDaniel CN, Mascarenhas JP (1980). Heat shock induced proteins in plant cells. Dev Gen 1:331-340.
16. Key JL, Lin CY, Chen YM (1981). Heat shock proteins of

higher plants. Proc Natl Acad Sci (USA) 78:3526-3530.
17. Hadwiger LA, Wagoner W (1983). Effect of heat shock on the mRNA-directed disease resistance response of peas. Plant Physiol 72:553-556.
18. Nover L, Scharf KD (1984). Synthesis, modification and structural binding of heat-shock proteins in tomato cell cultures. Eur J Biochem 139:303-313.
19. Nover L, Scharf KD, Neumann D (1983). Formation of cytoplasmic heat shock granules in tomato cell cultures and leaves. Mol and Cell Biol 3:1648-1655.
20. Pitto, L, Schiavo FL, Giuliano G, Terzi M (1983). Analysis of the heat-shock protein pattern during somatic embryogenesis of carrot. Plant Mol Biol 2:231-237.
21. Meyer Y, Chartier Y (1983). Long-lived and short-Lived heat-shock proteins in tobacco mesophyll protoplasts. Plant Physiol 72:26-32.
22. Chen Y-M, Kamisaka S, Masuda Y (1986). Enhancing effects of heat shock and gibberellic acid on the thermotolerance in etiolated Vigna radiata I. Physiological aspects of thermotolerance. Physiol Plant 66:595-601.
23. Belanger FC, Brodl MR, Ho T-HD (1986). Heat shock causes destabilization of specific mRNAs and destruction of endoplasmic reticulum in barley aleurone cells. Proc Natl Acad Sci (USA) 83:1354-1358.
24. Xiao C-M, Mascarenhas JP (1985). High temperature-induced thermotolerance in pollen tubes of Tradescantia and heat-shock proteins. Plant Physiol 78:887-890.
25. Ginzburg C, Salamon R (1986). The effect of dormancy on the heat shock response in Gladiolus cormels. Plant Physiol 81:259-267.
26. Hong-Qi A, Croes AF, Linskens HF (1984). Qualitative changes in protein synthesis in germinating pollen of Lilium longiflorum after a heat shock. Plant Cell and Environ 7:689-691.
27. Baszczynski CL, Walden DB, Atkinson BG (1983). Regulation of gene expression in corn (Zea mays L.) by heat shock. II. In vitro analysis of RNAs from heat-shocked seedlings. Can J Biochem Cell Biol 61:395-403.
28. Bewley DJ, Larsen KM, Papp JT (1983). Water stress induced changes in the pattern of protein synthesis in maize seedling mesocotyls: A comparison with the effects of heat shock. J Expt Bot 34:1126-1133.
29. Cooper P, Ho T-HD (1983). Heat shock proteins in maize. Plant Physiol 71:215-222.

30. Cooper P, Ho -T-HD (1984). Tissue specificity of the heat shock response in maize. Plant Physiol 75:431-441.
31. Burke JJ, Hatfield JL, Klein RR, Mullet JE (1985). Accumulation of heat shock proteins in field-grown cotton. Plant Physiol 78:394-398.
32. Key JL, Czarnecka E, Lin CY, Kimpel J, Mothershed C, Schoffl F (1983). A comparative analysis of the heat shock response in crop plants. In Randall DD, Blevins DG, Larson RL, Rapp BJ, (eds): "Current Topics in Plant Biochemistry and Physiology, Vol 2," Columbia: Univ Missouri, p 107.
33. Schoffl F, Baumann G (1985). Thermo-induced transcripts of a soybean heat shock gene after transfer into sunflower using a Ti plasmid vector. EMBO J 4:1119-1124.
34. Ougham HJ, Stoddart JL (1986). Synthesis of heat-shock protein and acquisition of thermotolerance in high-temperature tolerant and high-temperature susceptible lines of sorghum. Plant Sci 44:163-167.
35. Lin CY, Roberts JK, Key JL (1984). Acquisition of thermotolerance in soybean seedlings. Plant Physiol 74:152-160.
36. Damerval C, de Vienne D, Zivy M, Thiellement H (1986). Technical improvements in two dimensional electrophoresis increase the level of genetic variation detected in wheat seedling proteins. Electrophoresis 7:52-54.
37. Laemmli UK (1970). Cleavage of structural proteins during the assembly of the head of bacteriophage T4. Nature (London) 227:680-685.
38. Christiansen MN (1978). The physiology of plant tolerance to temperature extremes. In Jung GA (ed): "Crop tolerance to sub-optimal land conditions," Wisconsin: Am Soc Agron, p 173.
39. Smith FE (1951). Tetrazolium salt. Sci 113:751.
40. Leigh ET, Mazur P (1975). Studies on the reduction of 2,3,5-triphenyl tetrazolium chloride as a viability assay for plant tissue cultures. Can J Bot 53:1097-1102.
41. Blum A, Ebercon Adelina (1981). Cell membrane stability as a measure of drought and heat tolerance in wheat. Crop Sci 21:43-47.
42. Landry J, Bermier D, Gretien P, Nicole LM, Tanquay RM, Marceau N (1982). Synthesis and degradation of heat shock proteins during development and decay of thermotolerance. Cancer Res 42:2457-61.

Plant Gene Systems and Their Biology, pages 121–129

THE LACTATE DEHYDROGENASE SYSTEM OF BARLEY[1]

A.D. Hanson, N.E. Hoffman[2], D. Hondred, A.H.D. Brown, and D.C. Alexander

MSU-DOE Plant Research Laboratory, Michigan State University East Lansing MI 48824 (A.D.H., N.E.H., D.H.), CSIRO Division of Plant Industry, Canberra, ACT 2601, Australia (A.H.D.B.), ARCO Plant Cell Research Institute, Dublin CA 94568 (D.C.A.)

ABSTRACT Plant roots and seed tissues have the capacity for lactate glycolysis, as well as for ethanol glycolysis. Balanced operation of these two pathways is required for plant survival during O_2-deprivation. Plant lactate dehydrogenase (LDH) enzymes are tetramers and strongly resemble vertebrate and prokaryote L-LDHs. Genetic analysis of barley LDH isozyme variants indicates that there are two closely linked LDH loci, whose products associate randomly to generate a set of five isozymes. During O_2-deprivation, LDH activity increases up to 20-fold in barley root and seed tissues; there are parallel rises in LDH protein synthesis and in the level of translatable LDH mRNA.

In hypoxic aleurone tissues, induction of LDH activity shows a sharp optimum at 2-5% O_2, whereas alcohol dehydrogenase (ADH) induction is greatest at 0% O_2. At least in this tissue, regulation of the LDH system may therefore differ from that of the ADH system.

[1]The work of A.D.H., N.E.H. and D.H. was supported by Department of Energy Contract DE-AC02-76ERO-1338.
[2]Present address: Plant Science Institute, University of Pennsylvania, Philadelphia, PA 19104.

INTRODUCTION

L-Lactate dehydrogenases have been known for some time to be widespread in higher plants, but the role of LDH in plant anaerobic metabolism has received little attention. This is perhaps surprizing in the light (a) of extensive research on ADHs and ethanol glycolysis in plants (1,2), and (b) of the vast literature on the function, structure and evolution of vertebrate (3,4) and prokaryote (5) LDHs. However, two recent developments have focused interest on plant LDHs in relation to anaerobiosis: first, the experimental confirmation (6) of Davies' hypothesis (7) concerning the interdependence of lactate and ethanol glycolysis; and second, the demonstration that LDHs are strongly induced during hypoxia (8). This article reviews and discusses these developments, and summarizes available data on the protein structure and genetics of plant LDHs.

BIOCHEMICAL BACKGROUND: THE DAVIES-ROBERTS HYPOTHESIS

For a plant tissue with access to a supply of fermentable substrate, survival of O_2 deficit imposes three major requirements (9,10): (a) ATP generation; (b) maintenance of redox balance between NAD^+ and NADH; and (c) maintenance of cytosolic pH. Ethanol and lactate glycolysis meet the first two requirements equally well, but whereas ethanol is a neutral end product, lactate accumulation causes cytosolic acidosis which, when excessive, can lead to cell death (11,12). Davies *et al*.(7) proposed that plants avoid excessive acidosis through an *in vivo* pH-stat mechanism set by the pH requirements of the enzymes pyruvate decarboxylase (PDC) and LDH, which compete for pyruvate, thus:

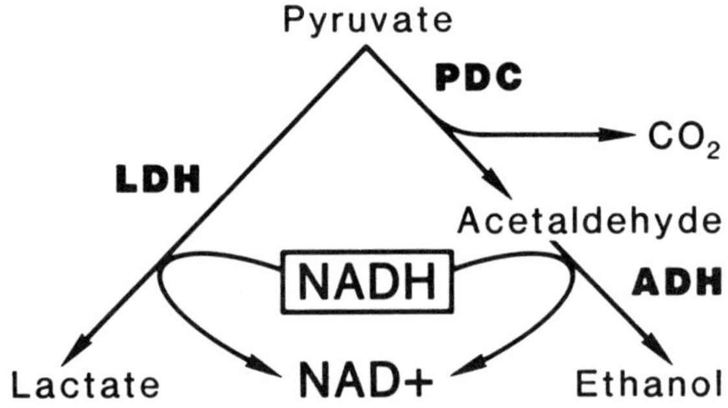

Simply put, this hypothesis holds that when hypoxia begins, the cytosolic pH is close to 7 so that LDH (with a near-neutral pH optimum) outcompetes PDC (with a very acid pH optimum) for pyruvate. As lactate builds up and cytosolic pH falls, LDH competes less and less effectively with PDC so that almost all glycolytic flux eventually switches to the ethanol route. Early data from an *in vitro* glycolyzing system (7) and later results of *in vivo* ^{14}C-labeling studies (13) fitted Davies' hypothesis, but were not conclusive. Unequivocal evidence for an early burst of lactate glycolysis and cytosolic acidification, followed by pH stabilization and a shift to long-term ethanol glycolysis, has come from *in vivo* NMR studies of maize root tips by Roberts *et al*. (6). Experiments with an *Adh-1* null demonstrated an uncontrolled cytosolic acidification, just as predicted by Davies' hypothesis (6). There is thus strong evidence that LDH serves an indispensable function in hypoxic plant tissues. In addition to mediating a transient burst of lactate synthesis associated with cytosolic pH regulation, LDH may in certain cases catalyze some continuous production of lactate, perhaps destined for secretion to the medium (14,15). There is as yet no evidence bearing on the possibility that plants can use lactate to export an O_2 debt from roots to shoots, in a manner analogous to glycolyzing muscle (8).

PLANT LDH ENZYMES

LDHs have been reported in a range of plant tissues. These include: developing and germinating seeds (15-19); roots (8,20) and root nodules (21); tubers (22,23); tissue cultures (24); and leaves (25,26). Certain of these reports are rigorous demonstrations that L-LDH is present, but caution is necessary in interpreting those based only on staining gels for LDH activity, since glycolate oxidase gives a positive reaction with the LDH stain (25). Also, coupled action of PDC and ADH can give apparent "LDH" activity in crude plant extracts if LDH is assayed spectrophotometrically in the pyruvate → lactate direction (7). At least in barley root extracts, LDH activity is low and quite unstable unless BSA, borate and DTT are added to the extraction buffer (8). Choice of extraction buffer may therefore explain reports of root extracts with undetectable (27) or low (20) LDH activity.

In all plant LDHs studied, the holoenzyme is a tetramer with subunits of 36-40 kD (19,22,28), suggesting a common evolutionary origin with vertebrate LDHs (22,29). Likewise, substrate specificities, kinetic properties and inhibitors of plant LDHs generally resemble those of vertebrate enzymes (17,22,23,28,30), although ATP inhibition of plant LDHs may not be strictly competitive with respect to NADH (22), and L-alanine may inhibit plant LDHs (31). Available primary structure data for the substrate-binding region of potato LDH also strongly support a common evolutionary origin of animal and plant LDH genes, as well as prokaryote L-LDH genes (32).

The best characterized plant LDH is that from potato tubers. In this case there are five isozymes; their protein chemistry indicates that they arise from random association of the polypeptide products of two related LDH genes (22,26,29), again conforming to the vertebrate pattern. Barley roots also express five LDH isozymes (8), and here there is genetic evidence consistent with a two-gene system (see below). However, not all tissues have five LDH isozymes. For example, only one major band has been found in soybean seeds (19), and in leaves of *Capsella* (33), lettuce (25), and potato (26). The distinct LDH isozyme profiles of potato leaves and tubers imply tissue-specific expression of plant LDH genes, which is the rule with vertebrate LDH genes. Thus, in mammals the LDH A gene predominates in muscle tissue, LDH B predominates in heart, and the LDH C gene is expressed only in testes and sperm cells (3).

ANAEROBIC INDUCTION OF LDH

Modest increases (∿1.5 to 5-fold) in extractable LDH activity have been reported for bean cotyledons (34), and for rhizomes (35) and root tips (20) of marsh plants given hypoxic treatments. However, none of these studies systematically varied the degree and duration of hypoxia. We tested these factors for barley aleurone layers (15) and for roots of hydroponically grown barley plants (8). In roots, LDH activity increased 20-fold (to 2 IU/g fr. wt.) over a 4 day period when O_2 levels in solution were equivalent to an atmosphere of 0.1-0.2% O_2; higher O_2 levels gave progressively less LDH induction. Upon resumption of aeration, induced LDH activity declined with an apparent half-life of

2 days. Inducible LDH activity was also present in roots of wheat, rye and maize.

In barley, immunoblot analysis established that LDH protein levels rose in parallel with enzyme activity (28). *In vivo* labeling of roots with [^{35}S]methionine followed by immunoprecipitation indicated that this was due to an increase in the rate of LDH protein synthesis (28). Data from *in vitro* translation experiments with mRNA from induced and control roots suggest a corresponding increase in functional LDH message level. LDH thus fits the description of an "anaerobic protein" (ANP), i.e. one of a fairly small group of proteins selectively synthesized in anaerobic roots as a result of anaerobically-induced increases in mRNA level and anaerobic-specific translation (36). It is therefore reasonable to speculate (a) that LDH genes are components of a control network which includes other known ANPs such as ADH (37) and PDC (27), and (b) that their shared property of anaerobic inducibility refects a shared mechanism of gene regulation. However, there is some reason to treat this idea of a "co-adapted gene battery" (36) with caution. In aleurone tissue of barley, ADH and LDH appear to be independently regulated: ADH activity increases steadily as the O_2 level is lowered to 0%, whereas LDH activity peaks at 2-5% O_2 (15). The increase in ADH enzyme activity in aleurone cells has already been shown to coincide with a 100-fold increase in ADH mRNA (38); it will be interesting to compare this with LDH mRNA levels when probes become available.

GENETIC BASIS OF LDH ISOZYMES

The LDH systems of several vertebrate species are well characterized genetically (3). Although various numbers of LDH isozymes have been found in a range of plant tissues (see above), formal genetic studies of plant LDHs are lacking. As the best-defined plant LDHs have either five isozymes or a single isozyme, it is reasonable at this point to suppose that other numbers of LDH bands (e.g. refs. 24,39) are derived from a basic one- or five-banded pattern, by differential expression or stability of two monomer types in a two-gene system (3), by proteolysis *in vivo* or *in vitro* (22,32), or by gel artifacts (40).

To investigate the genetic basis of the five-banded pattern in barley, we sought electrophoretic LDH variants among a group of 30 diverse *Hordeum spontaneum* lines. Three variants were found. In two variants the five isozymes were displaced as a group (fast or slow) relative to the normal set, whereas in the third there was a single main isozyme plus up to three minor bands. All variants bred true for LDH phenotype. In crosses among variants, the F_1 banding patterns were precisely those predicted by a model in which the products of two LDH loci associate freely in all possible combinations. However, no recombinants were found in the F_2 progenies, and F_2 ratios of parental to F_1-hybrid phenotypes did not deviate significantly from 1:2:1. These results can be explained by supposing barley to have two closely linked LDH genes, each with several alleles which have been fixed in particular combinations in natural *H. spontaneum* populations.

REFERENCES

1. Freeling M, Birchler JA (1981). Mutants and variants of the alcohol dehydrogenase-1 gene in maize. Genet Eng 3:223.
2. ap Rees T (1980). Integration of pathways of synthesis and degradation of hexose phosphates. In Preiss J (ed): "The Biochemistry of Plants," vol. 3, New York: Academic Press, p 1.
3. Markert CL, Shaklee JB, Whitt GS (1975). Evolution of a gene. Science 189:102.
4. Li SS-L, Tiano HF, Fukasawa KM, Yagi K, Shimizu M, Sharief FS, Nakashima Y, Pan Y-CE (1985). Protein structure and gene organization of mouse lactate dehydrogenase-A isozyme. Eur J Biochem 149:215.
5. Clarke AR, Wigley DB, Chia WN, Barstow D, Atkinson T, Holbrook JJ (1986). Site-directed mutagenesis reveals role of mobile arginine residue in lactate dehydrogenase catalysis. Nature 324:699.
6. Roberts JKM, Callis J, Wemmer D, Walbot V, Jardetzky O (1984). Mechanism of cytoplasmic pH regulation in hypoxic maize root tips and its role in survival under hypoxia. Proc Natl Acad Sci USA 81:3379.
7. Davies DD, Grego S, Kenworthy P (1974). The control of the production of lactate and ethanol by higher plants. Planta 118:297.
8. Hoffman N, Bent AF, Hanson AD (1986). Induction of lactate dehydrogenase isozymes by oxygen deficit in barley root tissue. Plant Physiol 82:658.

9. Hochachka PW (1981). "Living Without Oxygen." Cambridge: Harvard University Press, p 1.
10. Hochachka PW, Mommsen TP (1983). Protons and anaerobiosis. Science 219:1391.
11. Roberts JKM, Callis J, Jardetzky O, Walbot V, Freeling M (1984). Cytoplasmic acidosis as a determinant of flooding intolerance in plants. Proc Natl Acad Sci USA 81:6029.
12. Roberts JKM, Andrade FH, Anderson IC (1985). Further evidence that cytoplasmic acidosis is a determinant of flooding intolerance in plants. Plant Physiol 77:492.
13. Smith AM, ap Rees T (1979). Effects of anaerobiosis on carbohydrate oxidation by roots of *Pisum sativum*. Phytochemistry 18:1453.
14. Hook DD, Brown CL, Kormanik PP (1971). Inductive flood tolerance in swamp tupelo (*Nyassa sylvatica* var. *biflora* (Walt.) Sarg.). J Exp Bot 22:78.
15. Hanson AD, Jacobsen JV (1984). Control of lactate glycolysis, lactate dehydrogenase, and α-amylase by O_2 deficit in barley aleurone layers. Plant Physiol 75:566.
16. Leblova S, Zima J, Perglerova E (1976). Conversion of pyruvate under natural and artificial anaerobiosis in maize. Aust J Plant Physiol 3:755.
17. Barthova J, Borvak J, Leblova S (1976). Isolation and properties of lactate dehydrogenase from germinating pea plants. Phytochemistry 15:75.
18. Boyle SA, Yeung EC (1983). Embryogeny of *Phaseolus*: developmental pattern of lactate and alcohol dehydrogenases. Phytochemistry 22:2413.
19. Jervis L, Robertson ER, Schmidt CNG (1981). Affinity purification and subunit structure of soya bean lactate dehydrogenase. Phytochemistry 20:2117.
20. Smith AM, ap Rees T (1979). Pathways of carbohydrate fermentation in the roots of marsh plants. Planta 146:327.
21. Fottrell PF (1966). Dehydrogenase isoenzymes from legume root nodules. Nature 210:198.
22. Poerio E, Davies DD (1980). A comparison of potato and vertebrate lactate dehydrogenases. Biochem J 191:341.
23. Oba K, Murakami S, Uritani I (1977). Partial purification and characterization of L-lactate dehydrogenase isozymes from sweet potato roots. J Biochem 81:1193.
24. Gleba YY, Hoffmann F (1978). Hybrid cell lines *Arabidopsis thaliana* + *Brassica campestris*: no evidence for specific chromosome elimination. Mol Gen Genet 165:257.

25. Betsche T, Bosbach K, Gerhardt B (1979). L-Lactate dehydrogenase from leaves of Capsella bursa-pastoris (L.) Med. Planta 146:567.
26. Asker H, Davies DD (1984). The physiological role of the isoenzymes of lactate dehydrogenase in potatoes. Planta 161:272.
27. Laszlo A, St Lawrence P (1983). Parallel induction and synthesis of PDC and ADH in anoxic maize roots. Mol Gen Genet 192:110.
28. Hoffman NE, Hanson AD (1986). Purification and properties of hypoxically induced lactate dehydrogenase from barley roots. Plant Physiol 82:664.
29. Jervis L (1981). Characterization of potato (Solanum tuberosum) lactate dehydrogenase isoenzymes by affinity chromatography and hybridization. Biochem J 197:755.
30. Betsche T (1981). L-Lactate dehydrogenase from leaves of higher plants. Biochem J 195:615.
31. Jacob J-L, d'Auzac J (1968). La l-alanine, inhibiteur allosterique de la lactate deshydrogenase du latex d'Hevea brasiliensis. Influence du pH. C R Acad Sci Paris Ser D 266:631.
32. Mayr U, Hensel R, Kandler O (1982). Subunit composition and substrate binding region of potato L-lactate dehydrogenase. Phytochemistry 21:627.
33. Bosbach K, Hurka H (1981). Biosystematic studies on Capsella bursa-pastoris (Brassicaceae): enzyme polymorphism in natural populations. Pl Syst Evol 137:73.
34. Sherwin T, Simon EW (1969). The appearance of lactic acid in Phaseolus seeds germinating under wet conditions. J Exp Bot 20:776.
35. Monk LS, Crawford RMM, Brandle R (1984). Fermentation rates and ethanol accumulation in relation to flooding tolerance in rhizomes of monocotyledonous species. J Exp Bot 35:738.
36. Sachs MM, Freeling M, Okimoto R (1980). The anaerobic proteins of maize. Cell 20:761.
37. Dennis ES, Sachs MM, Gerlach WL, Finnegan EJ, Peacock WJ (1985). Molecular analysis of the alcohol dehydrogenase 2 (Adh2) gene of maize. Nucl Acids Res 13:727.
38. Hanson AD, Jacobsen JV, Zwar JA (1984). Regulated expression of three alcohol dehydrogenase genes in barley aleurone layers. Plant Physiol 75:573.

39. Tsay RC, Taylor IEP (1978). Isoenzyme complexes as indicators of genetic diversity in white spruce, *Picea glauca*, in southern Ontario and the Yukon Territory. Formic, glutamic, and lactic dehydrogenases and cationic peroxidases. Can J Bot 56:80.
40. Fritz PJ, Jacobson KB (1965). Multiple molecular forms of lactic dehydrogenase. Biochemistry 4:282.

Plant Gene Systems and Their Biology, pages 131–138

REGULATION OF PROTEIN GENE EXPRESSION IN ROOTS DURING SALINITY STRESS

Subbanaidu Ramagopal

USDA-ARS, Experiment Station
Hawaiian Sugar Planters' Association
P. O. Box 1057, Aiea, HI 96701

Pathways and Genetics of Na^+ and Cl^- Transport in Roots

Root system plays a pivotal role in ion uptake and regulation. Sodium chloride, the major compound leading to salinity of soils, presumably enters root cells by apoplastic and symplastic pathways (1). Compartmentation of NaCl into vacuoles is preceded by the symplastic pathway in the cytoplasm. Quite independent mechanisms are implicated in the import of Na^+, a nonessential element beneficial to some plants and Cl^-, an essential micronutrient, into cytoplasm and vacuoles; similarly, different mechanisms may be involved for the exit of each of these ions out of the cellular compartments (1, 2). Once in the stelar region, transport of Na^+ and Cl^- to upper parts of the plant occurs in xylem vessels. The compartmentation and recirculation of Na^+ and Cl^- in various parts of the shoot may follow mechanisms similar to those of the root system. Both ions are also transported, along with photosynthetic assimilates, back to the root system via the symplastic network of cells in the phloem tissue. The final exclusion of excess ions is supposed to take place again in the upper region of roots. It is evident, therefore, that root cells are crucial for both the entry and export of Na^+ and Cl^- ions. None of the biochemical and molecular details of these transport processes in roots are understood presently, although an involvement of specific, membrane carrier proteins has been postulated (1-3).

Ion uptake, transport, and exclusion all may be regulated by genetic mechanisms as suggested by earlier studies (4, 5). These studies employed primarily intraspecific variants. To my knowledge, only one study--that of Abel (6)--has been often quoted in this regard (1, 2, 5). He showed that the regulation of Cl^- transport to shoot by root xylem parenchyma which may be controlled by a single gene pair was heritable in two soybean cultivars (6). It was suggested that these alleles were dominant in the salt-tolerant genotype but were recessive in the salt-sensitive cultivar. Further genetic characterization of this sytem has apparently not been undertaken beyond these earlier findings. The above, or other investigations, have not addressed the genetics of Na^+ transport. Nevertheless, Abel's study is important in that it indicates the potential of utilizing genotypes of contrasting responses to salinity in investigating the genetic and molecular mechanisms of ion transport until well-defined transport mutants become available in plants.

Barley System and the Objective

Barley (*Hordeum vulgare*) is a good model plant to investigate the mechanisms of salt-tolerance. It is one of the most salt-tolerant crops known and cultivars showing differences in salt-tolerance have been identified (7). Our laboratory is currently concentrating on two such barley cultivars, California Mariout 72 (CM72), a salt-tolerant variety, and Prato, a salt-sensitive variety (8). Biochemically, the two varieties are quite homogenous, in terms of their protein patterns and mRNAs (8). Variation in gene expression during salinity stress is being examined with the hope that such an approach will ultimately lead to the identification of genes responsible for uptake and regulation of Na^+ and Cl^- ions in the salt-tolerant and salt-sensitive cultivars. In this paper, I summarize our recent findings (9, 11; unpublished data) on protein gene expression in roots of the above two barley cultivars.

Protein Synthesis in Seedling Roots

Table I shows data for protein synthesis in young seedlings of CM72 and Prato treated for various periods with NaCl. Incorporation of ^{35}S-methionine was progressively reduced by increasing NaCl concentration in both genotypes. However, the magnitude of inhibition was somewhat different in the two genotypes. The decrease in protein synthesis was similar in CM72 whether it was exposed to NaCl for 4 h or 24 h, whereas, in Prato, the inhibition appeared to be considerably less after 24 h of exposure, particularly, at 1% and 2% NaCl. Moreover, protein synthesis seemed less inhibited in Prato than in CM72 at all NaCl levels.

TABLE I.
SALINITY EFFECTS ON ^{35}S-METHIONINE INCORPORATION IN SEEDLING ROOTS.

NaCl	CM72 4 h		CM72 24 h		Prato 4 h		Prato 24 h	
	%							
0	100	(12.9)*	100	(10.6)*	100	(6.3)*	100	(6.6)*
1	31.8		32.1		46.0		87.8	
2	20.2		19.8		38.1		62.1	
3	10.1		9.7		20.6		22.7	

* Actual radioactivity in cpm x 10^{-6}.

Three-day old barley seedlings were treated with the indicated level of NaCl for 4 h and 24 h and 10 root tips from each treatment were pulse-labeled with ^{35}S-methionine for 2 h (9). Incorporation into proteins was measured as trichloroacetic acid-insoluble radioactivity and was determined as described (9).

The above incorporation data probably give a rough estimate of the protein synthesis status and may or may not reflect the absolute rate differences between the two genotypes. For example, elsewhere it will be reported (9) that there were intrinsic differences in uptake of ^{35}S-methionine among CM72 and Prato and the uptake was also differentially affected by salinity. In addition, the absolute growth rates of roots were also different in these barley cultivars (8). Therefore, at the moment, these factors confound an incisive analysis of the quantitative aspects of protein synthesis. An extensive study would be necessary in future to understand the quantitative parameters involved in the regulation of protein synthesis in these two barley genotypes, which should consider the levels of actual precursor pools and the various limiting factors (many yet to be characterized in plant systems) that control the global rates of protein synthesis (10).

One of our primary objectives of the above experiment (Table 1) was to identify the proteins altered by salinity stress. Proteins from roots of CM72 and Prato were resolved by gel electrophoresis. Staining of proteins with Commassie blue or silver revealed no salinity-induced differences. However, specific changes in protein patterns were discernible by a brief incubation of roots with ^{35}S-methionine and analysis of radioactively-labeled polypeptides by fluorography. Salinity either induced the synthesis of new proteins, or, increased or decreased the relative amounts of a few other proteins (Table 2). These protein changes were identical in both genotypes, except that the amount of one protein apparently increased in Prato and that of another decreased in CM72. Proteins were also separated by 2-D gels (9), and again, no genotypic differences were apparent. Moreover, when stress was released, the salinity-induced protein changes disappeared in both genotypes in the same manner (9).

TABLE 2.
SPECIFIC CHANGES IN NEWLY-SYNTHESIZED PROTEINS DURING SALINITY STRESS.

	Induced	Increased	Decreased
		(kDa)	
Proteins in:			
A. Seedling root	31.0 26.8	42.5 (P) 37.5	53.5 (CM) 51 38.5 25.6 19.8
B. Root culture	97 31.8 26.7 (P)	65	29.5 29 (P)
NaCl level causing response:			
A. Seedling root	2%	2%	2%
B. Root culture	1% 2% (97)	1% (P) 1.5% (CM)	1% (29.5) 2% (29)

Proteins were labeled with ^{35}S-methionine as described in Tables 1 and 3 and were resolved by SDS-PAGE and fluorography (12). Proteins under the column "induced" were synthesized only in salt-treated roots; those listed under other columns were made in both untreated and salt-treated roots.

If a protein change occurred in only one genotype, the name of that genotype is indicated by letters (P=Prato; CM=CM72) in parenthesis. Numbers in parenthesis following % NaCl refer to Mr of proteins affected at that salt-level.

Protein Synthesis in Root Meristem Cultures

Although salinity did induce specific protein changes in seedling roots, it was puzzling to us why the induced changes were so alike in the salt-tolerant and salt-sensitive cultivars. There may be several possible explanations. Perhaps the induced proteins may not be directly related to salt tolerance or sensitivity; however, only a detailed characterization of the individual, salinity altered proteins from each of the varieties would resolve that issue. We considered another possiblity. The expression of salinity-tolerance or sensitivity reportedly is closely linked to various processes of plant growth and development (13, 14). Perhaps the continuous and simultaneous occurrence of these processes in the seedling roots might have masked the detection of subtle, genotype-specific, salinity-induced protein differences between the two barley cultivars. At present, there are no simple experimental strategies to dissect such complex interactions of biological processes. Selection of mutants in growth or development and analysis of their responses towards saliity would be the best strategy. However, creation of such mutants in higher plants is currently out of our reach.

Root meristem cultures offer a convenient system to simplify the complex processes involved in growth and development. Here, I define growth as the process allowing a cell to divide (and enlarge) and development as a process which allows that cell to differentiate as well (and form specific tissue). Root cells can be manipulated by a choice of tissue culture media, particularly hormones, to undergo either division or differentiation (15, 16). As shown by others (15, 17), we have found that 2,4-D in the tissue culture medium allowed only cell division (callus formation) in apical root segments of both the salt-tolerant and salt-sensitive barley cultivars. These callus-forming root cultures permit analysis of the effects of salinity specific to the cell division process without the complication of cellulur differentiation. Furthermore, since only isolated root tissues were studied, possible complex interactions between the root and shoot cells, which occurred in seedling roots are avoided.

Root meristems from CM72 and Prato were exposed to NaCl and callus growth was monitored. Growth reduction was similar in both varieties under salinity stress (Table 3). No callus growth occurred above 2% NaCl. Salinity treatment also inhibited protein synthesis in both varieties. The magnitude of inhibition was similar in both varieties, except that protein synthesis in Prato was affected somewhat less than in CM72 at NaCl levels above 1.5% (Table 3). Again, these quantitative differences need to be critically evaluated as pointed out above in studies with seedling roots.

TABLE 3.
SALINITY EFFECTS ON GROWTH AND ^{35}S-METHIONINE INCORPORATION IN ROOT MERISTEM CULTURES.

NaCl (%)	Growth		Incorporation			
	CM72	Prato	CM72		Prato	
			2-Wk	4-Wk	2-Wk	4-Wk
0	100 (460)[a]	100 (250)[a]	100 (14.7)[b]	100 (7.3)[b]	100 (14.3)[b]	100 (8.0)[b]
1	24.7	35.6	26.5	23.3	25.2	30.0
1.5	16.8	18.4	13.6	8.2	13.3	15.0
2	5.0	3.6	0.2	0.3	14.7	2.5

[a] Actual protein content in µg per 10 root tips.
[b] Actual radioactivity in cpm x 10^{-6}.

Root tips were cultured on agar-solidified Murashige and Skoog medium with and without NaCl (12). Callus growth was monitored by changes in total protein content. Protein synthesis was determined after 2 and 4 weeks of culture as described in Table 1 except that the labeling period was 3 h.

Altogether, six of the newly-synthesized proteins showed notable changes during salinity stress in root cultures (Table 2). Based on molecular mass, only two of the six proteins were identical to those detected in seedling roots; all others were dissimilar. Therefore, salinity apparently regulates a distinct set of proteins in dividing root cells. Whereas the threshold NaCl eliciting the different protein changes was about 2% in seedling roots, it varied for different proteins in root cultures.

Furthermore, it is also evident, in contrast to seedling roots that not all salinity-induced protein changes in root cultures of CM72 and Prato were identical (Table 2 and unpublished data). All six proteins were regulated by salinity in Prato but only four of those in CM72. Although the 97 kDa protein was induced in both varieties, it was synthesized in lesser amounts in Prato. The threshold NaCl level leading to the increased synthesis of 65 kDa protein was 1% in Prato while it was 1.5% in CM72. Additional genotypic differences were also observed by following changes in the salinity-induced proteins after the withdrawl of NaCl: for example, the 31.8 kDa protein was continued to be made in CM72 but was not synthesized in Prato; the synthesis of 29.5 kDa protein was resumed in Prato but not in CM72. Possibly other genotypic variations may be apparent by extending these analyses by 2-D gels.

Concluding Remarks

The above findings demonstrate that a select group of proteins are regulated during salinity stress in barley roots. This study emphasizes the need to investigate the effects of salinity at simplified stages of root cell growth and development for unravelling the intrinsic varietal responses in gene expression. Salinity regulated a distinct set of proteins in root cultures where cells primarily undergo division, and another set in seedling roots in which cells undergo both division and differentiation. The newly-induced proteins, as well as the ones whose synthesis was elevated by salinity treatment, may be particularly important for ion regulation within root cells. Root cultures distinguished the differential regulation of proteins in the salt-tolerant and salt-sensitive cultivars and this system deserves further exploration. Additional gene expression studies on seedling roots has demonstrated the existence of genotype-specific differences at the mRNA level (11). Overall, these preliminary findings on gene expression in barley roots might help explain the underlying genetic mechanisms of salt-tolerance and sensitivity in plants. Additional research is warranted to establish the nature of the root proteins regulated during salinity stress and to understand their role in transport and regulation of Na^+ and Cl^- ions.

Acknowledgements

I thank Andrew Maretzki and Paul Moore for reading and Stella Uehara for typing the manuscript.

References

1. Marschner H (1986). "Mineral Nutrition in Higher Plants." New York: Academic Press.
2. Lüttge U (1983). Import and export of mineral nutrients in plant roots. In Läuchli A, Bielski RL (eds) "Encyclopedia of Plant Physiology, New Series Volume 15B: Inorganic Plant Nutrition," Berlin: Springer-Verlag, pp 181-211.
3. Leonard RT (1984). Membrane-associated ATPases and nutrient absorption by roots. In Tinker PB, Läuchli A (eds) Advances in Plant Nutrition, Volume 1, "New York: Praeger, pp 209-240.
4. Gerloff CC, Gableman WH (1983). Genetic basis of inorganic plant nutrition. In Läuchli A, Bielski RL (eds) "Encyclopedia of Plant Physiology, New Series Volume 15B: Inorganic Plant Nutrition," Berlin: Springer-Verlag, pp 453-480.

5. Jennings DH (1986). Salt relations of cells, tissues, and roots. In Steward FC (ed) "Plant Physiology, A Treatise, Volume IX: Water and Solutes in Plants." New York: Academic Press, pp 225-379.
6. Abel GH (1969). Inheritance of the capacity for chloride inclusion and chloride exclusion by soybeans. Crop Sci 9: 697-698.
7. Epstein E, Norlyn JD, Rush DW, Kingsbury RW, Kelley DB, Cunningham GA, Wrona AF (1980). Saline culture of crops: a genetic approach. Sci 210: 399-404.
8. Ramagopal S (1987). Molecular biology of salinity stress: Preliminary studies and perspectives. In Bruening G, Kosuge T, Harada J, Hollaender A (eds) "Tailoring Genes for Crop Improvement: An Agricultural Perspective." New York: Plenum Press, pp 111-119.
9. Ramagopal S (1987). Salinity stress induced tissue-specific proteins in barley seedlings. Plant Physiol (in press).
10. Moldave K (1985). Eukaryotic protein synthesis. Ann Rev Biochem 54: 1109-1149.
11. Ramagopal S (1987). Differential mRNA transcription during salinity stress in barley. Proc Natl Acad Sci USA 84: 94-98.
12. Ramagopal S (1986). Protein synthesis in a maize callus exposed to NaCl and mannitol. Plant Cell Rep 5: 430-434.
13. Bernstein L, Hayward HE (1958). Physiology of salt tolerance. Ann Rev Plant Physiol 9: 25-46.
14. Tal M (1985). Genetics of salt-tolerance in higher plants: theoretical and practical considerations. Plant and Soil 89: 199-226.
15. Dale PH, Deambrogio E (1974). A comparison of callus induction and plant regeneration from different explants of *Hordeum vulgare*. Z. Pflanzenphysiol. 94: 65-77.
16. Solomon M. Gedalvoich E, Mayer AM, Poljakoff-Mayber A (1986). Changes induced by salinity to the anatomy and morphology of excised pea roots in culture. Ann Bot 57: 811-818.
17. Cheng TY, Smith HH (1975). Organogenesis from callus cultures of *Hordeum vulgare*. Planta 123: 307-310.

Plant Gene Systems and Their Biology, pages 139–151

ANALYSIS OF FUNGAL ELICITOR- AND UV LIGHT-INDUCED GENE EXPRESSION IN PARSLEY CELLS[1]

Carl J. Douglas, Jeffery L. Dangl,
Heidi Hoffmann, Susanne Lipphardt[2]
and Klaus Hahlbrock

Abteilung Biochemie and
Abteilung Genetische Grundlagen der
Pflanzenzüchtung[2], Max-Planck-Institut für
Züchtungsforschung, D-5000 Köln 30, FRG

ABSTRACT We have used suspension-cultured parsley cells to study the fungal elicitor- or UV-light-induced transcriptional activation of plant defense-related genes. The structure and expression of genes encoding 4-coumarate:CoA ligase (4CL), a key enzyme of general phenylpropanoid metabolism, were studied in detail. Two highly homologous 4CL genes are present in parsley and each appears to be regulated by both elicitor and UV light. The sequences of the two genes 5' to transcription initiation sites are compared. A transient expression system utilizing homologous parsley protoplasts is described which will allow rapid analysis of potential DNA regulatory elements upstream of 4CL and other genes.

INTRODUCTION

Complex genetic and biochemical factors regulate the ability of plants to successfully defend themselves against pathogens (1, 2, 3). It is probable that in most cases the sum of a number of defense mechanisms determines the

[1]This work was supported by Max-Planck Society and Fonds der Chemischen Industrie. JLD is supported by a NSF Postdoctoral Fellowhip.

outcome of interactions between potential pathogens and plants. Among these defense mechanisms are those which are induced in plants after infection by pathogens. These inducible defense responses include the synthesis of specific antimicrobial enzymes such as ß-glucanases and chitanases (4, 5), structural components of cell walls such as hydroxyproline-rich glycoproteins (6) and lignin (7), and secondary metabolites which function as phytoalexins (plant antimicrobial compounds; 1, 8).

Many inducible defense reactions are triggered not only by pathogens, but also by elicitors (small molecular weight compounds derived from fungal cell walls or culture filtrates). Plants or plant cell cultures treated with elicitors respond by producing phytoalexins and other defense-related compounds (8, 9). This has allowed the establishment of simplified systems for studying the molecular biology of these induced reactions in which large numbers of cells react uniformly to elicitor preparations. Our studies have focused on the use of parsley cell cultures treated with an elicitor preparation from *Phytophthora megasperma* f.sp. *glycinea* (Pmg) as a model system.

Elicitor-treated parsley cells exhibit marked changes in gene expression, as demonstrated by two-dimensional gel electrophoresis of proteins newly synthesized after addition of elicitor. In these experiments, transient increases in the rates of synthesis of a large number of proteins was observed (26). Subsequently, differential screening of a cDNA library prepared using RNA from elicitor-treated cells led to the identification of a number of non-cross hybridizing cDNA clones encoding parsley genes which are transcriptionally regulated by elicitor (Somssich et al., in preparation). Transcription of many of these genes is strongly, but transiently, induced, while the transcription of some is repressed. Although the function of most of these genes is unknown, their rapid induction or repression indicates that *Pmg* elicitor treatment leads to a switch in the cellular program of gene expression at the transcriptional level. Similar findings have been reported by Cramer et al. (10) in cultured bean cells. It is possible that these observed changes in gene expression reflect events that also take place at and around sites of pathogen infection of plants. This question is approachable using techniques of immunocytochemistry and *in situ* hybridization.

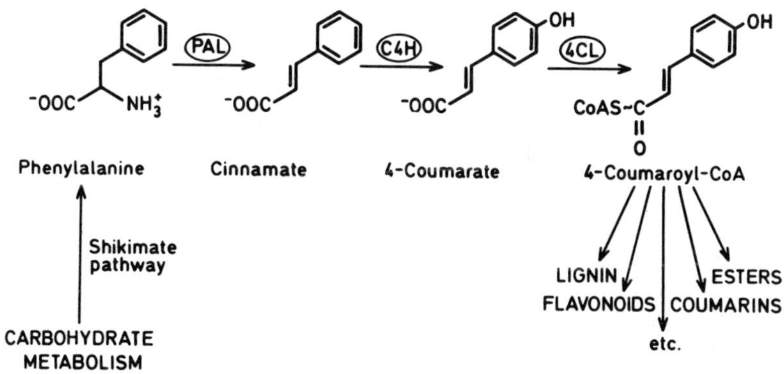

FIGURE 1. General phenylpropanoid metabolism, showing the key role played by 4CL in the synthesis of defense-related phenylpropanoid derivatives.

Our interests lie in the molecular mechanism(s) by which *Pmg* elicitor activates transcription of parsley genes. For this purpose, we have chosen to study the regulation of elicitor-induced genes of known function: those encoding 4-coumarate:CoA ligase (4CL), a key enzyme of general phenylpropanoid metabolism. Induction of phenylpropanoid metabolism (Figure 1), is the best characterized biochemical response of parsley cells to *Pmg* elicitor. Stimulation of general phenylpropanoid metabolism is a nearly ubiquitous response of plants to pathogen infection or elicitor treatment, presumably because of the role that various phenylpropanoid derivatives play as defense compounds (2). Genes encoding 4CL and phenylalanine ammonia-lyase (PAL), the first enzyme of general phenylpropanoid metabolism (Figure 1), are transcriptionally activated in parsley cells following elicitor treatment (11). The timing of this activation is very similar to that observed for many elicitor-activated genes of unknown function described above (Somssich et al., in preparation). This is followed by increases in 4CL and PAL mRNA abundance (12), enzyme activities (13, 14) and accumulation of furanocoumarins, considered to be the phytoalexins of parsley (15). Interestingly, PAL and 4CL are similarly regulated by UV irradiation, but in this case a different branch pathway is activated, leading, through chalcone synthase (CHS), to the synthesis of UV protective flavonoid glycosides (8).

Our initial studies show that 4CL genes are present at a low copy number in parsley (two copies per haploid genome). Genomic clones encoding two 4CL genes have been isolated and the genes sequenced. Analysis of the expression of these genes indicates that each is transcriptionally regulated by both elicitor and UV irradiation. The uncomplicated organization of the 4CL genes should facilitate molecular studies of regulatory elements 5' to the transcription initiation sites. The basis for such experiments has been established using parsley protoplasts, which respond to Pmg elicitor treatment and UV irradiation in the same way as suspension cultured parsley cells. We show that plasmid constructs can be introduced into the protoplasts and that transient gene expression from promotors on these constructs can be monitored.

RESULTS

Structure and Expression of 4CL Genes in Parsley

Lambda clones containing two separate 4CL genes were isolated from a parsley genomic library constructed in λ-EMBL-4 using a 450-bp 4CL cDNA (12). Restriction enzyme analysis and

Southern blots indicated that additional copies of 4CL are not present in the parsley genome. The complete nucleotide sequences of both parsley 4CL genes, designated Pc4CL-1 and Pc4CL-2, were determined and the structure of the genes deduced by comparing the genomic sequences with the cDNA sequences of near full-length 4CL cDNAs (H. Hoffmann et al., in preparation). Transcription start sites were determined by S1 nuclease protection and primer extension assays. Both genes contain 5 exons, separated by introns of varying lengths (Figure 2). The two genes are

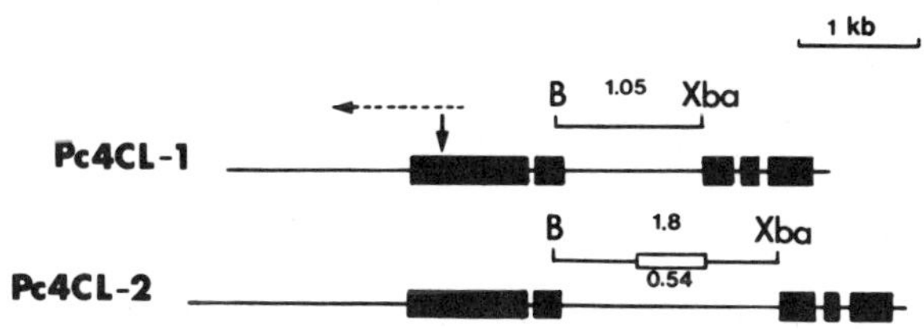

FIGURE 2. Structure of Pc4CL-1 and Pc4CL-2 (5' ends to the left), showing exons (filled-in boxes), location of run-off transcription probes derived from the second introns, and locations of the probe used in S1 nuclease protection (horizontal dashed arrow) and the oligonucleotide used in primer extension (vertical arrow) assays to determine the 5' transcription start sites.

highly homologous both in coding regions and 3' and 5' untranslated portions; the only major difference is in the size of the second large intron.

Because the high degree of homology between the two 4CL genes precluded the use of hybridization probes specific to mRNA transcripts from each gene, we used two alternative approaches to assess the expression of the genes after elicitor treatment or UV irradiation. Individual base pair differences in the 3' untranslated regions of the two genes allowed transcripts from the genes to be distinguished on the sequence level. Thus, the sequences of these regions from cDNA clones obtained from libraries made with RNA from elicitor-treated or UV-irradiated cells were compared to sequences of Pc4CL-1 and Pc4CL-2. As shown in Table 1, cDNA

TABLE 1
NUMBER OF cDNA CLONES DERIVED FROM Pc4CL-1 AND Pc4CL-2

	cDNA Source	
	elicitor-induced cells	UV-irradiated cells
Pc4CL-1	4	2
Pc4CL-2	4	7

clones corresponding to each gene were found in both types of libraries, suggesting that both genes are induced by both stimuli.

As a second approach, we assayed the level of transcription specific to Pc4CL-2 using a 0.54-kb DNA fragment unique to the second intron of this gene (see Figure 2). Run-off transcripts were hybridized to this fragment, as well as to DNA fragments containing sequences homologous to both genes (the Bam-Xba 1.8-kb fragment from the second intron and a 4CL cDNA). Transcripts specific to the Pc4CL-2 0.54-kb fragment were induced in nuclei isolated from both elicitor-treated and UV-irradiated cells, indicating that Pc4CL-2 is transcriptionally induced by both these stimuli. This induction was 2-3 fold lower than the induction of transcripts homologous to both 4CL genes. Taken together, the experiments indicate that each 4CL gene is induced following elicitor treatment or UV irradiation.

5' DNA Sequences of Pc4CL-1 and Pc4CL-2

A schematic representation of the nucleotide sequences of Pc4CL-1 and Pc4CL-2 5' to the transcription initiation sites is shown in Figure 3. The sequences are almost identical

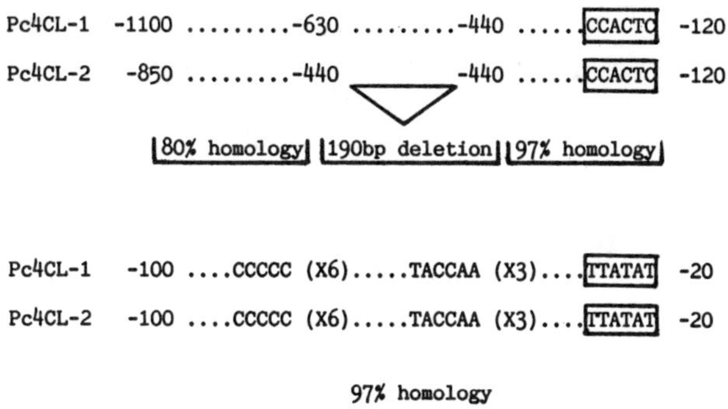

FIGURE 3. Schematic representation of the sequences of Pc4CL-1 and Pc4CL-2 upstream of the transcription initiation sites, comparing the two sequences and showing potential regulatory elements.

(97 % homologous) up to -440 bp; further upstream there is more sequence divergence and beyond about -1100 and -850 bp detectable homology ceases. Sequences homologous to CAAT (-120 bp) and TATA (-20 bp) boxes are found in the genes. Between these two elements are a three-fold repetition of the sequence TACCAA and a 6-fold repetition of the sequence CCCCC. The role that these elements may play in 4CL expression is being investigated using gene transfer techniques.

Gene Transfer into Parsley Protoplasts

In order to define the regulatory elements of the 4CL genes responsible for elicitor and UV induction, we developed a gene transfer system for parsley protoplasts. Protoplasts are easily prepared from parsley cell suspension cultures, and these protoplasts will rapidly regenerate new suspension cultures. Most importantly, there is no background activation due to protoplast preparation of either 4CL or of other genes encoding enzymes of phenylpropanoid metabolism such as PAL and chalcone synthase (CHS). Additionally, we have characterized the response of protoplasts to elicitor treatment and UV irradiation. We find that induction of 4CL and CHS mRNA accumulation is nearly identical, in timing and in amount, to the induction of these mRNAs in suspension cultured cells. Freshly isolated protoplasts and 3 day old regenerating protoplasts are indistinguishable in their responses. In the experiment shown in Figure 4,

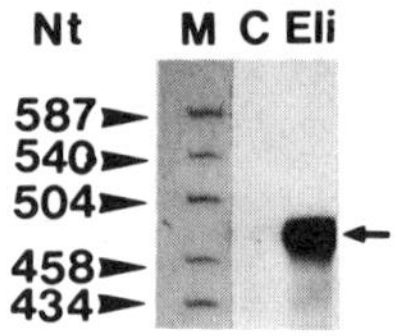

FIGURE 4. Quantitative S1 nuclease protection assay of RNA extracted from untreated 3 day old protoplasts (C) and from 3 day old protoplasts treated with 50 μg/ml Pmg elicitor (Eli), with the sizes of marker fragments (M) given in nucleotides (Nt). The arrow indicates the expected size of protected fragments.

4CL mRNA levels in untreated 3 day old protoplasts or in the same protoplasts treated 5 h with *Pmg* elicitor were determined by quantitative S1 nuclease protection assays. Elicitor treatment induced the accumulation of 4CL mRNA by about 20-fold above background, identical to the induction observed in suspension cultures. The quantitative and qualitative accumulation of furanocoumarin and flavonoid glycoside end products of elicitor and UV treatments is also maintained in protoplasts. We feel, therefore, that transient expression of newly introduced DNA in parsley protoplasts will provide a valid system in which to define regulatory elements of 4CL and other genes whose expression is inducible.

We introduced plasmid DNA into freshly isolated protoplasts by the PEG-Ca$(NO_3)_2$ method (16, 17). Our test vector contained the bacterial neomycin phosphotransferaseII (NPTII) coding sequence driven by the 35S promotor from cauliflower mosaic virus (CaMV) and flanked by the 35S polyadenylation signals (pRT100neo, a kind gift of R. Töpfer, Abt. Schell, MPI). Protoplasts were harvested and protein extracts analyzed for NPTII activity using the native gel analysis technique (18, 19). These experiments allowed us to establish the time course of transient expression of foreign DNA from a constitutive promotor, as well as the optimal amount of plasmid DNA required for expression. Results are shown in Figure 5. When 80 μg of

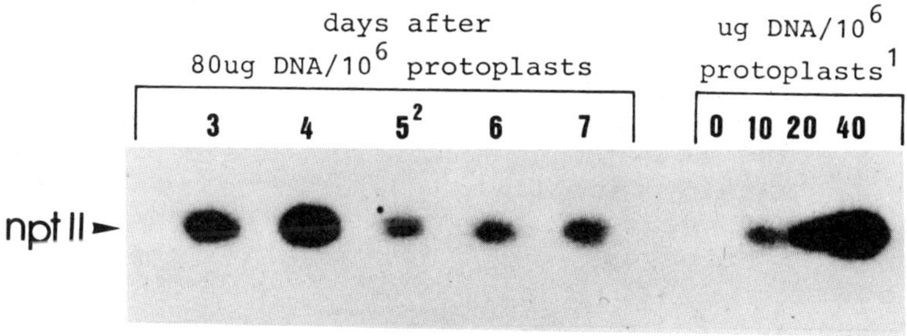

[1] 3 days after transformation
[2] The amount of protein loaded was one third of that in other samples

FIGURE 5. Transient expression of NPTII in parsley protoplasts (see text for details).

supercoiled plasmid DNA per 10^6 protoplasts was used, a maximum of NPTII expression was observed between 3 and 5 days after transfer. DNA titration revealed that expression is detectable 3 days after transformation with as little as 10 μg DNA per 10^6 protoplasts, and that 20-40 μg/10^6 protoplasts is probably optimal. We can detect NPTII activity with as little as 2.5 μg DNA/10^6 protoplasts (not shown). We are currently developing methods to analyze transient gene expression in elicitor- and UV light-treated parsley protoplasts.

DISCUSSION

Our studies show that genes encoding 4CL are arranged in a simple manner in the parsley genome. Two 4CL genes are present, and each appears to be regulated by both fungal elicitor and UV irradiation. In further studies aimed at determining the DNA sequences involved in the transcriptional regulation of 4CL, it will be of interest to establish whether the same or distinct DNA regulatory elements are involved in regulation by elicitor and UV irradiation. There are precedents for a single promotor being responsive to distinct stimuli, such as those of metallothionein (20) and heat shock (21) genes in animal systems. However, the nature of plant regulatory elements controlling the response to such diverse signals as elicitor and UV irradiation remains unknown. The low copy number of 4CL should facilitate these studies, as the transcriptional activation of 4CL is the sum of the activation of only two highly homologous genes. Thus, the complications of a multigene family, in which different members may be differently regulated (22), is avoided.

Recent studies analyzing sequences involved in regulation of inducible plant genes, mostly those regulated by light (23, 24, 25), have made use of transgenic plants or calli transformed using *Agrobacterium tumefaciens*. We have decided to develop a transient gene expression system utilizing homologous parsley protoplasts to delineate DNA elements involved in elicitor and UV light inducible expression of 4CL. We show that freshly isolated or 3 day old regenerating protoplasts respond to elicitor treatment and UV irradiation in a manner identical to suspension cultured cells. We also characterized the expression from a constitutive promotor, CaMV 35S, introduced into these cells. We now plan to use this system to study the inducible expression of constructions containing 4CL sequences 5' to the transcription initiation sites. The use of a transient assay system has the obvious advantage of rapidity, allowing analysis of a large number of promotor constructions. Results obtained from transient experiments can be confirmed in stably transformed cells. The same assay system could also be used to rapidly assess the properties of promotors from individual genes belonging to more complex multigene families. The information obtained from analysis of the regulatory regions of 4CL should aid in elucidating the molecular mechanisms involved in transcriptional activation of 4CL and other genes by fungal elicitor or UV light.

REFERENCES

1. Bailey JA (1983). Biological perspectives of host-pathogen interactions. In Bailey JA, Deverall BJ (eds): "The Dynamics of Host Defense", New York: Academic Press, p 1.
2. Hahlbrock K and Scheel D (1987). Biochemical responses of plants to pathogens. In Chet I (ed): "Non-Conventional Approaches to Plant Disease Control", in press.
3. Ellingboe AH (1984). Genetics of host-parasite relations: an essay. Adv Plant Pathol 2:131.
4. Bell AA (1981). Biochemical mechanisms of disease resistance. Ann Rev Plant Physiol 32:21.
5. Boller T (1985). Induction of hydrolases as a defense reaction against pathogens. In Key JL Kosuge T (eds): "Cellular and Molecular Biology of Plant Stress," New York: Alan Liss, p 247.
6. Showalter AM, Bell JN, Cramer CL, Bailey JA, Varner JE and Lamb CJ (1985). Accumulation of hydroxyproline-rich glycoprotein mRNAs in response to fungal elicitor and infection. Proc Natn Acad Sci USA 82:6551.
7. Ride JP (1983). Cell walls and other structural barriers in defense. In Callow TA (ed): "Biochemical Plant Pathology", New

York: John Wiley and Sons, p 215.

8. Hahlbrock K, Kreuzaler F, Ragg H, Fautz E. and Kuhn DN (1982). Regulation of flavonoid and phytoalexin accumulation through mRNA and enzyme induction in cultured plant cells. In Jaenicke L (ed): "Biochemistry of Differentiation and Morphogenesis", Berlin: Springer-Verlag, p 34.
9. Dixon RA (1986). The phytoalexin response: elicititation, signalling and control of host gene expression. Biol Rev 61:239.
10. Cramer CL, Ryder TB, Bell, JN and Lamb CJ (1985) Rapid switching of plant gene expression induced by fungal elicitor. Science 227:1240.
11. Chappell J and Hahlbrock K (1984). Transcription of plant defense genes in response to UV light or fungal elicitor. Nature 311:76.
12. Kuhn DN, Chappel J, Boudet A and Hahlbrock K (1984) Induction of phytoalexin ammonia-lyase and 4-coumarate:CoA ligase mRNAs in cultured plant cells by UV light or fungal elicitor. Proc Natl Acad Sci USA 81:1102.
13. Hahlbrock K, Lamb CJ, Purwin C, Ebel J, Fautz E and Schäfer E (1981) Rapid response of suspension-cultured parsley cells to the elicitor from *Phytophthora megasperma* var. *sojae*. Plant Physiol 67:768.
14. Ragg H, Kuhn DN and Hahlbrock K (1981) Coordinated regulation of 4-coumarate:CoA ligase and phenylalanine ammonia-lyase mRNAs in cultured plant cells. J Biol Chem 256:10061.
15. Knogge W, Kombrink E, Schmelzer E and Hahlbrock K (1987) Occurrence of phytoalexins and other putative defense-related substances in uninfected parsley plants. Planta, in press.
16. Krens FA, Molendijk L, Wullems, GJ and Schilperoort RA (1982) *In vitro* transformation of plant protoplasts with Ti-plasmid DNA. Nature 296:72.
17. Hain R, Stabel P, Czernilofsky AP, Steinbiß HH, Herrera-Estrella L and Schell J (1985)

Uptake, integration, expression and genetic transmission of a selectable chimeric gene gy plant protoplasts. Molec Gen Genet 199:161.

18. Reiss B, Sprengel R, Will H and Schaller H (1984). A new sensitive method for qualitative and quantitative analysis of NPT in crude cell extracts. Gene 30:211.

19. Schreier P, Seftor EA, Schell J and Bohnert HJ (1985) The use of nuclear sequences to direct the light-regulated synthesis and transport of a foreign gene into plant chloroplasts. EMBO J 4:25.

20. Karin M, Haslinger A, Holtgreve H, Richards RI, Krauter P, Westphal HM and Beato M (1984) Characterization of DNA sequences through with cadmium and glucocorticoid hormones induce human metallothionein-II_A gene. Nature 308:513.

21. Wu BJ, Kingston RE and Morimoto RI (1986). Human HSP70 promoter contains at least two distinct regulatory domains. Proc Natn Acad Sci USA 83:629

22. Fluhr R, Moses P, Morelli G, Coruzzi G and Chua N-H (1986). Expression dynamics of the peal *rbc*S multigene family and organ distribuation of the transcripts. EMBO J 5:2063.

23. Fluhr R, Kuhlemeier C, Nagy F and Chua H-H (1986). Organ-specific and light-induced expression of plant genes. Science 232:1106.

24. Simpson J, Van Montagu M and Herrera-Estrella L (1986) Photosynthesis-associated gene families: differences in response to tissue-specific and environmental factors. Science 233:34.

25. Kaulen H, Schell J and Kreuzaler F (1986). Light-induced expression of the chimeric chalcone synthase-NPTII gene in tobacco cells. EMBO J 5:1.

26. Kombrink E, Bollmann J, Hauffe KD, Knogge W, Scheel D, Schmelzer E, Somssich I and Hahlbrock K (1986). Biochemical responses of non-host plant cells to fungi and fungal elicitors. In Bailey J (ed): "Biology and Molecular Biology of Plant-Pathogen Interactions", Berlin: Springer-Verlag, p 253.

Plant Gene Systems and Their Biology, pages 153–160

ISOLATION OF cDNA CLONE ENCODING WHEAT GERM AGGLUTININ

Natasha V. Raikhel[1], Thea A. Wilkins[1] and W. Michael Ainley[2]

[1]MSU-DOE Plant Research Laboratory, Michigan State University, East Lansing, MI 48824-1312 and [2]Botany Department, University of Georgia, Athens, GA 30602

ABSTRACT. Two sets of synthetic oligonucleotides coding for amino acids in the amino- and carboxy-termini of the wheat germ agglutinin polypeptide (Wright *et al.* 1984) were synthesized. Labelled oligonucleotides were used as hybridization probes to screen cDNA libraries derived from developing wheat embryos. The nucleotide sequence for a cDNA clone recovered from the cDNA libraries was determined by dideoxy sequencing in M13. The amino acid sequence deduced from the DNA sequence confirmed that the cDNA clone encodes wheat germ agglutinin.

Introduction

Wheat germ agglutinin (WGA) was the first cereal lectin characterized in detail at the biochemical and structural level. It is a 36,000 Da protein composed of two identical, non-covalently associated protomers of Mr 18,000 Da (1,2). The protomers are particularly rich in half-cystine (18%) and glycine (23%), lack covalently-bound sugar residues, and possess two independent saccharide-binding sites specific for N-acetylglucosamine oligomers (3). X-ray crystallography studies of the tertiary structure of native WGA (4) indicate that the protomers are closely associated and centered around the crystallographic 2-fold axes. Each protomer consists of four structurally homologous and spatially distinct domains designated A, B, C and D. Each

This work was supported by National Science Foundation Grant DMB 83-14374 and by Department of Energy Grant DE-AC02-76ERO-1338 to N.V.R.

domain contains 41 amino acid residues folded in an irregular configuration with no ordered secondary structure. Domains within each protomer have limited contact, whereas contact between domains of different protomers is extensive.

Three structurally and functionally closely related isolectins of WGA exist in wheat and are presumably due to the allopolyploid nature of wheat (5,6). The isolectins associate as homodimeric and heterodimeric forms. The homology among the isolectins enables an exchange of protomers *in vitro* (5). The high degree of stability displayed by WGA *in vivo* is probably related to the presence of disulfide bonds and blocked amino termini identified in two isolectins (7,8).

WGA is localized in the surface cell layer of adventitious roots of adult plants and the radicle and coleoptile of embryos (9,10). On the other hand, WGA is distributed throughout the root caps of adult plants and the coleorhiza and epiblast of embryos (9,11,12, 13). At the subcellular level, WGA is sequestered in protein bodies of the embryo and in vacuoles of adventitious roots (9,14). Accumulation of WGA in embryos is shown to be temporally regulated and tissue-specific. The radicle and coleorhiza accumulate WGA around 10 days post anthesis (DPA). However, WGA is not found in the epiblast and coleoptile until 15 and 20 DPA, respectively (13).

While these biochemical, immunological and microscopic studies have helped to define and characterize the configuration, composition and distribution of WGA, genes encoding WGA have not been isolated. We report here the isolation of a cDNA clone encoding WGA and present a comparative analysis of the amino acid sequence obtained by peptide sequencing (7) and the amino acid sequence deduced from the nucleotide sequence.

MATERIALS AND METHODS

Materials. Two cDNA libraries, derived from mRNA isolated from developing wheat grains at 3- and 4- weeks post-anthesis, were kindly provided by Dr. Chris Brinegar (ARCO Plant Research Institute). Two sets of degenerate synthetic oligonucleotides were prepared (Fig.1, probe 1 and 2). Each oligonucleotide consisted of 20 nucleotides complementary to the 5'or 3'ends of the mRNA for WGA as predicted from the amino acid sequence of the protein (7).

The oligonucleotide mixtures shown in Table 1 were synthesized on an Applied Biosystem 380 DNA synthesizer by a solid phase method (15) and separated by electrophoresis in a 20% polyacrylamide gel containing 8 M urea in TBE pH 8.3 (0.89 M Tris, 0.089 M boric acid, 2.7 mM EDTA). Oligonucleotides were eluted in 0.5 M ammonium acetate, 10 mM magnesium acetate, 0.1% SDS, and end-labelled with ^{32}P using polynucleotide kinase (16).

Isolation and screening of cDNA clones. The cDNA libraries, in *E. coli* strain DH5α (17), were plated directly onto nitrocellulose filters (18) and hybridized with the oligonucleotide probes described above. Replicate filters were prehybridized in 6X SSC, 0.25% non-fat milk and hybridized in the same buffer containing the labelled oligonucleotide probes and 0.1% SDS at 36°C (probe 1) and 38°C (probe 2). Following hybridization, filters were washed three times in 6X SSC, 0.25% non-fat milk, 0.1% SDS at room temperature, followed by 2 minutes at 46°C (probe 1) or at 48°C (probe 2). Filters were dried and autoradiographed for 16 to 18 h. Colonies that produced positive signals were selected and rescreened using the same probes under the same conditions.

DNA Sequence Determination. Inserts from recombinant plasmids were purified by electrophoresis in low-melting point agarose. Excised cDNA inserts or appropriate restriction fragments were then subcloned into M13mp18 or M13mp19. Dideoxy sequencing from single-stranded M13 templates was accomplished using a Bethesda Research Laboratory M13 sequencing kit.

RESULTS

Isolation of cDNA clones. Two sets of synthetic oligonucleotides were used for isolation of cDNA clones specific for WGA. Each oligonucleotide consisted of 20 nucleotides (Fig. 1), complementary to the 5' or 3' end of the WGA mRNA (7). Due to codon degeneracy, probe 1 is a mixture of 64 oligonucleotides and probe 2 is a mixture of 128 oligonucleotides (Fig. 1).

```
                          9    10   11   12  13   14  15

Amino acid sequence      Asn  Met  Glu  Cys Pro  Asn Asn

                           U         A    U   U    U
mRNA                  5' AA   AUG  GA   UG  CCC  AA  AA  3'
                           C         G    C   A    C
                                              G

                           A         T    A   A    A
Probe 1               3' TT   TAC  CT   AC  GGG  TT  TT  5'
                           G         C    G   T    G
                                              C

                         141  142  143  144 145  146 147

Amino acid sequence      Cys  Thr  Asn  Asn Tyr  Cys Cys

                           U    U    U    U   U    U
mRNA                  5' UG   ACC  AA   AA  UA   UG  UG  3'
                           C    A    C    C   C    C
                                G

                           A    A    A    A   A    A
Probe 2               3' AC   TGG  TT   TT  AT   AC  AC  5'
                           G    T    G    G   G    G
                                C
```

Figure 1. The amino acid sequence (7) and corresponding possible mRNA sequences for WGA that were used to prepare oligonucleotide probes 1 and 2 are shown.

Six cDNA clones, designated pNVR1 through pNVR6, were selected by hybridization to one or both probes. Only one cDNA clone, pNVR1, was recognized by both probes, suggesting that this insert contains sequences spanning the coding region delimited by the oligonucleotide probes. The cDNA insert of pNVR1 was subcloned into M13mp18 and M13mp19, and its nucleotide sequence was determined as described in Materials and Methods.

Amino acid sequences. A comparison of the deduced amino acid sequence of WGA cDNA clone pNVR1 with the published amino acid sequence of WGA is shown in Fig.2.

```
pNVR1     Q R C G E Q G S[G]M E C P N N L C C S Q
aa 1-20   Q R C G E Q G S[N]M E C P N N L C C S Q

pNVR1     Y G Y C G M G G D Y C G K G C Q[N]G A C
 21-40    Y G Y C G M G G D Y C G K G C Q[D]G A C

pNVR1     W T S K R C G S Q A G G[K]T C P N N H C
 41-60    W T S K R C G S Q A G G[A]T C P N N H C

pNVR1     C S Q Y G H C G F G A E Y C G A G C Q G
 61-80    C S Q Y G H C G F G A E Y C G A G C Q G

pNVR1     G P C R A D I K C G S Q[A]G G K L C P N
 81-100   G P C R A D I K C G S Q[S]G G K L C P N

pNVR1     N L C C S Q W G[Y]C G L G S E F C G[E]G
101-120   N L C C S Q W G[S]C G L G S E F C G[G]G

pNVR1     C Q[N]G A C S T D K P C G K D A G G R V
121-140   C Q[S]G A C S T D K P C G K D A G G R V

pNVR1     C T N N Y C C S K W G S C G I G P G Y C
141-160   C T N N Y C C S K W G S C G I G P G Y C

pNVR1     G A G C Q S G G C D[G]V F A F A I A T N
161-180   G A G C Q S G G C D[A]

pNVR1     S T L L A E termination codon
```

Fig.2. The amino acid sequence deduced from pNVR1 cDNA sequence coding for WGA is represented by the single letter code of amino acids. The differences between the deduced amino acid sequence and published amino acid sequence are designated by a box.

DISCUSSION

In this paper we report the isolation and characterization of a cDNA clone for WGA. The identity of this clone has been verified by comparison of the deduced amino acid sequence with sequence obtained by direct amino acid sequencing of the purified protein (7,8).

Wheat germ agglutinin in hexaploid wheat consists of three similar isolectins. Amino acid sequence studies on isolectins 1 and 2 indicate that they differ distinctly in their histidine content: two histidines in isolectin 2 and one histidine in isolectin 1 (8). The deduced amino acid sequence from isolated cDNA clone pNVR1 sequenced here confirms strong homology with isolectin 2, but suggests variability at 8 sequence positions. In addition, the DNA sequence of pNVR1 encodes a protein which extends 15 amino acids beyond the COOH-terminus of the published amino acid sequence. Amino acids 1 through 171 of the predicted sequence displays a 90% homology to the published amino acid sequence.

The length of the polypeptide derived from the deduced amino acid sequence is 186 amino acids and the calculated molecular weight based on the sequence is 18,754 Da. We do not know if the mature protein consists of 171 amino acids such that the COOH-terminal residues are lost during proteolytic processing. On the other hand this modification may occur during the purification of WGA and the mature protein actually contains 186 amino acids *in vivo*.

The internal homology among four domains of WGA was reported earlier and is probably important for the integrity of the molecule (7,8). The deduced amino acid sequence confirmed extensive interdomain homology. There is a short repeated peptide of 7 amino acids that has sequence the GCQNGAC. This peptide begins at amino acid 34 and extends through amino acid 40. A perfect repeat of the same sequence lies between amino acid 120 and 126. The short stretches of YCG, GCQ, CCS or CGG are found throughout the cDNA clone.

REFERENCES

1. Nagata Y, Burger MM (1974). Wheat Germ Agglutinin. Molecular characteristics and specificity for sugar binding. J Biol Chem 249:3116.

2. Allen AK, Neuberger A, Sharon N (1973). The purification, composition and specificity of wheat-germ agglutinin. Biochem J 131:155.

3. Goldstein IJ, Hayes CE (1978). The lectins: carbohydrate binding proteins of plants and animals. In Tipson RT, Horton D (eds): "Advances in Carbohydrate Chemistry and Biochemistry," New York: Academic Press, p 127.

4. Wright CS (1981). Histidine determination in wheat germ agglutinin isolectin by x-ray diffraction analysis. J Cell Biol 145:453.

5. Rice RH, Etzler ME (1975). Chemical modification and hybridization of wheat germ agglutinin. Biochem 14:4093.

6. Rice RH (1976). Wheat germ agglutinin. Evidence for a genetic basis of multiple forms. Biochem et Biophys Acta 444:175.

7. Wright CS, Gavilanes F, Peterson DL (1984). Primary structure of wheat germ agglutinin isolectin 2. Peptide order deduced from x-ray structure. Biochem 23:280.

8. Wright CS, Olafsdottir S (1986). Strucutral differences in the two major wheat germ agglutinin isolectins. J Biol Chem 261:7191.

9. Mishkind M, Raikhel NV, Palevitz BA, Keegstra K (1982) Immunocytochemical localization of wheat germ agglutinin in wheat. J Cell Biol 92:753.

10. Raikhel NV, Mishkind ML, Palevitz BA (1984). Characterization of a wheat germ agglutinin-like lectin from adult wheat plants. Planta 162:55.

11. Miskind ML, Raikhel NV, Palevitz BA, Keegstra K (1983). The cell biology of wheat germ agglutinin and related lectins. In Goldstein IJ, Etzler ME (eds): "Chemical Taxonomy, Molecular Biology, and Function of Plant Lectins," Alan R. Liss, p 163.

12. Raikhel NV, Palevitz BA, Quatrano RS (1986). Pattern of wheat germ agglutinin accumulation in different genotypes of wheat. In Bog-Hansen TC, van Driessche E (eds): "Lectins," Vol. V, p 75.

13. Raikhel NV, Quatrano RS (1986). Localization of wheat-germ agglutinin in developing wheat embryos and those cultured in abscisic acid. Planta 168:433.

14. Raikhel NV, Mishkind M, Palevitz BA (1984). Immunocytochemistry in plants with colloidal gold conjugates. Protoplasma 121:25.

15. Matteucci MD, Caruthers MH (1980). The synthesis of oligodeoxypyrimidines on a polystyrene support. Tetrahedron Lett 21:719.

16. Maniatis T, Fritsch EF, Sambrook J (1982). Molecular cloning. Cold Spring Harbor, NY: Cold Spring Harbor Laboratory.

17. Hanahan D (1985). Techniques for transformation of E. coli. In Glover (ed): "DNA cloning", Oxford: IRL Press, Vol. 1, p. 109.

18. Hanahan D, Meselson M (1983). Plasmid screening at high colony density. In Wu et al (eds): "Methods in Enzymology," New York: Academic Press, Vol 100, p 333.

Plant Gene Systems and Their Biology, pages 161–170

DISEASE RESISTANCE RESPONSE GENES: INDUCTION IN PEAS BY FUNGAL AND BACTERIAL WALL COMPONENTS AND BLOCKAGE BY HEAT SHOCK

C. H. Daniels, L. A. Hadwiger, Y. S. Cody and D. F. Kendra

Department of Plant Pathology, Washington State University, Pullman, WA 99164-6430

Introduction

The plants surviving throughout evolution have developed various systems for resisting challenges by invading microorganisms. Of the multitude of organisms which contact plants only a few can make the necessary inroads that lead to proliferation and sporulation within infected tissue. The definition of a single general process by which plant systems defend against various pathogens, each possessing unique functions and cellular features, has not been forthcoming. However, host defense is often associated with an induced host response (1).

The newly exposed endocarp surface of an immature pea pod can be used to assay both for the induction of a host response subsequent to challenge by a variety of organisms or chemicals and for the presence of elicitors released from plant pathogenic organisms which are placed in contact with the pod (2,3,4,5,6). Induction of the host resistance response has been studied primarily using the non-host pathogen *Fusarium solani* f. sp. *phaseoli* (Fsph) while *F. solani* f. sp. *pisi* (Fsp), a true pea pathogen, has been used to study the host response in a susceptible interaction. Macrocondia of the bean pathogen Fsph germinate on the pea endocarp surface but stop growing in 4-6 hours (2). This period of rapid resistance expression by the host is associated with the enhanced synthesis and accumulation of numerous mRNA species (5,7) which when translated result in the accumulation of at least 20 major plant proteins (7). Some of the proteins which increase include: i) Phenylalanine ammonia lyase (8) a key enzyme in phenylpropanoid metabolism which is necessary for the subsequent increases in lignin and pisatin (an isoflavonoid phytoalexin) production and ii) chitinase and β 1,3-glucanase (9,10), plant enzymes capable of degrading the major wall components in the fungus. In the process of developing a cDNA library of fungal-induced pea mRNA species (6), we discovered some 0.6 to 0.9 Kb messenger RNAs whose accumulation profiles are remarkably synchronized with the period in which resistance to Fsph is maximally expressed, and whose copy numbers are subsequently decreased prior to the onset of rapid proliferation by the true pea pathogen Fsp (5).

These same two fungi have been studied in relation to the release of elicitors. We have found that a minor fungal wall component chitosan, when applied to healthy pea endocarp, can mimic the bean pathogen's induction of the mRNA and proteins thus far described (11), including the induction of complete immunity against subsequent inoculation with the pea pathogen Fsp (4)(12). Thus it appears that as the cell wall of the bean pathogen contacts the viable, intact endocarp cells, the plant enzymes chitinase and β-glucanase attack the major compounds chitin and β-glucan (11) in the fungal walls releasing the minor wall component chitosan along with other wall fragments. Immunochemical and histochemical marking techniques have shown that chitosan is taken up by both *Fusarium solani forma speciales* and pea host cytoplasm within 20 minutes following inoculation (4). Chitosan acts in the fungal cell by inhibiting RNA synthesis (13) and thus growth. In the plant, chitosan is hypothesized to induce nuclear genes and can render the plant resistant (14).

Although many components contributing to resistance are in place following inoculation by either the bean or pea pathogen, after a brief period (4-12 hours) of suppressed growth the pea pathogen commences rapid growth at 20-24 hours after inoculation. This growth occurs in spite of the presence of continually increasing accumulations of chitinase (9,10), β-glucanase (9,10), lignin (Hadwiger, L. A. and Victory, S., unpublished) and pisatin (2). The accumulation of some unique mRNA species [whose corresponding genes are termed "disease resistance response genes" (DRRG)](5) and major disease resistance-associated proteins (15) are maintained in the resistance response but are depleted prior to the onset of susceptibility in the susceptible reaction (16). Therefore, we propose that resistance or the maintenance of resistance is dependent on some or all of the DRRG plant genes being transcribed or translated.

In this paper we will both summarize and relate additional information concerning *F. solani* cell wall elicitors. We will also relate some of our findings on the induction in peas of the DRRG by live cells of three races of the bacterial pathogen *Pseudomonas syringae* pv. *pisi*. In addition we will describe a number of heat shock experiments designed to test for suppression of the DRRG.

INDUCTION

F. solani cell wall elicitors.

The ability of cell wall components to induce phytoalexins has been observed in many host-parasite systems (17) and thus the production of the phytoalexin, pisatin, in peas is utilized as a simple assay to screen for elicitors. In the pea-*Fusarium solani* system, chitosan appears to be a major elicitor both of phytoalexin production and cellular immunity. In comparing the chitin- derived chitosan extractable from Fsph with that from Fsp, we found that there are differences in the concentrations required to elicit host immunity, to induce pisatin in the pea, and to inhibit growth in the two *forma speciales*. The Fsp-derived chitosan was composed of a much higher proportion of small oligomers and required 125 μg/ml compared to 20 μg/ml of that derived from Fsph to inhibit fungal growth, and 100 μg/ml compound to 10 μg/ml respectively to induce immunity in the host. The compounds chitin and β-glucan, also present in *Fusarium* walls are ineffective in inducing host immunity or inhibiting fungal growth at these levels (4). The optimal size of chitosan oligomer (shrimp shell derived) required for fungal inhibition or pisatin elicitation is 7 or more glucosamine residues (18). Interestingly both medium and high MW polygalactosamine polymers which comprise the flocculating component (19) of the fungal strain *Paecilomyces* sp. I-1 are ineffective in inducing immunity in peas to Fsp and only induce low levels of pisatin (Hadwiger, unpublished). Thus the structure of the hexosamine polymer inducing these responses is important beyond simply being a polycationic polymer. However, both of the hexosamine polymers, chitosan and polygalactosamine, can inhibit fungal growth at somewhat comparable concentrations. Fsp is completely inhibited by 8 μg/ml of shrimp shell chitosan or 62 μg/ml of polygalactosamine while Fsph is inhibited by 4 μg/ml or 31 μg/ml, respectively. These fungal inhibition values differ from those given earlier in this section as the latter values were obtained using shrimp shell chitosan rather than chitosan derived from fungal cell walls. At this point it is worth remarking that chitosan from different origins possesses differing amounts of biological activity.

As a polycation, chitosan would be expected to influence membrane structure and viability of both plant and fungal cells. The extent of membrane leakage and cell death following chitosan application was evaluated both with cellular vital stains and conductivity measurements of components exuding from the tissue (18). The disease resistance response, which suppresses growth of Fsph and can be induced by physiologically significant levels of chitosan within 6 hours is not associated with either the enhanced release of electrolytes or measurable plant cell death (18). Both chitosan treated and *F. solani* - inoculated pea tissue take up the vital stain phenosafranin within 24 hours which indicates membrane alterations do indeed occur but may not be required for initiating the resistance response.

Chitosan mode of action.

The mode of action of chitosan in inhibiting fungal growth could relate to its action as a polycationic polymer since other polycationic molecules (20) can also inhibit fungal growth. However, in the pea plant chitosan activates immunity, while other cations tested do not. Since the induction of genes associated with disease resistance in peas is mimicked by treatments such as long U.V. light-activated xanthotoxin and low levels of actinomycin D (15), which affect DNA specifically, it has been proposed that chitosan also acts on the host cellular DNA. We established that the complexing of DNA by chitosan *in vitro* can alter melting temperature curves of DNA (4), influence specific restriction enzyme digestion, and influence circular dichroism spectra of poly G-poly C rich DNA segments (14). Such conformational perterbations may influence the transcription of pea genes by alterations of pea chromatin structure.

The biosynthesis and availability of chitosan.

The route of chitosan biosynthesis in *Fusarium* is not known. In *Mucor rouxii* the biosynthesis of chitosan is *not* via the direct incorporation of UDP-glucosamine into a polymeric chain (21). Rather, UDP-N-acetyl glucosamine is first incorporated into chitin by chitin synthetase which in tandem with chitin deacetylase forms chitosan. Davis and Bartnicki-Garcia (22) show that chitosan yields are increased by adding the detergent digitonin to "chitosomes." Digitonin presumably improves chitosan yields by dissociating chitin synthetase from the "chitosome" complex. Any condition which enhances chitin synthesis or interferes with chitin fibril deposition may potentially enhance chitin deacetylation thus increasing the amount of chitosan. We observed two-fold enhancement of chitosan accumulation in *F. solani* 6 hrs following the addition of digitonin (12 μg/ml) to 2 day old shake cultured mycelia. This digitonin concentration also inhibits fungal growth. In light of the digitonin enhanced chitin/chitosan conversions observed in *Mucor* (22), the steroid/carbohydrate hybrid digitonin may suppress fungal growth in part by increases in chitosan. Along these same lines, other functionally similar plant compounds might then influence or modify infection by manipulation of regulatory compounds inherent to the fungal pathogen.

In summation, we would like to make the point that although chitosan is present in all *Fusarium* walls, its size, accumulation and availability to both host and fungal cells is dependent on multiple factors and therefore the degree of accumulation is variable depending on the specific interaction (4).

Elicitation of disease resistance response genes by bacterial races and cell wall components.

Part of the "non-host" response of pea endocarp tissue to *Fusarium solani* f. sp. *phaseoli* is similar to the resistance response which occurs when pea cultivars are challenged by incompatible races of *Pseudomonas syringae* pv. *pisi* (Psp). In this race-specific system, resistance is expressed in hosts possessing single Mendelian genes (23). Also, in the incompatible response the pea accumulates higher levels of pisatin than when the same host is challenged by a compatible race of Psp (24). When the behavior of the DRRG is monitored during an infection time course we see that some of the specific mRNAs homologous to the cDNA probes are present in high levels in incompatible interactions and at very low levels in compatible interactions (25). From this data we have proposed that there may be a common response mechanism in the host which can be triggered by more than one type of pathogenic organism. Since these pea differentials each possess single dominant Mendelian gene(s) which determine resistance, yet we show that at least five DRRG are active in correlation with phenotypic resistance, it is reasonable to postulate that the single gene is either a regulatory element, or an essential first step in a cascade pathway. Another possibility is that the gene codes for a receptor which in some fashion allows the plant to recognize the bacterium and initiate a resistance response.

This last hypothesis was tested in two types of experiments, the first entailed running 2-D PAGE separations of _in vitro_ translated proteins of each pea cultivar (Wagoner and Hadwiger, unpublished). It was not possible to detect the enhanced synthesis of individual proteins which corresponded to the different dominant genes specific in cultivar/race interactions. The inability to detect protein products of the Mendelian genes for disease resistance may indicate that there are none, or only reflect the lack of sensitivity of these techniques to resolve such products. The second type of experimentation was designed to determine if these Mendelian traits code for a receptor which could initiate a host response following contact with an inducing substance(s) in the bacterial wall. Ten hours following inoculation with bacterial races the DRRG specific mRNAs were induced over the level of water control (in approximately 80% of the experiments). Although the RNA levels in the incompatible reactions were higher and more sustained throughout the time course than RNA levels in compatible reactions, the fact remained that the DRRG were induced in nearly all reactions by 10 h after inoculation. The finding that DRRG were induced to some degree in both compatible and incompatible interactions does not support the hypothesis that the single gene is a receptor which allows the plant to recognize all but a compatible bacterial race.

Since the fungal elicitor chitosan is not a normal component of bacterial walls the host response to bacterial pathogens must rely on other types of elicitors. Bacterial cell wall extracts from races of _P. syringae_ pv. _pisi_ are capable of eliciting pisatin non-specifically in pea cultivars. However, isolated components of the wall, such as lipopolysaccharides, membrane proteins and extracellular polysaccharides either do not induce or are low level inducers of pisatin, even when applied at high concentrations to those pea cultivars (Cody and Hadwiger, unpublished).

In an effort to determine if the Psp races differed in cell wall composition, isolated cell walls of each of the three Psp races were subjected to IR spectroscopy and the results compared (Daniels and Hadwiger, unpublished). No differences, within the limitations of the procedures, could be detected between the races. Although the constituents of the cell wall may not differ between the races, the arrangement and sizes of wall macromolecules could well differ. Whether this is the mechanism behind compatibility vs. incompatibility in this system remains to be proven.

An experiment identical to the one using live Psp cells was performed using isolated bacterial walls as inoculum. Preliminary data suggest that the expression of the DRRG in a race-specific manner is dependent upon the presence of live Psp cells. When live cells were included in the cell wall inoculum at 10^6 cfu/ml (as opposed to 10^9 cfu/ml in the live cell experiment), results closely mirrored those of the live intact cell-challenged pea tissue. That is, each of the five strongest DRRG showed a strong pattern of mRNA accumulation over the time course that correlated with plant expressed resistance to one or more Psp races. Isolated cell walls without the live bacteria did induce the DRRG genes but not in a race-specific fashion. From this data we propose that the specificity may be based on the biosynthesis, accumulation, degradation, processing or localization of unknown elicitors, occurring only in the presence of the living bacterial cell.

SUPPRESSION

Previous work (16) has demonstrated that when the pea tissue is given a 40°C heat shock, then challenged with Fsph, resistance is broken and the fungus readily colonizes the pea pod unless a 9 h recovery period is allowed between heat treatment and challenge. Proteins labelled in an *in vitro* translation system using mRNA from heat shocked, heat shocked then challenged, or just challenged pea pods each showed a characteristic constellation of proteins when subjected to 2-D PAGE (16). It was concluded that tissue that is heat shocked does not produce proteins known to be associated with the resistance response unless the tissue is allowed to recover at least 9 h following heat treatment.

When cDNAs of DRRG are used as probes against Northern blots containing RNA from heat shocked tissue we see a marked decline in the accumulation of homologous mRNA versus those levels in control tissue. Densitometric quantitations of autoradiograms indicate that in heat shocked tissue the specific mRNA level generally drops an order of magnitude compared to levels in induced tissue, while falling to levels approximately 20% of water control tissue. If the tissue in challenged following a 9 h recovery from heat shock the homologous mRNA again accumulates. This renewed potential to accumulate specific DRRG mRNAs correlates well with the restoration of the resistance response of the pea pod to the Fsph spores.

SUMMARY

In this paper we have presented and reviewed evidence of gene activity in the pea host, following challenge by both a bean and a pea pathogen. Many aspects of the bean pathogen-pea endocarp interaction have been well characterized, most notably the action of chitosan as an elicitor. Some aspects of the pea pathogen-pea interaction remain to be clarified, primarily the means by which the pathogen can turn off the DRRG and successfully colonize the tissue. We have also shown that the same DRRG are active in a race-specific manner in certain pea differential lines when the host is challenged by the live Psp races but are activated non-specifically when challenged by isolated Psp cell walls alone. In addition to demonstrating that the DRRG are active during resistance responses, we have also shown that in susceptible reactions they become inactive or less active. Relatively low expression of the DRRG in the susceptible reaction of the pea-pea pathogen at time periods when fungal growth rapidly accelerates, and the lower copy numbers of those mRNAs following resistance-breaking heat treatments both verify this observation.

REFERENCES

1. Hadwiger LA, Schwochau ME (1969). Host resistance responses - an induction hypothesis. Phytopathol 59:223.
2. Teasdale J, Daniels D, Davis WC, Eddy R, Hadwiger LA (1974). Physiological and cytological similarities between disease resistance and cellular incompatibility responses. Plant Physiol 54:690.
3. Hadwiger LA, Loschke DC (1981). Molecular communication in host-parasite interactions. Hexosamine polymers (chitosan) as regulator compounds in race-specific and other interactions. Phytopathol 71:756.
4. Hadwiger LA, Beckman JM (1980). Chitosan as a component of pea-F. solani interactions. Plant Physiol 66:205.

5. Fristensky B, Riggleman RC, Wagoner W, Hadwiger LA (1985). Gene expression in susceptible and disease resistant interactions of peas induced with Fusarium solani pathogens and chitosan. Physiol Plant Pathol 27:15.
6. Riggleman R, Fristensky BW, Hadwiger LA (1985). The disease resistance response in pea is associated with increased transcription of specific mRNAs. Plant Molecular Biology 4:81.
7. Hadwiger LA, Wagoner W. (1983). Electrophoretic patterns of pea and Fusarium solani proteins synthesized in vitro which characterize the compatible and incompatible interactions. Physiol Plant Pathol 23:153.
8. Loschke DC, Hadwiger LA, Schroder J, Hahlbrock K (1981). Effects of pathogenic and nonpathogenic races of Fusarium solani and of light in peas: Rate of synthesis and activity of phenylalanine ammonia-lyase. Plant Physiol 68:680.
9. Nichols J, Beckman JM, Hadwiger LA (1980). Glycosidic enzyme activity in pea tissue and pea-Fusarium solani interactions. Plant Physiol 66:199.
10. Mauch F, Hadwiger LA, Boller T (1984). Ethylene: symptom not signal for the induction of chitinase and β-1,3-glucanase in pea pods by pathogens and elicitors. Plant Physiol 76:607.
11. Hadwiger LA, Daniels C, Fristensky BW, Kendra DF, Wagoner W (1986). Pea genes associated with the non-host resistance to Fusarium solani are also induced by chitosan and in race-specific resistance by Pseudomonas syringae. In Bailey JA (ed): "Biology and molecular biology of plant-pathogen interactions," Berlin: Springer-Verlag, p 263.
12. Kendra DF, Hadwiger LA (1984). Characterization of the smallest chitosan oligomer that is maximally antifungal to Fusarium solani and elicits pisatin formation in Pisum sativum. Exp. Mycol. 8:276.
13. Hadwiger LA, Kendra DF, Fristensky BW, Wagoner W (1986). Chitosan both activates genes in plants and inhibits RNA synthesis in fungi. In Muzzarelli R, Jeunianx C, Gooday GW (eds): "Chitin in nature and technology," New York: Plenum Press, p 209.
14. Kendra DF, Fristensky BW, Daniels CH, Hadwiger LA (1987). Disease resistance response genes in plants: expression and proposed mechanisms of induction. In Arntzen C, Ryan CA (eds): "Molecular strategies for crop protection," New York: Alan R. Liss, Inc. (in press).

15. Loschke DC, Hadwiger LA, Wagoner W (1983). Comparison of mRNA populations coding for phenylalanine ammonia lyase and other peptides from pea tissue treated with biotic and abiotic phytoalexin inducers. Physiol Plant Pathol 23:163.
16. Hadwiger LA, Wagoner W (1983). Effect of heat shock on the mRNA-directed disease resistance response of peas. Plant Physiol 72:553.
17. Bailey JA (1982). "Phytoalexins." New York: John Wiley and Sons, p 289.
18. Kendra DF, Hadwiger LA (1986). Cell death and membrane leakage are not associated with the induction of disease resistance in peas by chitosan. Phytopathol (In Press).
19. Takagi H, Kadowaki, K (1986). Polygalactosamine produced by a microorganism. In Muzzarelli R, Jeuniaux C, Gooday GW (eds): "Chitin in nature and technology," New York: Plenum Press, p 121.
20. Hadwiger LA, Loschke DC, Teasdale JR (1977). An evaluation of pea histones as disease resistance factors. Phytopathol 67:755.
21. Davis LL, Bartnicki-Garcia S (1984). Chitosan synthesis by the tandem action of chitin synthetase and chitin deacetylase from Mucor rouxii. Biochem. 23:1065.
22. Davis LL, Bartnick-Garcia S (1984). The co-ordination of chitosan and chitin synthesis in Mucor rouxii. J Gen Microb 130:2095.
23. Taylor JD, Vivian A, Malik A (1984). Race structure in Pseudomonas syringae pv. pisi. In ISPP Committee on Phytopathogens Bacteria (ed.): "Proceedings of the 2nd Working Group on Pseudomonas syringae Pathovars," Athens: The Hellenic Phytopathol Soc, p 90.
24. Hadwiger LA, Webster DM (1984). Phytoalexin production in five cultivars of peas differentially resistant to three races of Pseudomonas syringae pv. pisi. Phytopathol 74:1312.
25. Daniels CH, Fristensky BW, Hadwiger LA (1987). The genes associated with non-host disease resistance to Fusarium are also active in race-specific disease resistance to Pseudomonas. Plant Mol Biol (in press).

Plant Gene Systems and Their Biology, pages 171–181

ENGINEERING IMPROVED CROPS FOR AGRICULTURE : PROTECTION FROM INSECTS AND RESISTANCE TO HERBICIDES

M. Vaeck, J. Botterman, A. Reynaerts, M. De Block, H. Hofte, J. Leemans

Plant Genetic Systems N.V., J. Plateaustraat 22, 9000 Gent, Belgium.

I. ABSTRACT

Using the *Agrobacterium* mediated transformation system, foreign genes can be transferred and expressed in plants.

A crystal protein gene (*bt2*) has been cloned from *Bacillus thuringiensis berliner* 1715. In *E. coli* *bt2* directs the synthesis of a 130 kDa protein which is highly toxic to Lepidoptera insect larvae. Tobacco plants transformed with modified genes, derived from *bt2*, express active insecticidal protein and are protected from insect feeding damage.

A gene which confers resistance to phosphinothricin, a new nonselective contact herbicide, has been cloned from *Streptomyces hygroscopicus*. The gene product is an enzyme which acetylates the herbicide. Expression of this detoxifying enzyme in transgenic tobacco, tomato and potato plants confers complete resistance to the herbicide. Both the insect protection and the herbicide resistance trait are stably inherited throughout the next generations.

II. INTRODUCTION

Several procedures have been developed to introduce foreign DNA in plant cells. The most successful one is the Ti-plasmid mediated gene transfer by *Agrobacterium* (1). Its development into an efficient vector system has opened the way to genetically engineer plants. This technology is now applied to transfer agronomically important traits to plants in order to obtain improved crop varieties. We report here two examples: transgenic plants protected from insect damage and herbicide resistant plants.

Bacillus thuringiensis (*B.t.*) has been used for many years as a safe biological insecticide. Its toxicity towards insects is due to the production of a protein, called delta-endotoxin, which is selectively toxic to larvae of insects. Different *B.t.* toxins with activity against

Lepidoptera, Diptera and Coleoptera insects have been identified and partially characterized (2,3). We have cloned a gene from *B.t.* strain *berliner* 1715 which encodes a protein, highly toxic against larvae of Lepidoptera. This gene has been used to engineer resistance against such insects in tobacco plants.

The use of herbicides to reduce loss in crop yield due to weed has become an integral part of modern agriculture. In all agronomically important crops adequate weed control programs are needed and new herbicides which are highly effective and safe for the environment are searched for. Several new herbicides act by inhibiting specific amino acid biosynthesis pathways in plants, but most do not distinguish between weeds and crops (4). Hence, engineering plants to become resistant to such broad-spectrum herbicides would allow their selective use for crop protection. We cloned a gene from the herbicide producing strain *Streptomyces hygroscopicus*, which encodes a herbicide detoxifying enzyme to prevent auto-toxicity (5, 6). Expression of this gene in plants yields plants resistant to the nonselective herbicide (7).

III. RESULTS AND DISCUSSION

1. Protection from insect damage

1.1. Cloning of a B.t. gene encoding an insecticidal protein. Since different *B.t.* strains differ in their insecticidal spectrum, we analysed the specific toxicity of a number of randomly chosen strains towards two important Lepidoptera pest insects on tobacco: the tobacco hornworm, *Manduca sexta*, and the tobacco budworm, *Heliothis virescens*. Strain B.t. *berliner* 1715 was highly toxic against both these insects (table 1). Crystals produced by B.t. *berliner* 1715 contain two major polypeptides of 130 kDa and 140 kDa respectively. These can be distinguished on the basis of their antigenic structure. The gene encoding the 130 kDa protein is plasmid localized and has been cloned and expressed in *E. coli* (8). This gene (*bt2*) encodes a protein of 1155 amino acids with a predicted M.W. of 130, 533 Da.

The recombinant protein, termed Bt2, is highly active against both *M. sexta* and *H. virescens* (table 1). It therefore represents a potentially interesting protein to provide resistance against those insects in tobacco plants. The amino acid sequence of Bt2 was compared to that of other B.t. toxins. Extensive sequence homology was detected with the 130 kDa crystal proteins of *B.t.* strains *kurstaki* HD1, *kurstaki* HD73 and *sotto* (8). The four proteins are even almost identical in the 300 N-terminal amino acids. Such high conservation suggests a critical role for this region in the toxin function.

TABLE 1
TOXICITY OF SOLUBILIZED B.T. CRYSTAL PROTEINS TOWARDS TOBACCO PEST INSECTS

B.t. strain	Protein	LC50 (in ng/cm^2) *M. sexta*	*H. virescens*
B.t. aizawai HD68	total crystals	35	300
B.t. aizawai HD127		13	45
B.t. entomocidus HD110		26	350
B.t. kenyae HD551		21	17
B.t. kurstaki HD73		10	12
B.t. berliner 1715		11	10
B.t. berliner 1715	recombinant Bt2	11	13

1.2. Bt2 is a protoxin which is processed towards a 60 kDa active toxin by insect proteases. The *B.t.* endotoxins are generally believed to be protoxins which, upon ingestion, are degraded in the insect midgut into smaller toxic polypeptides (9). We therefore investigated wether smaller fragments could be generated from Bt2 using insect midgut enzymes. Purified Bt2 protein was treated with midgut extracts from different Lepidoptera insect larvae: *M. sexta*, *H virescens*, *M. brassicae*. The resulting polypeptides were analysed on SDS-PAGE and their specific toxicity was evaluated in insect bioassays. Bt2 is rapidly degraded (10 min at 37°C) essentially towards a 60 kDa polypeptide by midgut enzymes from different Lepidoptera larvae. Treatment of Bt2 with other proteolytic enzymes such as trypsin or chymotrypsin also results in the generation of a similar 60 kDa fragment (Fig. 1).

Interestingly this polypeptide is relatively resistant to further degradation by the different proteolytic enzymes over a period of several hours at 37°C (8). Therefore it constitutes a protease-resistant core within the Bt2 protein. In addition, the 60 kDa polypeptide preparations generated by treatment with the different enzymes have retained full insecticidal activity and, on a molar basis, are as toxic as the intact Bt2 protein.

The N-terminal amino acid sequence of the trypsin generated 60 kDa fragment, determined by gas-phase sequencing, was identical to a deduced amino acid sequence of Bt2 starting at residue isoleucine 29 (8). Therefore the active toxic fragment could be unambigously located in the N-terminal half of Bt2.

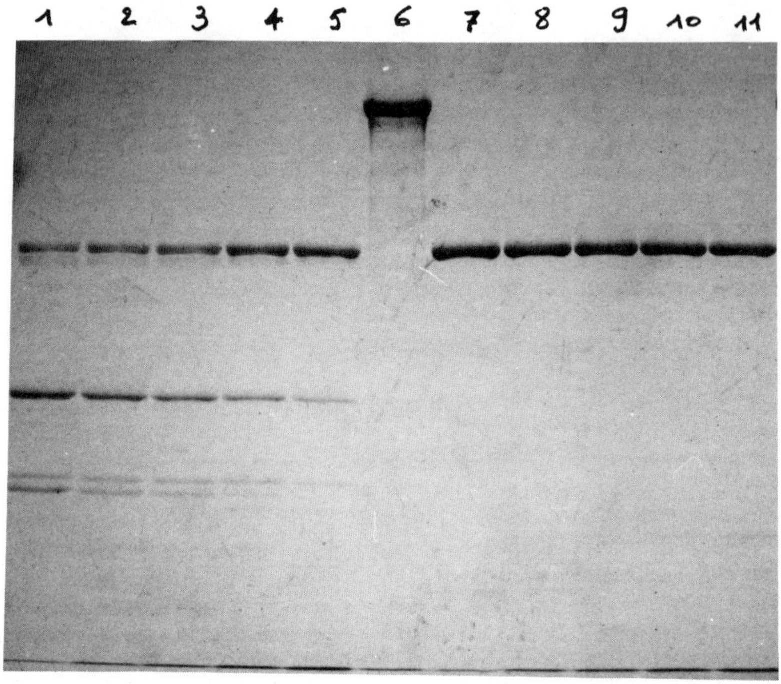

Fig. 1: SDS polyacrylamide gel of purified Bt2 protein (lane 6), Bt2 treated with trypsin for 10, 20, 40, 60 and 120 minutes (lanes 11-7), Bt2 treated with chymotrypsin for 10, 20, 40, 60 and 120 minutes (lanes 5-1).

1.3. Genetic mapping of the minimal toxic polypeptide fragment in Bt2. Since the previous data do not completely exclude the possibility that smaller, still actively toxic fragments, are contained within the 60 kDa protease resistant core, we performed genetic experiments to explore this possibility and to determine the minimal gene fragment which still encodes an active insecticidal protein.

Different deletions at the 5′ or the 3′end of the bt2 gene were constructed. The corresponding polypeptides were expressed in E. coli and analysed for insect toxicity (8).

The results from the experiments showed that the minimal toxic fragment is delineated by amino acid positions 29 and 607 on the Bt2 protein sequence. Deleting only a few amino acids from the N-terminus or the C-terminus of this fragment results in a completely nonfunctional polypeptide.

1.4. Construction of modified B.t. toxin genes. Since only a fragment of the bt2 gene encodes the active insecticidal protein, we designed different gene constructs to express the insect toxin in plants (11).

First, we used truncated genes encoding N-terminal fragments of the Bt2 protein, such as bt884 and bt820 (Fig. 2). Second, we constructed fusion genes between 5′ fragments of bt2 and the neomycin phosphotransferase (NPTII) gene, neo. In both constructs the B.t. coding sequence comprises the minimal fragment encoding an active toxin, contained between codon 29 and 607 (Fig. 2). Translational fusions to the 5′end of neo have been shown to confer kanamycin resistance in bacteria (12) and in plants (13). The fusion genes bt:neo860 and bt:neo23, when expressed in E. coli, confer resistance to kanamycin and give rise to stable fusion proteins, Bt:NPT860 and Bt:NPT23, which have the expected sizes of 106.000 d and 110.000 d respectively and react with anti-NPTII and anti-Bt2 antibodies. The enzymatic activity of the fusion proteins is comparable to that of wildtype NPTII. In addition insect bioassays revealed that their toxic activity towards insect larvae is identical to that of Bt2 protein (11).

We anticipated that plant cells transformed with these fusion genes would produce fusion proteins expressing NPTII activity. Selection for high levels of kanamycin resistance therefore would allow to select directly for transformants producing relatively high quantities of B.t. toxin.

Fig. 2

Toxin gene constructions used in plant transformation experiments.

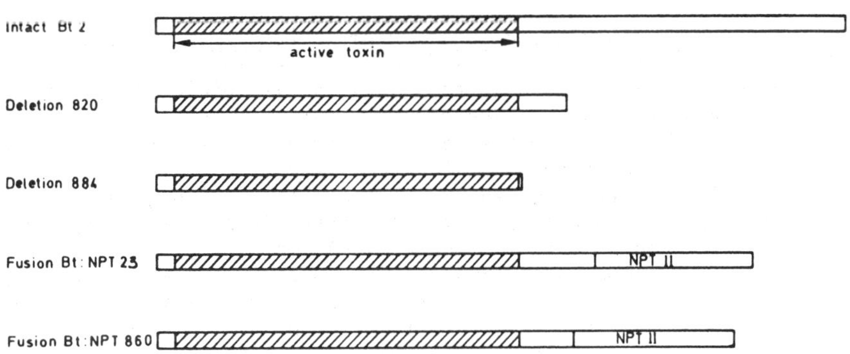

1.5. Chimeric genes and vectors for expression in plants. The intact bt2 gene, as well as the different modified toxin genes, were inserted between the T-DNA borders of Agrobacterium-based expression vectors pGSH150 or pGSH160 (14). In this way chimeric genes were created containing plant specific promotors and bt2 derived toxin genes. Different promotor sequences were used such as the promotor of the Agrobacterium TR 2′ gene of the Ti A6 plasmid (15) or the 35S promotor of cauliflower mosaic virus (16). The intermediate expression vectors were mobilized into Agrobacterium recipient C58C1Rif^R pGV2260, through homologous recombination with the pBR sequence on plasmid pGV2260 (a disarmed Ti plasmid).

1.6. Expression of B.t. toxin in transgenic tobacco plants. Transgenic tobacco plants were obtained using leaf disc infection of *Nicotiana tabacum* var. Petit Havana SR1. Kanamycin resistant shoots were obtained in all experiments, indicating that the fusion genes also confer NPTII activity to the transformed plant cells. Expression of B.t. protein in the leaves of greenhouse grown plants was determined using a quantitative immunoassay (ELISA) with monoclonal anti-Bt2 antibodies. Plants transformed with the bt:neo fusions and selected for high Km resistance (resistant to 500 mg/l Km) express *B.t.* protein levels of 20-200 ng per mg of total leaf protein. Similar levels of expression were detected in plants transformed with the truncated B.t. genes. In contrast, plants containing the intact bt2 gene behind the same promotors express levels that are at least ten times lower, between 2 and 5 ng per mg of total protein. A similar difference in expression between truncated and intact bt2 was observed at the level of RNA, when steady state levels of mRNA were measured. The reasons for this are not clear at present. Different factors such as RNA stability or translation efficiency might play a role.

1.7. Transgenic plants are protected from insect attack. Insect bioassays were performed to evaluate wether the *B.t.* protein in transgenic tobacco plants is biologically active, and expressed at levels sufficient to kill leaf feeding insects. Mortality rates of *M. sexta* larvae were monitored after 6 days of feeding on tobacco leaves.

Plants transformed with the two fusion genes bt:neo860 and bt:neo23 exhibit significant insecticidal activity. About half of these plants induce 100% mortality in 1st instar larvae. A comparable insecticidal activity is also detected in plants transformed with the truncated bt884 gene. In contrast plants containing the intact bt2 do not show insect killing above background levels (0-10%), obtained in untransformed control plants. This observation together with the data on expression levels of B.t. protein indicates that the intact bt2 gene does not give rise to efficient protein synthesis and that only truncated versions of this gene can generate insect killing levels in plants.

We then evaluated whether expression of modified bt2 toxin genes in plants would result in efficient protection against insect damage. Selected transgenic plants were grown in the greenhouse and were infested with freshly hatched larvae of *M. sexta*. Plants expressing levels of B.t. protein above 40 ng/mg showed complete resistance towards *M. sexta*: the larvae stopped feeding within 18 hours and all died within 3 days. Single larvae caused damage of only a few mm^2 areas. Control plants were severely damaged within 4-6 days and were entirely consumed after 10-12 days.

1.8. Inheritance of the new trait. A large number of transgenic tobacco plants, transformed with the different types of chimeric B.t. genes, and expressing various levels of active toxin, were grown in the greenhouse. All grew normally and were indistinguishable from controls with respect to morphology and growth vigor. The inheritance pattern of the kanamycin resistance was analysed. Most plants contained one or two kanamycin resistance loci. These numbers were confirmed by Southern blot analysis. Insect toxicity in the F1 progeny from transgenic plants was correlated with the presence of the kanamycin resistance marker. The insecticidal activity as well as B.t. protein levels were similar to those observed in the parental plants. Thus, the insect resistance trait was stably inherited to the next generation, following Mendelian rules of a dominant marker.

1.9. Conclusion. Clearly, the present experiments exemplify the feasibility of using genetic engineering techniques to generate plants which have the potential to defend themselves against certain predatory insects. Transfer of different B.t. genes controlling a wide variety of insect species into a range of crop species such as corn, cotton, tomato and brassica varieties, may provide a new and general method to control destructive insect pests.

2. Engineering herbicide resistance in crop plants

2.1. The herbicide. Phoshinothricin (PPT) is a new nonselective contact herbicide. It is obtained by chemical synthesis or naturally produced as part of a tripeptide antibiotic such as bialaphos which is composed of two L-alanine residues and PPT (Fig. 1). Upon removal of these residues, PPT is a potent inhibitor of glutamine synthetase (GS). This enzyme plays a central role in plant metabolism and inhibition by PPT causes rapid accumulation of ammonia which leads to death of the plant cell.

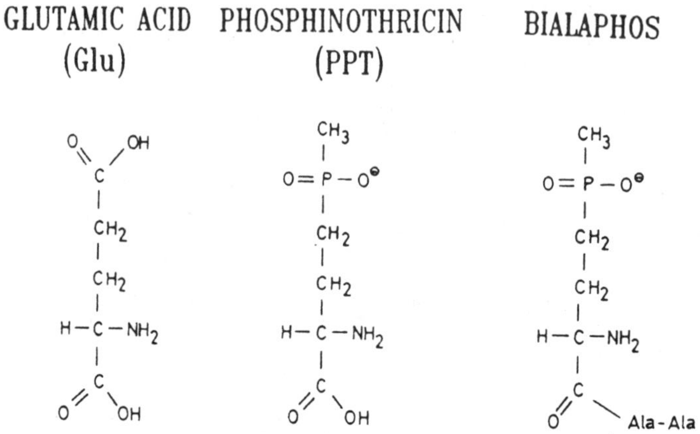

Fig. 3. Chemical structure of the herbicide compounds bialaphos and phosphinothricin. Their analogy with glutamic acid is shown.

2.2. Cloning of a herbicide resistance gene and characterization of the gene product. It is frequently noticed that Streptomyces which produce antibiotics have developed mechanisms to neutralize the toxicity of their own products. A genomic bank was made from total DNA of the bialaphos producing strain *Streptomyces hygroscopicus* ATCC21702 and transferred in the sensitive host *Streptomyces lividans* 66. Transformants resistant to bialaphos were selected. After subcloning and optimization of gene expression, it was shown by *in vitro* coupled transcription-translation, that the resistance gene (*bar*) encodes a 22 kDal polypeptide.

The complete sequence of the *bar* gene which encodes a phosphinothricin acetyl transferase (PAT) was determined. This enzyme transfers an acetyl group from acetyl-coA to the amino group from PPT yielding acetylated PPT (Fig. 4). PAT was purified from an *E. coli* overproducing strain and comparison of the specificity of the enzyme towards PPT and chemically related compounds showed that PAT has a very high substrate specificity.

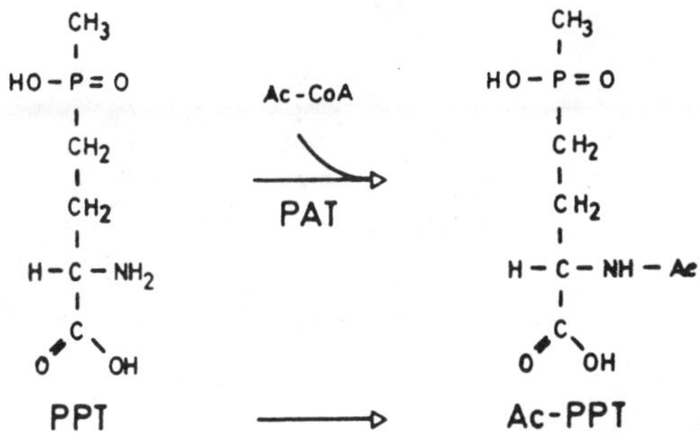

Fig. 4. Enzymatic reaction of the phosphinothricin acetyl transferase. An acetyl group from acetyl CoA is transferred on the amino group of PPT.

2.3. Expression of the resistance gene in plants. In order to guarantee proper translation initiation in plants, the GTG initiation codon was substituted for an ATG. The bar gene was placed under control of the CaMV 35S promoter. This chimeric gene was inserted between the T-DNA borders in an intermediate vector and cointegrated in the Agrobacterium strain C58C1 (pGV2260) (14). PAT was used as a selectable marker in tobacco protoplast cocultivation. Transformed calli were selected on 25 to 50 mg/l and were resistant up to 1 g/l PPT, whereas no growth was observed on control plants.

Resistant calli were selected after infection of tobacco (Nicotiana tabacum cv. Petit Havana SR1), tomato (Lycopersicum esculentum cv. Lucullus) and potato (Solanum tuberosum cv. Berolina) leaf discs with Agrobacterium. From these calli, transgenic tobacco, tomato and potato plants were regenerated and tested in spraytests for their resistance towards the commercial herbicides Basta[R] (Hoechst AG-active compound PPT) and Herbiace[R] (Meiji Seika, Ltd - active compound bialaphos). Transgenic plants expressing various levels of PAT were fully resistant to high doses. Normal applications for weed control vary from 2.5 to 7.5 l/ha. For example, Basta[R] effectively kills control tobacco plants at 1 l/ha while transgenic plants were fully resistant to repeated spraying with 20 l/ha. Treated resistant plants did not show any increase in NH_4^+ content and showed a complete protection of the plant glutamine synthetase from the action of the herbicide.

2.4. Inheritance of PPT resistance. Transgenic tobacco plants produced normal amounts of viable seed. Five week old plants from the F1 progeny were sprayed. Resistant and sensitive seedlings segregated in a 3 to 1 ratio (Fig. 5), which indicates Mendelian inheritance as a single dominant locus.

Fig. 5. Spraytest on F_1 seedlings of tobacco plants. Seedlings of a transgenic tobacco and a control nontransformed plant grew indistinguishable. After five weeks they were sprayed with a dose equivalent to 8 l/ha Basta^R. Seven days after treatment all controls were dead, while segregants from a transformed plant were resistant in a ratio 3 to 1.

2.5. CONCLUSION

This work illustrates a successful approach to obtain herbicide resistant plants. The transgenic plants are resistant to herbicide doses exceeding those used in regular field conditions. As apposed to other strategies, in which a nonsensitive mutant enzyme is produced or in which the target enzyme is overproduced, we introduced a detoxification pathway for the herbicide. This system is independant from the plant species, highly effective and has no effect on crop performance. This strategy can now be applied for different herbicides and to a variety of crop species.

REFERENCES

1. Gheysen, G. et al. (1985). in Genetic Flux in Plants. Advances in plant gene research, vol. 2. (eds. Hohn, B. and Dennis, E.S.) 11-47, (Springer Verlag, New York).
2. Dulmage, H.T. (1980) in "Microbial control of pests and plant diseases 1970-1980" (Burges, H.D. ed.) pp. 193-222, Ac. Press, London.
3. Krieg, V.A. et al. (1983). Angew. Ent., 96, 500-508.
4. La Rossa, R.A. and Falco, S.C. (1984) Trends in Biotechnology, 2, 158-161.
5. Thompson, C., et al. (1987) EMBO J., in press.
6. Murakami, T. et al. (1986) Mol. Gen. Genet. 205, 42-52.
7. De Block, M., et al. (1987). EMBO J., in press.
8. Höfte, H. et al. (1986). Eur. J. Biochem., 161, 273-280.
9. Lilley, M. et al. (1980). Gen. Microbiol., 118, 1-11.
10. Fast, P.G. and Martin, W.G. (1980). Biochem. Biophys. Res. Commun., 95, 1314-1320.
11. Vaeck, M. et al. (1987). Nature, in press.
12. Reiss, B. et al. (1984). EMBO J., 3, 3317-3322.
13. Schreier, P. et al. (1985). EMBO J., 4, 25-32.
14. Deblaere, R. et al. (1985) Nucleic Acid Res., 13, 4777-4788.
15. Velten, J. et al. (1984) EMBO J., 3, 2723-2730.
16. Odell, J.T. et al. (1985). Nature, 313, 810-812.

Plant Gene Systems and Their Biology, pages 183–204

MOLECULAR CLONING OF MAIZE GENES BY TRANSPOSON TAGGING WITH *MUTATOR*[1]

Jeffrey L. Bennetzen, April Cresse, Willis E. Brown, and Lisa Lee

Department of Biological Sciences, Purdue University, West Lafayette, Indiana 47907

ABSTRACT Transposon tagging is currently the most promising general technique for the molecular isolation of plant genes associated with a distinct phenotype but not yet associated with a specific gene product. The *Mutator* transposable element system of maize is particularly well suited to this approach due to its exceptional mutagenic activity. The *Mutator* system induces new mutations at all loci examined at frequencies as high as one in one thousand. These new mutations are generally associated with the insertion of a 1.4 kb *Mu1* element or closely related molecules. We have developed a routine procedure to optimize the recovery, molecular cloning, and confirmation of *Mutator*-induced mutations.

INTRODUCTION

Many of the most interesting genes in maize and other plant species have not yet been associated with a specific gene product. These include mapped single loci which grant resistance to microbial pathogens (1), regulate tissue-specific genes and gene families (2,3), or determine the developmental potential of the embryo (4,5). Lacking identified RNA transcripts or protein products, molecular cloning of any of these genes is not feasible by the standard cDNA technologies. The major interspersed repetitive DNA component and high complexity of most higher plant genomes plus the relatively low efficiency of plant cell transformation

[1]This research was supported by Grant 8400350, Competitive Research Grants Office, USDA, and by Eli Lilly and Company.

and recovery limit the applicability of such technologies as cloning by complementation or chromosome walking to highly specialized cases (6). An alternative approach is to use the presence of a known transposable element within a gene of interest as a source of homology for cloning. This technique, generally referred to as "transposon tagging," has been used extensively in bacteria and Drosophila (7,8) and recently in plants (9-19).

Several criteria must be met before a transposon tagging protocol may be used to clone a plant gene. First, the gene to be cloned must have a definitive and non-lethal phenotype associated with its partial or complete inactivation. Second, the gene you wish to tag should be reasonably stable in the absence of the transposable element system you will use in the tagging approach. Third, the plant species in which your desired gene is found needs to have a well characterized and mutagenically active transposable element system. To date, this last requirement limits the transposon tagging approach to *Zea mays* and *Antirrhinum majus*. Transposition of the maize *Ac* element in tobacco (20), however, suggests that transposon tagging may soon be possible in any transformable plant. If these minimal criteria are met, then a variety of considerations should be taken into account in order to maximize the efficiency of the tagging program. The most important of these considerations is the choice of a tagging module.

CHOICE OF A TRANSPOSABLE ELEMENT SYSTEM FOR TRANSPOSON TAGGING

Which transposable element system one will use for a gene tagging procedure should depend on the particular plant gene targetted. In maize, elements from each of the *Ac-Ds* (21,22), *En* (*Spm*) (23), and *Mutator* (24) transposable element systems have been cloned and employed successfully in tagging programs (9-11,13-19). Several other transposable elements from maize have been molecularly cloned (25-28) or soon will be (29-32). Many of these elements transpose very rarely (25-28) or seem to induce mutations at a very limited number of loci (29-32). For these reasons, the *Ac-Ds*, *En* (*Spm*), and *Mutator* transposable elements are the current systems of choice in maize.

Each of the three major maize tagging systems has its advantages and disadvantages. The autonomous *Ac* and *En* (*Spm*) elements are present in low genomic copy numbers (33) and are structurally different from the numerous non-autonomous elements of the same family present in all maize lines (9,15,16, 21,23,34). This facilitates cloning of the transposon tagged locus. Presumed *Ac*- or *En* (*Spm*)-induced mutants at a locus can be confirmed by demonstrating segregation of the mutation with the *trans* activational component of the autonomous element. Finally, the frequent somatic and germinal reversion of many *Ac*- and *En* (*Spm*)-induced mutations can be used to confirm the element-induced nature of a mutation. Molecular analysis of these germinal revertants also helps confirm that the final cloned gene is actually the locus that was targetted.

Compared to *Ac* and *En* (*Spm*), *Mutator* has one major technical advantage: mutation rate. *Mutator* lines produce new mutations at rates at least 30-fold higher than lines with *Ac* or *En* (*Spm*) (35,36). With the appropriate crossing program, this can mean rates as high as one in one thousand for a particular gene (37). For a mutational phenotype that must be detected at the seedling or whole plant level, the difference between screening over a million plants carrying *Ac* or *En* (*Spm*) versus scoring ten thousand with *Mutator* may be the deciding factor in whether transposon tagging is feasible.

The preferential transposition of *Ac* (38,39) to sites near the original insertion position can lead to significantly increased mutagenic frequencies at a nearby locus (15). The crosses to bring *Ac* or *En* (*Spm*) onto a specific chromosome arm are time consuming, however, and this approach requires previous mapping of the gene you wish to tag. One difficulty with using *Ac* or *En* (*Spm*) for a tagging procedure is the lack of information we have regarding insertion specificities for these systems. In the last forty years, despite the efforts of several laboratories, only twelve loci in maize have been associated with either an *Ac/Ds*- or *En* (*Spm*)-induced mutation (compiled in 32). In the last eight years, thousands of independent mutations induced by *Mutator* have been isolated, primarily by Robertson. We have performed allelism tests on twenty *de novo* seedling mutations (albino, yellow, yellow-green, pale green, glossy) and ten *de novo* kernel mutations (defective kernel, viviparous) which segregated out of our *Mutator* self cross populations and found that all were at separate loci. In addition, all twelve loci that have been specifically targetted with *Mutator* over the last eight years have produced mutations at the expected frequencies (16,27,40, 41). We can therefore state with complete confidence that *Mutator* will induce mutations at high frequencies at virtually any maize locus.

RECOVERY OF NEW MUTATIONS FROM *MUTATOR* LINES

In order to isolate a *Mutator*-induced mutation, one must first have at least 100 seeds of a mutagenically active *Mutator* stock in the appropriate genetic background. For the isolation of mutations in genes which are active in standard maize dent lines, the dihybrid *Mutator* stocks described by Robertson (35) may be suitable. We have found that the four inbreds which contribute to this line carry a high level of restriction fragment length polymorphisms, however, as is true of maize in general (42,43). In our experience, confirmation of a cloned locus containing a *Mu* element as the gene we targetted is complicated by difficulties in ascertaining the structure of the non-mutant, progenitor allele. This difficulty may be overcome by gel blot hybridization analysis or cloning of all four possible parental alleles (44), but this delays the tagging process considerably. Another shortcoming of using this standard line is that it has very few useful genetic markers.

Our general approach has been to cross an active *Mutator* line into a background which contains a dominant, functional allele of the gene we will target and then to homozygose for that allele (45). This stock may then be amplified by one or more generations of sibling crosses.

Since *Mutator* mutagenic activity can be lost in either a cross between two *Mutator* lines (46) or between a *Mutator* line and any other maize line (35,47), it is necessary to score any lines derived from active *Mutator* plants for the presence of *Mutator* mutagenic activity. The most accurate and direct test of mutagenic activity requires a self cross <u>and</u> outcross of a suspected *Mutator* plant and then self crosses of at least twenty progeny of the outcross. These twenty or more selfs of the original outcross are then scored for 1:3 segregation of kernel and seedling mutations not segregating in the original self cross progeny (35). An active *Mutator* line will generate new mutations in five to thirty percent of the self crosses (35,36,46). This crossing and scoring program is very arduous, requires two growing seasons, and only confirms the *Mutator* activity of a single starting seed.

We have tested to see whether *Mutator* mutagenic activity could be followed by a simpler procedure, relying on the somatic mutability of a *Mutator*-derived mutation, *bzMum4* (44, 48). The *bzMum4* mutation conditions an unstable inactivation of the *bronze* locus yielding a bronze kernel with purple, phenotypic revertant, sectors. This somatic instability is

TABLE 1
MUTABILITY OF *bzMum4* VERSUS MUTAGENIC ACTIVITY

bzMum4 parent	Mutability of *bzMum4* parent[a]	Mutability in F_1[b]	Frequency of new mutations in selfs of F_1[c]
A	sm	sm	12% (3/26)
B	sm	m	0% (0/21)
C	m-sm	m	8% (3/38)
D	m-sm	m	5% (2/38)
E	m-sm	wm	0% (0/25)
F	m	m	9% (2/22)
G	m	m	0% (0/27)
H	wm	wm	13% (4/30)
I	wm	vwm	6% (2/31)
J	wm	vwm	0% (0/29)
K	vwm	vwm	0% (0/26)
L	nm	nm	0% (0/22)
M	nm	nm	0% (0/23)

[a]Classification of the frequency of revertant sectors per kernel in these parental seeds is as described by Bennetzen (48). sm = over 200, m-sm = about 200, m = 100 to 200, wm = 20 to 100, vwm = 1 to 20, nm - 0.

[b]The mutable kernels were outcrossed to *bz sh* deletion tester 1302A (57) to create an F_1. This column gives the modal mutability level of this F_1 ear.

[c]At least twenty individual F_1 seeds from each cross were then selfed, and 20-40 kernels off each of these selfed ears were tested for 1:3 segregating seedling and kernel mutations. Any mutations also seen segregating 1:3 in selfs of the original *bzMum4* parent were not counted as new mutations. The numbers in parentheses record the number of new mutations over the total number of F_1 self ears scored.

occasionally lost on outcrossing (48). Analysis of the mutagenic activity of *Mutator*-derived lines carrying the *bzMum4* locus but varying in the frequency of phenotypic revertant sectors indicates that mutagenic activity is fairly well correlated with the level of mutability (Table 1). Lines that have lost or have very low levels of mutability yielded no new mutations in twenty or more selfs scored. In agreement with Robertson (49), we find that some lines with a high level of *Mutator*-associated mutability may also have fairly low mutagenic activities (Table 1, lines B, E, G). Since the populations tested were very small here, however, we can not conclude that any of these individual lines had totally lost *Mutator* mutagenic activity.

Another possible technique we have investigated for scoring *Mutator* activity is gel blot hybridization analysis of *Mutator* elements. The *Mu1* element or sequences that cross-hybridize with its central region (50) have been found associated with all but one of the fifteen *Mutator*-derived mutations, at four different loci, that have been carefully analyzed (10,17,24,40,44,51). Since sequences hybridizing to the center of *Mu1* are relatively rare in standard maize lines (52,53), the multiple *Mu* elements derived from a *Mutator* line can be specifically analyzed. In lines with high *Mutator* mutagenic activity, we have always seen multiple new *Mu* element bands in the genomes of outcross progeny (54 and Figure 1). In lines that have lost *Mutator* mutagenic activity, new *Mu* elements are rarely seen in outcross progeny (54).

Copy number determination of *Mu* elements, in and of itself, is not a good predicter of *Mutator* mutagenic activity (54) or *bzMum4* mutability (48). We have not seen, however, lines with a high mutagenic activity and a very low copy number of *Mu1*-like elements (fewer than ten elements per diploid genome) (unpublished observations).

Covalent DNA modification, apparently 5-methylation of multiple cytosine residues in the *Mu1* element, has been found to correlate with some losses of *Mutator* mutagenic activity (48,55). Loss of *Mutator*-derived mutability is also commonly associated with similar *Mu* element modifications (56,57). Gel blot hybridization analysis of genomic DNA from a *Mutator*-derived plant with 5-methylcytosine-sensitive restriction enzymes can serve as a simple and rapid screen for the degree of *Mu* element modification (Figure 2). All lines we have investigated that contain extensive *Mu* element modification also have lost *Mutator* mutagenic activity. Lines with low levels of *Mu* element modification tend to lose *Mutator* activity in subsequent generations and are thus best avoided (57). However, some lines which have lost *Mutator* mutagenic activity have fully unmodified *Mu* elements (55). Since some sites in the *Mu1*-like elements are more likely to be modified than others in a *Mutator*-loss line (55), the choice of restriction enzymes used for this analysis is significant. In particular, the restriction enzyme *Hin*fI fully digests some otherwise modified *Mu* elements (48), while it does not digest some others (56,57). In all cases where *Mu* elements become modified and *Mutator* mutagenic activity is lost, we have seen a

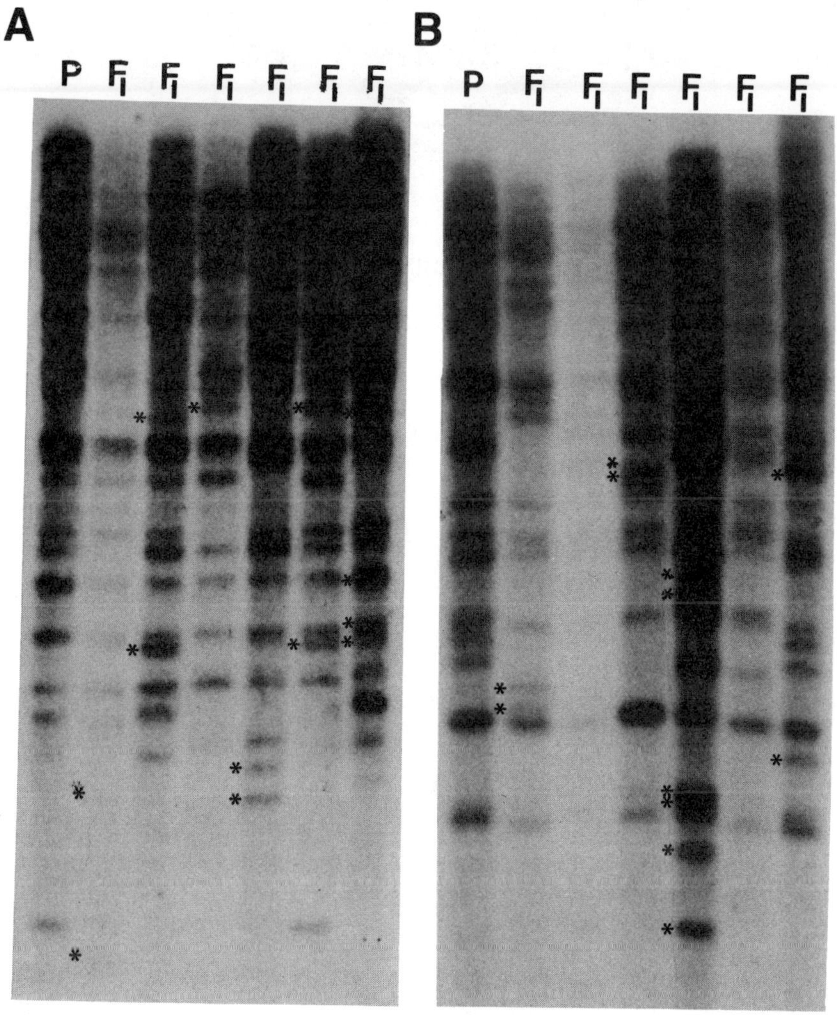

FIGURE 1. Multiple new *Mu* element band positions observed in gel blot hybridization analysis of the progeny of active *Mutator* plants. Approximately 10 µg of DNA from a *Mutator* parent ("P") or the F_1 progeny ("F_1") of this same plant crossed to non-*Mutator* line 1302A (57) was digested with *Hin*dIII and the fragments separated by horizontal gel electrophoresis through 0.7% agarose. The resultant slab gel was filter replicated, hybridized to the *Mu1* central *Tth*111I fragment (50), and washed as previously published (52,54). A = *Mutator* parent A and its progeny. B = *Mutator* parent B and its progeny. The asterisks mark the most obvious new *Mu* element bands in the F_1 progeny.

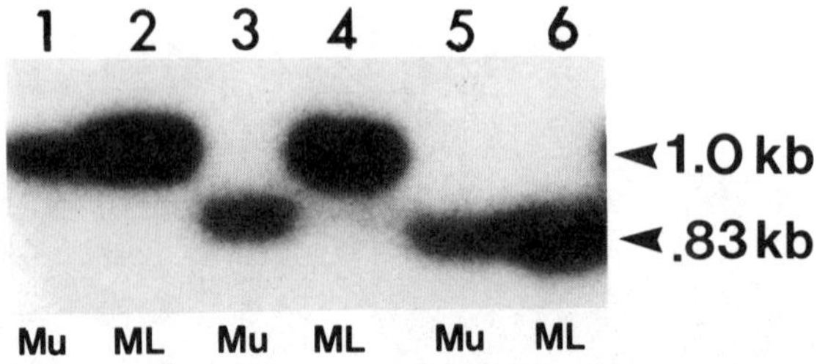

FIGURE 2. Internal modification of *Mu* elements in a *Mu*-loss maize line. Approximately five micrograms of DNA from either an active *Mutator* plant ("Mu") or a plant which had lost *Mutator* mutagenic activity ("ML") was digested with *Tth*111I (lanes 1,2), *Tth*111I, and *Eco*RII (lanes 3,4), or *Tth*111I and *Bst*NI (lanes 5,6). The digested DNA was resolved on an 0.8% agarose gel, filter replicated, hybridized to the 1.0 kb central *Tth*111I fragment of *Mu1* (50), and washed as previously described (52,54). Restriction enzyme *Tth*111I digests all *Mu1*-like elements from active or inactive *Mutator*-derived lines, to a fragment size of 1.0 kb (50,52,55). The enzyme *Bst*NI is similarly insensitive to *Mu* element modification and cuts once in *Mu1* to generate fragments of 0.83 and 0.17 kb (50). *Eco*RII recognizes and cleaves the same DNA sequence as *Bst*NI but will not cut DNA which is cytosine methylated in the sequence 5'-CNG-3' (55). Hence, lanes 3 and 4 demonstrate that the active *Mu* line contains unmodified *Mu* elements, while the *Mu* loss line has modified *Mu* elements.

concomitant loss of genomic *Mu* element digestibility with *Eco*RII. In all gel blot hybridization analyses screening for *Mu* element modification, it is essential to rehybridize the filters with a second probe [e.g., ribosomal or organellar DNA (55)] to guarantee that the enzymatic digestions have gone to completion. An incomplete digestion artifact would be mistaken for *Mu* element modification.

To summarize, loss of *Mutator* mutagenic activity by a particular maize line is well correlated with [1] loss of mutability by a *Mutator*-derived reported allele, [2] very low copy numbers of *Mu* elements, [3] modification of *Mu* elements (55), and/or [4] loss of frequent *Mu* element transpositions as detected by gel blot hybridization (54). Only the last of these tests, numerous *Mu* element band transpositions [and the basic *Mutator* mutagenicity test (35)], serves as a dependable positive correlate for the retention of high *Mutator* mutagenic activity.

Once *Mutator* has been crossed into a particular background and the retention of *Mutator* mutagenic activity confirmed in the resulting line, increases in the mutagenicity of this stock can be generated by a careful sibling cross program. Intercrossing of active *Mutator* lines can increase mutagenic activity of the derived lines at least two fold (46). Continuing this intercrossing procedure leads to eventual inactivation of the *Mutator* trait (46) associated with *Mu* element modification events (55). Crosses between *Mutator* lines, in some backgrounds, can increase *Mu* element copy numbers to as high as 70 to 100 per diploid genome without *Mu* element modification or loss of *Mutator* activity (54,55). Continuing such an intercrossing program, until *Mu* element modification is noted, should optimize *Mu* element copy number and *Mutator* mutagenic activity.

Mutator activity appears to be maintained at different efficiencies in different genetic backgrounds. Robertson has reported (47) that *Mutator* mutagenic activity is lost within four generations of backcrosses into five of the six dent inbreds he employed. Some backgrounds also appear to maintain high *Mu* element copy numbers without loss of *Mutator* activity and *Mu* element modification better than others (55,56). We have initiated studies to investigate the retention of *Mutator* mutagenic activity and *bzMum4* mutability in several inbred and hybrid backgrounds. To date, these preliminary studies indicate that *bzMum4* mutability is efficiently maintained in H95, H99, Ms71, and Oh43 through at least two generations of backcrosses. Mutability was lost in two or fewer generations in A188, B73, P8, and Q66 backgrounds. Multiple separate introductions into the same background and extension of these results will be needed to substantiate these preliminary findings. For the retention of *Mutator* activity, based on these studies and those of Robertson (47), it might be best to avoid crossing *Mutator* into A188, Mo17, B73, B76, P8, or Q66 inbred backgrounds and maintain them in A632, Ms71, Oh43, H95, H99, or [preferably] hybrid backgrounds.

ISOLATION OF A *MUTATOR*-INDUCED MUTATION

With an active *Mutator* line in the appropriate background, a *Mutator*-induced mutation may be simply detected by a cross between a female *Mutator* plant and a male recessive tester. The *Mutator* line is used as the female to avoid the possible interpretation of a pollen contaminant as a *Mutator*-induced mutation. Figure 3 outlines the crossing scheme we used to

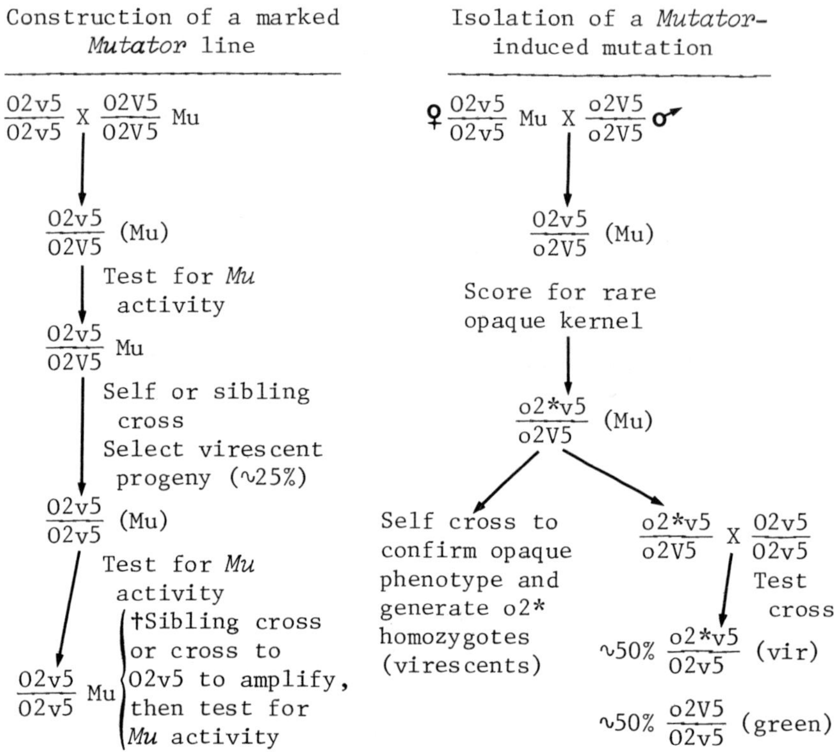

FIGURE 3. Construction and use of a marked *Mutator* line for the isolation of a *Mutator*-derived mutation at *opaque2*. Tests for *Mu* activity were performed after each construction generation (see text). The crossing program may be continued before the data on retention of *Mutator* activity has been assembled, as long as the identical plants are used for the continued crosses as for the *Mutator* tests and accurate records are kept. Lines which have lost *Mu* activity may then be removed from the more advanced generations. o2* = a presumed *Mutator*-induced, mutant allele of *opaque2*. † = a further generation of inbreeding may increase or decrease *Mutator* activity (see text). Mu = active *Mutator* line. (Mu) = possible active *Mutator* line. vir = virescent.

isolate *Mutator*-derived mutations at the *opaque 2* (*o2*) locus. The use of a flanking marker, in this case the *v5* locus, also helps insure against a contamination artifact and aids in the later identification of a *Mu* element at the mutated locus.

This approach generated two novel, confirmed *o2* mutations (*o2**) out of approximately 30,000 kernels screened.

Confirmation of the origin of a putative *Mutator*-derived mutation can be made by comparing the frequency of mutant production in your *Mutator* background versus a standard maize background. For most maize genes, "natural" mutation rates are around 10^{-6}, although some maize genes mutate to an inactive form at rates over a thousand fold more frequently (58). The standard *Mutator* mutagenic rates of 10^{-3} to 10^{-5}, depending upon the locus targetted and the activity of the *Mutator* line used (37,41,45), would be significant confirmation of mutant induction at a locus normally stable to 10^{-6}. For this confirmation, the number of individual seed or seedlings screened must be accurately noted, and a study of "natural" mutation rates at the locus of interest should be undertaken.

A second technique for confirming the *Mutator* origin of a mutation isolated from a *Mutator* line is possible with that subset of *Mu*-induced mutations that show somatic or germinal mutability (35,37,41,44). Almost all unstable *Mutator*-induced mutations show a very similar pattern of somatic reversion: uniform small sectors (37,59). This indicates a relatively precise and late developmental timing of phenotypic reversion events. New mutations isolated from a *Mutator* line with this uniform small sector reversion phenotype have, to date, always been due to the insertion of a *Mu1*-related element (10,17,44,51). For many genes, however, somatic reversion sectors would not be detectable. Germinal reversion events do occur with some *Mutator*-induced lesions. We have seen germinal reversions of *bzMum4* and *bzMum3* at rates of approximately 10^{-4}, apparently by a *Mu* element excision event (unpublished observations). Robertson has observed germinal *Waxy* revertants of *wxMum2* at frequencies ranging from 6×10^{-2} to 3×10^{-4} (60). Many *Mutator*-induced mutations are not mutable (35), however, so this confirmation is not universally applicable. In some cases, stable *Mutator*-derived mutations can be partially or wholly reactivated to a mutable phenotype by crossing to an active *Mutator* line (57,61), thus extending the potential use of this *Mutator* confirmation approach to some originally stable mutations.

THE MOLECULAR CLONING OF A *MUTATOR* TAGGED GENE

Once a *Mutator*-induced mutation has been isolated and confirmed, two complementary approaches may be taken to cloning the mutated gene. The first method requires identification of a *Mu* element which segregates with the mutated gene's phenotype (or that of a flanking marker), followed by cloning of that specific *Mu* element. The second approach involves the molecular isolation of most or all of the *Mu* elements from the plant carrying the *Mutator*-induced mutation, with subsequent identification of the clone carrying the gene of interest.

The *o2** mutations identified in the procedure delineated in Figure 3 were induced in *Mutator* plants containing a homozygous *v5* marker. Crossing plants derived from the *opaque* mutant seed to an *O2 v5* tester yielded a 1:1 ratio of virescent (*v5/v5*) to green (*V5/v5*) plants. Since *v5* is tightly lined to *o2* (8 map units), one expects that 92% of the virescent plants from this outcross will carry the *Mu* element band at *o2** (*o2* v5/O2 v5*) and 92% of the green plants will not (*o2 V5/O2 v5*). Digestion of DNA from green and virescent plants with a restriction enzyme that cuts in the DNA flanking the *Mu* elements, followed by gel blot hybridization to a *Mu* element probe, should allow the identification of a particular *Mu* element band at the *o2** locus due to its segregation with *v5*. However, since *Mutator* lines contain in the range of 10 to 70 *Mu1*-related elements (52), the overlap and coincidental comigration of these numerous *Mu* element bands makes identification of an *o2**- or *v5*-linked band difficult (Figure 4). This is especially true if none of the restriction enzymes chosen for this analysis happens, by chance, to generate a relatively low molecular weight DNA fragment carrying the tagged gene. Only such a small fragment can be easily discerned from the bulk of higher molecular weight bands (Figure 4). Unfortunately, even if a *Mu* element band is seen to segregate with *v5* or *o2** at a statistically significant level, this only indicates linkage and does not prove the *Mu* element identified is actually at *o2**.

With the *Ac* and *En* (*Spm*) systems, the unique structure of the autonomous element [a lack of DNA modification in the active element (16,34)], can be used to clone the single autonomous element in your mutant line and at the tagged locus (9,15,16). Since all or most of the many *Mu* elements present in an active *Mutator* line are unmodified (48,56,57), this approach can not be used to identify a particular *Mu* element at the tagged locus.

The second approach for the molecular cloning of a *Mutator*-induced mutation requires the isolation of representatives of most of the *Mu1*-like elements from your mutant

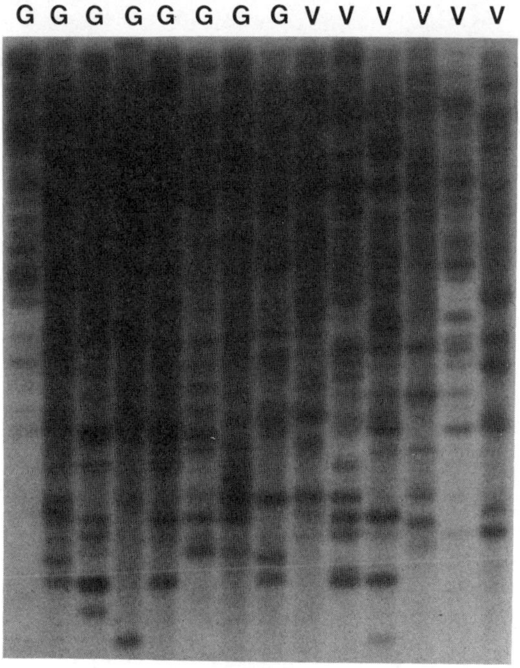

FIGURE 4. Segregation of *Mu* elements in an o2* v5/o2 V5 outcross to a tester (O2 v5/O2 v5). Approximately eight micrograms of DNA from either green ("G") or virescent ("V") plants were digested with *Bam*HI and *Bcl*I, resolved, filter blotted, and hybridized to a *Mu1* central probe as described (52,54). The green and virescent plants that were the source of these DNAs are from the approximately 50% green, 50% virescent progeny of the cross in the lower right corner of Figure 3. Nearly all virescent plants should have an o2* *Mu* band which would be lacking from the green plant genome.

plant. The DNA flanking the *Mu* element in these individual clones is then tested for the properties expected of your particular locus (see below). Although it may appear both arduous and artifact-prone, this method has been used successfully in two *Mutator* tagging programs (10,17).

Both the techniques for cloning *Mu* element associated mutations mentioned above would be simplified if the genomic *Mu* element copy number in the mutant line could be reduced. Crossing a maize line that has lost *Mutator* activity to a non-*Mutator* plant (i.e., a standard inbred, for example) leads to an approximate two-fold dilution in *Mu1*-like element copy number (54). Such a *Mutator*-loss line is easy to obtain, since *Mutator* activity is lost by about 10% of the progeny of a cross between standard *Mutator* and non-*Mutator* plants (35) and at much higher rates in some backgrounds (47) or *Mutator* lineages (48). The *Mutator* loss event may be detected by loss of mutability in a *Mu*-induced reported allele, loss of mutagenic activity, acquired modification of *Mu* elements, or loss of frequent *Mu* element transpositions, all detected as described above. Once a *Mutator*-loss lineage of your *Mu*-induced mutant plant is identified, repetitive outcrosses can dilute the *Mu* element number to more manageable levels.

CONFIRMATION OF THE MOLECULAR CLONING OF A *MUTATOR* TAGGED LOCUS

Once a *Mu* element inserted into genomic maize DNA has been molecularly cloned, several methods exist to confirm or determine whether the DNA flanking the *Mu* element contains the locus originally targetted in the transposon tagging program. All of these approaches are facilitated by a rapid technique for specifically labelling the DNA next to a *Mu* element. The method we employ (suggested by S. Dellaporta) is to use a short oligonucleotide homologous to the extreme termini of the *Mu1* inverted repeats (50) as a primer for synthesizing a high specific activity probe (62) directed outside the *Mu* element. Since *Mu* elements insert primarily into low copy number DNA (59 and unpubl. obs.), the radioactive probe generated will commonly hybridize to a single genomic DNA band on gel blot hybridization. Before labelling, the *Mu* element associated clone is generally digested with a four base pair specificity restriction enzyme that generates blunt ends (e.g., *Alu*I or *Hae*III). This guarantees that the probe will be fairly small and unlikely to extend into the middle and highly repetitive DNA sequences that flank single copy maize genes (63).

If the DNA flanking a cloned *Mu* element is the locus targetted in a *Mutator* mutagenesis program, then any major structural alterations in that gene should be detected by gel blot hybridization and should segregate with the targetted locus. A maize line with a deletion of the tagged gene should show no hybridization to DNA flanking the appropriate *Mu*-containing clone (9). Alterations in restriction fragment lengths associated with insertion of a transposable element, including the original *Mutator*-induced deficiency, should be detectable in comparisons of the progenitor and mutant alleles. For example, a hybridizational probe homologous to the *opaque2* locus in our *o2* tagging program will hybridize to a different size restriction fragment in the *O2 v5/O2 v5* and *o2* v5/o2* v5* line (Figure 3). The *o2 V5/o2 V5* tester undoubtedly will have unique size *opaque2* hybridizing restriction fragments, as well, allowing a test for segregation of the presumed *o2** restriction fragment length with the flanking *v5* phenotype in the tester outcross. Since restriction fragment length polymorphisms are so common in maize (42,43), it is likely that any segregating allelic variations of unrelated or unknown origin at the *o2* locus will also segregate a restriction fragment length polymorphism. Specific germinal reversions of the *Mu* element-induced mutation should have a gel blot hybridization restriction map like that of the parental *O2 v5* allele (unpubl. obs.), and this can also be tested with the *Mu*-flanking probe.

Another technique for identifying and/or confirming a *Mu* element-containing clone as a specific, targetted gene requires cross-hybridization of the DNA flanking your *Mu* element clone to a second library of transposable element-containing clones isolated from a maize line with a second transposable element-induced mutation at the locus of interest (10). If the two mutations have very different origins, then only the DNA flanking the *Mu* element at the targetted locus will hybridize to one of the second set of clones (10). If both mutations were induced by *Mutator*, however, there is a good chance that *Mu* elements in identical genomic positions in the progenitors of these two mutants will also lead to a crosshybridization result. This technique also has the disadvantage that all or nearly all of the appropriate transposable elements from each line must be cloned before the crosshybridization program can be initiated with confidence.

A final mode in which a *Mu* element-containing clone may be tested for its genetic content requires some understanding of the physiology of the gene targetted. If the gene of interest regulates, for instance, flower development, it is more likely that the targetted gene will be transcriptionally active in flowers than in leaves or roots. In some cases, it may be possible to predict not only the specific organ but the specific time in development at which a gene will be active. A radioactive probe homologous to the targetted gene should hybridize only to RNA synthesized at the appropriate time and in the appropriate tissues. This is not a particularly good technique for identifying a *Mu* element associated

I. Isolate a *Mutator*-associated mutation (Figure 3).
II. Confirm the *Mutator* origin of the mutation.
  A. Frequency.
  B. Somatic mutability.
  C. Germinal mutability.
III. Clone the *Mutator*-induced mutant allele.
  A. Clone representatives of most or all *Mu* elements from the mutant line. Identify the specific *Mu* element at the targetted locus by the confirmation techniques outlined below.
  B. Identify a *Mu* element restriction fragment that segregates with the mutation. Clone this fragment specifically.
IV. Confirm the *Mu* element flanking DNA as the targetted locus.
  A. Absence from a deletion mutant.
  B. Predicted restriction fragment length polymorphisms associated with various alleles, including transposable element insertions and excisions.
  C. Segregation of a restriction fragment length polymorphism with the targetted allele.
  D. Homology to DNA in an independently derived, putative clone of the locus.
  E. Expression of the DNA at the RNA level in the appropriate tissue(s) and time(s) in development expected of the targetted gene.
V. Clone the fully functional allele of the targetted locus using the *Mu* element flanking DNA as a probe.

FIGURE 5. Outline of an approach for the molecular cloning of a maize gene by *Mutator* transposon tagging. See text for a detailed description of each step.

clone as a targetted gene because [1] it begs the question on the molecular behavior of the gene you wish to study and [2] *Mu* elements tend to insert in single copy, transcriptionally active DNA in most cases (59 and unpubl. obs.), giving rise to the potential for numerous false positives. This technique is useful, however, for final confirmation of a clone that is already firmly associated with the targetted locus by other criteria.

Once the *Mutator*-induced allele of the targetted gene has been cloned, identified, and confirmed, then the single copy DNA flanking the *Mu* insertion can be used as a hybridizational probe to molecularly clone a fully functional version of the desired gene from a standard maize background. In the not-too-distant future, absolute confirmation of this clone will be made by transformation and complementation of a recessive mutation at the locus.

SUMMARY

Due primarily to its high mutagenic activity, the *Mutator* system has been and will continue to be a useful system for transposon tagging in maize. We have assembled a routine protocol for the isolation and molecular cloning of *Mutator*-induced mutations (outlined in Figure 5). The key steps in this procedure are isolation of a *Mutator*-induced mutation from a certified *Mutator* maize line and confirmation of a *Mu* element containing clone as the targetted allele. Recent advances in our understanding of the *Mutator* system have greatly simplified the tagging/cloning process. Depending on the degree of line construction needed, the genetics program to isolate and confirm a *Mutator*-induced mutation at a given locus requires two to five plant generations. The molecular cloning of a *Mutator*-induced mutation should then require four or more months, depending on the experience and enthusiasm of the researcher involved.

ACKNOWLEDGMENTS

We wish to thank S. Dellaporta and V. Walbot for permission to reference unpublished data. We also wish to thank D. Robertson and P. Starlinger for permission to cite their observations in the *Maize Genetics Cooperative Newsletter*. The expert technical assistance of R. Fracasso and C. Carter and the clerical expertise of V. Livingston are greatly appreciated.

REFERENCES

1. Person C, Sidhu G (1971) Genetics of host-parasite interrelationships. In "Mutation Breeding for Disease Resistance," Vienna: Int At Energy Agency, p 31.
2. Coe EH Jr, Neuffer MG (1977) The genetics of corn. In Sprague GF (ed): "Corn and Corn Improvement," Madison, WI: Am Soc Agron, p 111.

3. Nelson OE (1980) Genetic control of polysaccharide and storage protein synthesis in the endosperm of barley, maize and sorghum. In Pomerantz Y (ed): "Advances in Cereal Science and Technology," Vol 3, St Paul, MN: Am Assoc Cereal Chemists, p 41.
4. Sheridan WF, Neuffer MG (1982) Maize developmental mutants: Embryos unable to form leaf primordia. J. Hered 73:318.
5. Poethig RS (1985) Homeotic mutations in maize. UCLA Symp Mol Cell Biol 35:33.
6. Bennetzen JL (1984) Genetic engineering for improved crop disease resistance. In Collins GB, Petolino F (eds): "Applications of Genetic Engineering to Crop Improvement," Dordrecht, The Netherlands: Nijhof/Junk, p 491.
7. Kleckner N, Roth J, Botstein D (1977) Genetic engineering *in vivo* using translocatable drug-resistance elements. J Mol Biol 116:125.
8. Bingham PM, Levis R, Rubin GM (1981) Cloning of DNA sequences from the *white* locus of *D. melanogaster* by a novel and general method. Cell 25:693.
9. Fedoroff NV, Furtek DB, Nelson OE Jr (1984) Cloning of the *bronze* locus in maize by a simple and generalizable procedure using the transposable controlling element *Activator* (*Ac*). Proc Natl Acad Sci USA 81:3825.
10. O'Reilly C, Shepherd NS, Pereira A, Schwarz-Sommer Zs, Bertram I, Robertson DS, Peterson PA, Saedler H (1985) Molecular cloning of the *a1* locus of *Zea mays* using the transposable elements *En* and *Mu1*. EMBO J 4:877.
11. Dooner HK, Weck E, Adams S, Ralston E, Farreau M, English J (1985) A molecular genetic analysis of insertions in the bronze locus of maize. Mol Gen Genet 200:240.
12. Martin C, Carpenter R, Sommer H, Saedler H, Coen ES (1985) Molecular analysis of instability in flower pigmentation of *Antirrhinum majus* following isolation of the *pallida* locus by transposon tagging. EMBO J 4:1625.
13. Paz-Ares J, Wienand U, Peterson PA, Saedler H (1986) Molecular cloning of the *C* locus of *Zea mays*: a locus regulating the anthocyanin pathway. EMBO J 5:829.
14. Wienand U, Weydemann U, Niesbach-Kloesgen U, Peterson PA, Saedler H (1986) Molecular cloning of the *C2* locus of *Zea mays* - the gene coding for chalcone synthase. Mol Gen Genet 203:202.
15. Dellaporta SL, Greenblatt I, Kermicle J, Hicks JB, Wessler S (1987) Molecular cloning and structural characterization of the *R* locus of maize. EMBO J, in press.

16. Cone KC, Burr FA, Burr B (1986) Molecular analysis of the maize anthocyanin regulatory locus *C1*. Proc Natl Acad Sci USA 83:9631.
17. McLaughlin M, Walbot V (1987) Cloning of the *bronze2* locus of maize by transposon tagging with *Mutator*. Manuscript in preparation.
18. Theres K, Starlinger P (1986) Molecular cloning of *bz2-m*. Maize Genetics Coop Newsl 60:40.
19. Döring H-P, Starlinger P (1987) Molecular genetics of transposable elements in plants. Ann Rev Genet 20:175.
20. Baker B, Schell J, Lörz H, Fedoroff N (1986) Transposition of maize controlling element *Activator* in tobacco. Proc Natl Acad Sci USA 83:4844.
21. Fedoroff N, Wessler S, Shure M (1983) Isolation of transposable maize controlling elements *Ac* and *Ds*. Cell 35: 235.
22. Behrens U, Fedoroff N, Laird A, Müller-Neumann M, Starlinger P, Yoder J (1984) Cloning of the *Zea mays* controlling element *Ac* from the *wx-m7* aliele. Mol Gen Genet 194:346.
23. Pereira A, Schwarz-Sommer Zs, Gierl A, Bertram I, Peterson PA, Saedler H (1985) Genetic and molecular analysis of the Enhancer (*En*) transposable element system of *Zea mays*. EMBO J 4:17.
24. Bennetzen JL, Swanson J, Taylor WC, Freeling M (1984) DNA insertion in the first intron of maize *Adh1* affects message levels: Cloning of progenitor and mutant *Adh1* alleles. Proc Natl Acad Sci USA 81:4125.
25. Shepherd NS, Schwarz-Sommer Zs, Wienand U, Sommer H, Deumling B, Peterson PA, Saedler H (1982) Cloning of a genomic fragment carrying the insertion element *Cin1* of *Zea mays*. Mol Gen Genet 188:266.
26. Shepherd NS, Schwarz-Sommer Z, Blumberg vel Spalve J, Gupta M, Wienand U, Saedler H (1984) Similarities of the *Cin1* repetitive family of *Zea mays* to eucaryotic transposable elements. Nature 307:185.
27. Dellaporta SL, Chomet PS, Mottinger JP, Wood J, Yu SM, Hicks JB (1984) Endogenous transposable elements associated with virus infection in maize. Cold Spring Harbor Symp Quant Biol 49:321.
28. Johns MA, Mottinger J, Freeling M (1985) A low copy number, *copia*-like transposon in maize. EMBO J 4:1093.
29. Rhoades MM (1938) Effect of the *Dt* gene on the mutability of the a_1 allele in maize. Genetics 23:377.
30. Ashman RB (1960) Stippled aleurone in maize. Genetics 45:19.

31. Salamini F (1980) Controlling elements at the *Opaque-2* locus of maize: their involvement in the origin of spontaneous mutation. Cold Spring Harbor Symp Quant Biol 45:467.
32. Nevers P, Shepherd NS, Saedler H (1986) Plant transposable elements. Adv Bot Res 12:103.
33. McClintock B (1950) The origin and behavior of mutable loci in maize. Proc Natl Acad Sci USA 36:344.
34. Dellaporta SL, Chomet PS (1985) The activation of maize controlling elements. In Hohn B, Dennis ES (eds): "Genetic Flux in Plants," Wien/New York: Springer Verlag, p 169.
35. Robertson DS (1978) Characterization of a Mutator system in maize. Mutat Res 51:21.
36. Robertson DS, Mascia PN (1981) Tests of four controlling element systems of maize for mutator activity and their interaction with *Mu* mutator. Mutat Res 84:283.
37. Robertson DS (1985) Differential activity of the maize mutator *Mu* at different loci and in different cell lineages. Mol Gen Genet 200:9.
38. Greenblatt IM, Brink RA (1962) Twin mutations in medium variegated pericarp maize. Genetics 47:489.
39. Greenblatt IM (1984) A chromosomal replication pattern deduced from pericarp phenotypes resulting from movements of the transposable element, *Modulator*, in maize. Genetics 108:471.
40. Strommer JN, Hake S, Bennetzen J, Taylor WC, Freeling M (1982) Regulatory mutants of the maize *Adh1* gene caused by DNA insertions. Nature 300:542.
41. Walbot V, Briggs CP, Chandler V (1986) Properties of mutable alleles recovered from Mutator stocks of *Zea mays* L. Stadler Genet Symp, in press.
42. Rivin CJ, Zimmer EA, Cullis CA, Walbot V, Huynh T, Davis RW (1983) Evaluation of genomic variability at the nucleic acid level. Plant Mol Biol Rptr 1:9.
43. Helentjaris T, King G, Slocum M, Siedenstrang C, Wegman S (1985) Restriction fragment polymorphisms as probes for plant diversity and their development as tools for applied plant breeding. Plant Mol Biol 5:109.
44. Brown WE, Robertson DS, Bennetzen JL (1987) Molecular analysis of multiple *Mutator*-derived mutations at the *bronze* locus in maize. Manuscript in preparation.
45. Ellingboe AE, Bennetzen JL (1987) Genetic properties of *Mutator*-associated mutations at the *Rp1* locus of maize. Manuscript in preparation.

46. Robertson DS (1983) A possible dose-dependent inactivation of Mutator (*Mu*) in maize. Mol Gen Genet 191:86.
47. Robertson DS (1986) Loss of *Mu* mutator activity when active *Mu* systems are transferred to inbred lines. Maize Genetics Coop Newsl 60:10.
48. Bennetzen JL (1985) The regulation of *Mutator* function and *Mu1* transposition. UCLA Symp Mol Cell Biol 35:343.
49. Robertson DS (1986) Evidence bearing on the relationship between germinal instability (mutator activity) and somatic instability (excision) in *Mu*-induced mutants. Maize Genetics Coop Newsl 60:8.
50. Barker RF, Thompson DV, Talbot DR, Swanson J, Bennetzen JL (1984) Nucleotide sequence of the maize transposable element *Mu1*. Nucl Acids Res 12:5955.
51. Taylor LP, Chandler VL, Walbot V (1986) Insertion of 1.4 kb and 1.7 kb *Mu* elements into the *Bronze-1* gene of *Zea mays* L. Maydica 31:31.
52. Bennetzen JL (1984) Transposable element *Mu1* is found in multiple copies only in Robertson's *Mutator* maize lines. J Mol Appl Genet 2:519.
53. Chandler V, Rivin C, Walbot V (1986) Stable non-Mutator stocks of maize have sequences homologous to the *Mu1* transposable element. Genetics 114:1007.
54. Bennetzen JL, Fracasso RP, Morris DW, Robertson DS, Skogen-Hagenson MJ (1987) Concomitant regulation of *Mu1* transposition and *Mutator* activity in maize. Mol Gen Genet, in press.
55. Bennetzen JL (1987) Covalent DNA modification and the regulation of *Mutator* element transposition in maize. Submitted.
56. Chandler V, Walbot V (1986) DNA modification of a maize transposable element correlates with loss of activity. Proc Natl Acad Sci USA 83:1767.
57. Bennetzen JL (1987) Inactivation and reactivation of a *Mutator*-derived *bronze* mutable allele in maize. Submitted.
58. Bennetzen JL, Ellingboe AE (1987) Allele-specific instability at the *Rp1* locus of maize. Manuscript in preparation.
59. Bennetzen JL (1987) The molecular genetics of plant transposable elements. Intl Rev Cytol, in press.
60. Robertson DS (1986) Studies of pollen and somatic instability in *Mu*-induced waxy mutants. Maize Genetics Coop Newsl 60:7.

61. Robertson DS (1985) Studies on the nature of the loss of somatic mutability for two *Mu*-induced mutable *a1* mutants. Maize Genetics Coop Newsl 59:11.
62. Feinberg A, Vogelstein B (1983) A technique for radiolabelling DNA restriction endonuclease fragments to high specific activity. Anal Biochem 132:6.
63. Bennetzen JL, Schrick KS, Hake S (1987) Active, single copy maize genes are flanked by diverse highly repetitive DNA sequences. Manuscript in preparation.

IV. ENZYME–GENE SYSTEMS IN METABOLIC REGULATION

Plant Gene Systems and Their Biology, pages 207–216

THE PHOSPHOENOLPYRUVATE CARBOXYLASE GENE FAMILY OF MAIZE[1]

John W. Grula and Richard L. Hudspeth

Phytogen, 101 Waverly Drive
Pasadena, California 91105

ABSTRACT We have isolated clones from a maize (var.B73) genomic library that contain four members of the phosphoenolpyruvate (PEP) carboxylase gene family. Three of the four genes share substantial sequence homology as revealed by restriction map similarities and by using the clones to probe genome blots and Northern blots of RNA from green leaves, etiolated leaves, and roots. These three more closely related genes are divergent from the PEP carboxylase mRNA in green leaves and exhibit comparatively strong reactions to the PEP carboxylase mRNA in roots. In contrast, a fourth gene hybridizes intensely with PEP carboxylase message in green leaves, but weakly to the PEP carboxylase mRNA in etiolated leaves and roots. Thus, this gene probably encodes the form of PEP carboxylase involved in C_4 photosynthesis, and it appears to have diverged from the other members of the family. Finally, the blot data also indicate the presence of at least one other PEP carboxylase gene in maize that is divergent from the four we have cloned.

INTRODUCTION

The enzyme phosphoenolpyruvate (PEP) carboxylase catalyzes the vital reaction of atmospheric CO_2 fixation in plants utilizing C_4 and crassulacean acid metabolism (CAM) photosynthesis (reviewed in 1). In several species of C_4 plants, PEP carboxylase has been shown to be abundantly expressed in a light-inducible fashion in leaves to fulfill

[1]This work was supported by Eli Lilly and Company (Indianapolis, IN) and the J. G. Boswell Company (Los Angeles, CA)

this function (2-5). However, this enzyme is also present in lesser amounts in nonphotosynthetic cells of C_4 plants (6), as well as in various tissues of C_3 plants (7), and procaryotic cells (8,9). Depending on the cell type in which PEP carboxylase has been detected, several nonphotosynthetic roles for the organic acids that derive from this enzyme's activity have been proposed. These include the maintenance of intracellular ionic and pH balance (1), providing carbon skeletons for the process of nitrogen fixation (10), and replenishing the levels of TCA cycle and amino acid biosynthetic intermediates (10,11). High levels of PEP carboxylase activity found in some fruits and seeds may also be functioning to refix CO_2 lost during periods of high respiration rates (1).

A C_4 plant such as maize (*Zea mays*) would be expected to have a PEP carboxylase gene family that encoded multiple forms of the enzyme needed for C_4 photosynthesis as well as some of the functions described above. Here we describe genomic clones that contain four different members of the maize PEP carboxylase gene family, including one that appears to encode the isozyme involved in C_4 photosynthesis.

RESULTS

PEP Carboxylase Genomic Clones and Relevant Sequence Information.

The isolation of the maize (var. B73) leaf PEP carboxylase cDNA clone pH1 has been previously described (12). We have determined that the sequence of pH1 is 99% homologous to the maize (var. indentata STURT.) PEP carboxylase cDNA clone reported by Izui et al. (13; data not shown). The restriction maps of four different genomic clones screened from a maize (var. B73) library with pH1 are shown in Fig. 1. We initially observed that pH1 hybridizes more intensely with H1λ14 than to the other three clones (for an example see Fig. 2, lanes 1, 2, 4 and 5). Subsequently, we have found the sequence of pH1 to be identical to that of H1λ14 over a region that includes 155 bp of protein-coding sequence and 234 bp of 3'-untranslated sequence (data not shown). Thus, the different intensities with which pH1 hybridizes to the four genomic clones in Fig. 1 can be explained by different degrees of sequence relatedness. In addition, the sequence data also indicate that pH1 derives from the mRNA encoded by the gene in H1λ14.

Although the restriction site maps of the four PEP carboxylase genes and their flanking regions are clearly different from each other, several sites appear to be held in common (Fig. 1). In particular, there is a cluster of two Sal I sites and one Bam HI site which maps in the vicinity of the 5'-end of all four genes. The clones designated 22, 11, and 4 also appear to share one-to-two EcoRI sites and an additional Bam HI site in the same region. These data suggest that the genes in clones 22, 11, and 4 may be more closely related to each other than to the gene in clone 14, but clearly more information is needed.

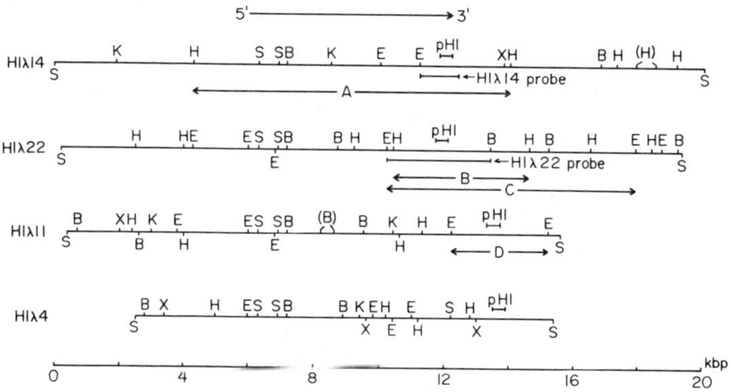

Figure 1. Restriction maps of genomic clones containing four different members of the maize (B73) PEP carboxylase gene family. The pH1 cDNA clone insert (12) is indicated just above the fragments to which it hybridizes in each genomic clone. The 5'-3' orientation indicated applies to all four genes. The end points of the arrow indicating gene orientation mark the transcription start and stop sites of the gene in H1λ14 (unpublished data). Fragments from H1λ14 and H1λ22 that were used as genome blot probes (Fig. 2) are indicated just below each of these clones. Hind III and EcoRI fragments in the maize (B73) genome which hybridize to these probes and pH1 are designated by double-headed arrows labeled A-D (see also Fig. 2). The clone designated H1λ22 is identical to the clone H1λ21 we previously reported (12), except it contains several kbp of additional sequence at both ends of the insert. Restriction sites: E, EcoRI; H, HindIII; S, Sal I; B, Bam HI; K, Kpn I; X, Xho I. Sites enclosed in parentheses map to one of the two positions indicated.

Genome Blots Probed with Different Gene Fragments.

The Hind III fragments in the maize (B73) genome that hybridize to the PEP carboxylase cDNA clone pH1 are shown in lane 3 of Fig. 2 (12). The reconstructions in lanes 1, 2, 4 and 5 of this figure indicate that the genes represented by H1λ14 and H1λ22 are present in 1-2 copies per haploid genome. The 9.5 kbp fragment labled A is present in H1λ14 and the 4.3 kbp fragment labeled B is present in H1λ22 (also see Fig. 1).

The perfect sequence homology between pH1 and H1λ14 that was described earlier is reflected in the hybridization intensities of the fragments observed in lane 6 of Fig. 2. In this case a Hind III digestion of maize (B73) DNA was probed with a 1.2 kbp EcoRI/HincII fragment from H1λ14 that includes the region homologous to pH1 plus approximately 500 bp of additional protein-coding sequence (see "H1λ14 probe" in Fig. 1). The size and reaction intensities of the fragments that hybridize with this genomic clone probe (lane 6) are essentially identical to that obtained with the pH1 probe (lane 3), with a strong reaction observed to a 9.5 kbp fragment and weak reactions to fragments 22, 5.4 and 4.3 kbp in size.

When a duplicate Hind III digestion of the same maize DNA is probed with a 3.3 kbp EcoR1/BamH1 fragment from H1λ22 (see Fig. 1), the same fragments that hybridize to the pH1 and H1λ14 probes are detected, but the intensities of the reactions are very different (Fig. 2, lane 7). In this case the 5.4 and 4.3 kbp fragments exhibit strong reactions, while the 9.5 kbp reaction is weak. Interestingly, the 22 kbp fragment reaction remains weak. Therefore, this fragment probably contains a member of the PEP carboxylase gene family divergent from both probes.

The strong reaction to the 4.3 kbp HindIII fragment exhibited by the H1λ22 probe is expected because this is the genomic fragment that corresponds to this probe. The strong reaction to the 5.4 kbp Hind III fragment (which could be double the strength of the 4.3 kbp reaction) indicates there is at least one other PEP carboxylase gene in the maize genome that at least in part has great similarity to the gene in H1λ22. This is verified by the result shown in lane 8 of Fig. 2, in which an EcoR1 digestion of the same maize DNA was probed with the H1λ22 fragment. Three strongly-reacting fragments are observed, one of which again corresponds to the H1λ22 probe (the 7.8kbp fragment labeled C in Figs. 1 and 2) and another which is in H1λ11 (the 3.0 kbp fragment labeled D in Figs. 1 and 2). The third strongly-reacting fragment, which measures 18 kbp, cannot be assigned to any of the genomic clones in Fig. 1. However, it indicates there are three PEP carboxylase genes in the maize (B73) genome that have close homology in at least a portion of their sequence. Two of these genes are contained in genomic clones H1λ22 and H1λ11.

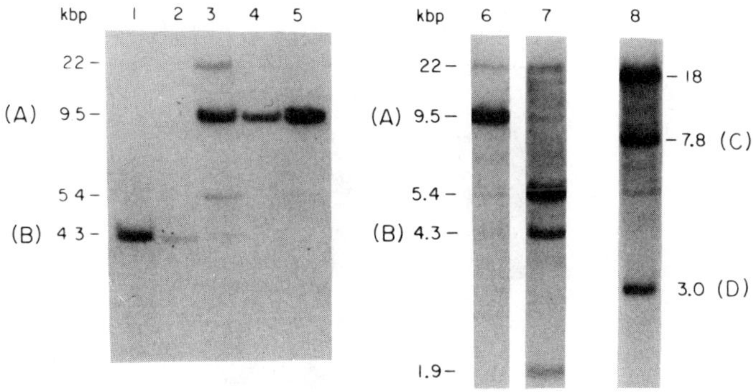

Figure 2. Southern blots of maize (B73) DNA probed with different PEP carboxylase gene fragments. Lanes 1-5 are from a previous report (12) and contain one-and three-copy reconstructions of H1λ21 (lanes 2 and 1; see also the Fig. 1 legend) and H1λ14 (lanes 4 and 5), and 10 µg of total maize DNA (lane 3) probed with the PEP carboxylase cDNA clone pH1. Total maize DNA (10µg) was probed with the H1λ14 probe (Fig. 1) in lane 6 and with the H1λ22 probe (Fig. 1) in lanes 7 and 8. In lanes 1-7 the DNAs were digested with HindIII; lane 8 is an EcoRI digestion. The fragments labeled A-D correspond to the fragments indicated in Fig. 1 with the same labels.

RNA from Different Maize Organs Probed with Genomic Clones.

Northern blots of RNA from maize (hybrid Golden Jubilee) green leaves, etiolated leaves, and roots were probed with the genomic clones in Fig. 1 to provide additional information about the protein-coding information and sequence relationships of the four genes contained in these clones (Fig. 3). As previously reported (12), the H1λ14 clone reacts intensely with a 3.5kb mRNA in green leaves and very weakly to mRNAs of the same size in etiolated leaves and roots. In contrast, the other three genomic clones exhibit relatively weak reactions to the green leaf mRNA and generally stronger reactions to the etiolated leaf and root mRNAs (Fig. 3). These results provide an additional indication that the DNA sequences of the PEP carboxylase genes in H1λ22, H1λ11, and H1λ4 are more closely related to each other than to the gene in H1λ14.

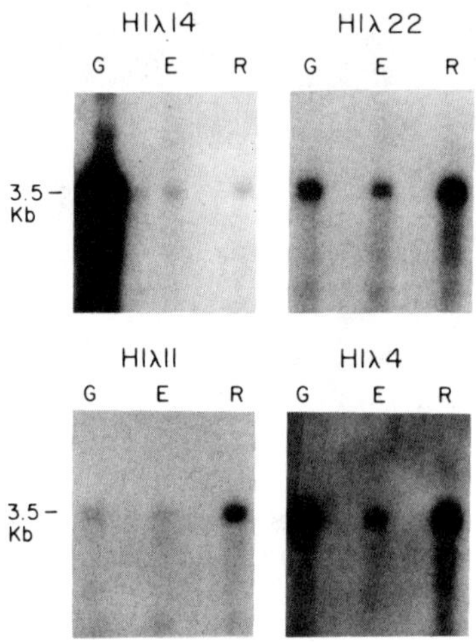

Figure 3. Northern blots of poly(A)$^+$ RNA (1μg) from green leaves(G), etiolated leaves(E), and roots(R) of maize probed with the PEP carboxylase genomic clones in Fig. 1. The size of the mRNAs (3.5kb) has been estimated using maize leaf rRNAs and λDNA digested with Hind III as markers.

Restriction Site Analysis of Additional PEP Carboxylase Genomic and cDNA Clones.

We have mapped eight additional genomic clone isolates that exhibit strong reactions with pH1 in the same manner as H1λ14. These analyses have shown that all eight of these clones overlap with H1λ14 to varying extents and we have been unable to find any differences among these clones in the six-base pair recognition sites indicated in Fig. 1. In addition, we have mapped 14 additional restriction sites in a 2.5 kbp subcloned region from five of these genomic clones (including H1λ14) that includes approximately 1.5 kbp of DNA flanking the 3'-end of the gene. All 14 of these sites mapped the same in each clone, including seven that are in the 3'-flanking region. These results support the view that H1λ14 and the other eight isolates represent one locus in the maize genome. We have no evidence for the existance of more than one gene that can hybridize intensely with pH1.

We have also mapped in varying degrees of detail 17 additional maize (B73) leaf PEP carboxylase cDNA clones, including several that are close to full-length (Fig. 4). These clones all appear to have the same maps, and the small amount of sequence we have obtained so far (from the 3'-untranslated region of the clone in Fig. 4) matches that of pH1 and H1λ14 (data not shown). Given that different mRNAs have not been differentially cloned, this would suggest there is only one form of abundant PEP carboxylase mRNA in green maize leaves that encodes the C_4 photosynthetic form of the enzyme. This conclusion is consistent with previous studies in maize and also sorghum (2-4,14-16), with the exception of one report by Harpster and Taylor (17).

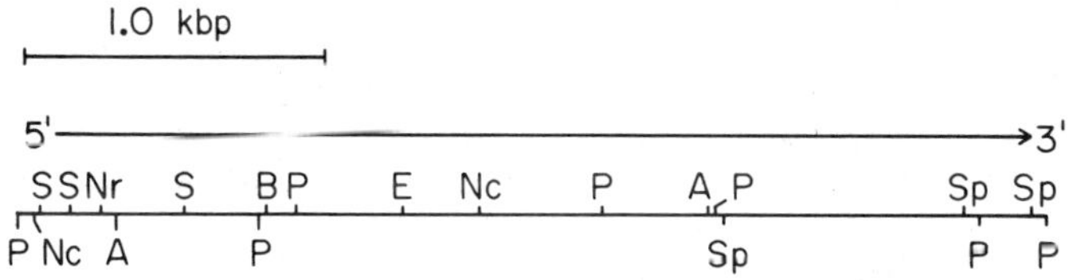

Figure 4. Restriction site map and orientation of a nearly full-length PEP carboxylase cDNA clone derived from maize (B73) green leaves. Partial sequence information and restriction site analysis indicate that this clone derives from the same mRNA as pH1, and therefore is encoded by the same gene (represented by genomic clone H1λ14). The restriction site map is also very similar to the map of the maize (indentata STURT.) leaf PEP carboxylase cDNA clone isolated by Izui et al (13). Restriction sites: S, Sal I; Nc, Nco I; Nr, Nru I; A, Apa I; P, Pst I; B, BamHI; E, EcoRI; Sp, Sph I.

DISCUSSION

We have cloned and partially characterized four members of the maize (B73) PEP carboxylase gene family. The genome blot data in Fig. 2 indicate that the total family size may be as small as five, but we cannot exclude the possibility that probes and/or conditions different from the ones we have used here may reveal additional members. Sequence data and restriction site analysis of genomic and cDNA clones indicate that the C_4 photosynthetic enzyme is probably encoded by one gene. Furthermore, our results show that the sequence of the gene encoding this isozyme (contained in clone H1λ14) is divergent from the sequences of the other three genes we have cloned. These latter three genes exhibit comparatively strong reactions to root mRNA and appear to have much more homology with each other than with the gene in H1λ14. A fifth gene detected in the genome blots (Fig. 2) is apparently divergent from all four of the others that we have cloned. Our results are consistent with studies of PEP carboxylase from C_3, C_3-C_4, and C_4 Flaveria and Atriplex species (18,19), which support the view that a unique form of this enzyme has evolved to function in the C_4 photosynthetic pathway.

Note added after manuscript submitted: The chromosomal location of several maize PEP carboxylase genes has been determined by Scott Wright and Tim Helentjaris of NPI in Salt Lake City, Utah, using restriction fragment length polymorphism analysis of monosomic lines (20) with the cDNA clone shown in Fig. 4 and the H1λ22 fragment indicated in Fig. 1 as probes. The results obtained with the cDNA clone probe reproduced those obtained previously with a maize PEP carboxylase probe provided by W. Taylor, and show that the gene encoding the C_4 photosynthetic isozyme maps to a locus near the centromere on the long arm of chromosome 9. With the H1λ22 probe it was possible to map two additional members of the gene family. One of these is located close to the distal end of the long arm of chromosome 4, and the other maps near the centromere on the long arm of chromosome 5. These latter loci fall within two regions that are one of several examples of chromosomal segment duplication in the maize genome (21).

ACKNOWLEDGMENTS

The authors thank Carlotta Glackin for preparation of the three RNAs used for the blots shown in Fig. 3. Drs. Eric Davidson, Robert Goldberg, David Anderson, and Richard Yenofsky provided criticism and information that was very useful for the completion of this work. We thank Mary Strother for preparation of the manuscript.

REFERENCES

1. Edwards G, Walker DA (1983). "C_3, C_4: Mechanisms, and Cellular and Environmental Regulation, of Photosynthesis." Berkeley: Univ. of California Press, pp 275-298, 479-493.
2. Hague DR, Sims TL (1980). Evidence for light-stimulated synthesis of phosphoenolpyruvate carboxylase in leaves of maize. Plant Physiol 66:505.
3. Hayakawa S, Matsunaga K, Sugiyama T (1981). Light induction of phosphoenolpyruvate carboxylase in etiolated maize leaf tissue. Plant Physiol 67:133.
4. Vidal J, Gadal P (1983). Influence of light on phosphoenolpyruvate carboxylase in sorghum leaves. I. Identification and properties of two isoforms. Physiol Plant 57:119.
5. Bassett CL, Rinehart CA, Rawson JRY (1985). Immunological determination of phosphoenolpyruvate carboxylase and the large and small subunits of ribulose 1,5-bisphosphate carboxylase in leaves of the C_4 plant pearl millet. Plant Physiol 77:828.
6. Ting IP (1968). CO_2 metabolism in corn roots. III. Inhibition of P-enolpyruvate carboxylase by 1-malate. Plant Physiol 43:1919.
7. O'Leary MH (1982). Phosphoenolpyruvate carboxylase: An enzymologist's view. Ann Rev Plant Physiol 33:297.
8. Colman B, Cheng KH, Ingle RK (1976). The relative activities of PEP carboxylase and RuDP carboxylase in blue-green algae. Plant Sci Lett 6:123.
9. Yoshinaga T, Teraoka H, Izui K, Katsuki H (1974). Molecular properties of phosphoenolpyruvate carboxylase of Escherichia coli. W J Biochem 75:913.
10. Schubert KR, (1986). Products of biological nitrogen fixation in higher plants:synthesis, transport, and metabolism. Ann Rev Plant Physiol 37: 539.

11. Latzko E, Kelly GJ (1983). The many-faceted function of phosphoenolpyruvate carboxylase in C_3 plants. Physiol Veg 21:805.
12. Hudspeth RL, Glackin CA, Bonner J, Grula JW (1986). Genomic and cDNA clones for maize phosphoenolpyruvate carboxylase and pyruvate, orthophosphate dikinase: Expression of different gene-family members in leaves and roots. Proc Natl Acad Sci USA 83:2884.
13. Izui K. Ishijima S, Yamaguchi Y, Katagiri F, Murata T, Shigesada K, Sugiyama T, Katsuki H (1986). Cloning and sequence analysis of cDNA encoding active phosphoenolpyruvate carboxylase of the C_4-pathway from maize. Nuc Acids Res 14:1615.
14. Ting IP, Osmond CB (1973). Multiple forms of plant phosphoenolpyruvate carboxylase associated with different metabolic pathways. Plant Physiol 51:448.
15. Mukerji SK (1977). Corn leaf phosphoenolpyruvate carboxylases: Purification and properties of two isozymes. Arch Biochem Biophys 182:343.
16. Stiborová M, Leblová S (1983). Isolation and partial characterisation of two phosphoenolpyruvate carboxylases from maize (*Zea mays* L). Photosynthetica 17:379.
17. Harpster MH, Taylor WC (1986). Maize phosphoenolpyruvate carboxylase. J Biol Chem 261:6132.
18. Adams CA, Leung F, Sun SSM (1986). Molecular properties of phosphoenolpyruvate carboxylase from C_3, C_3-C_4 intermediate, and C_4 *Flaveria* species. Planta 167:218.
19. Hatch MD, Osmond CB, Troughton JH, Björkman O (1971). Physiological and biochemical characteristics of C_3 and C_4 *Atriplex* species and hybrids in relation to the evolution of the C_4 pathway. Carnegie Inst Washington Yrbk 71:135.
20. Helentjaris T, Weber DF, Wright S (1986). Use of monosomics to map cloned DNA fragments in maize. Proc. Natl Acad Sci USA 83:6035.
21. Helentjaris, T, Weber D, Wright S (1987). Duplicate sequences in maize and identification of their genomic locations through restriction fragment length polymorphisms. Genetics, in press.

Plant Gene Systems and Their Biology, pages 217–226

MOLECULAR ANALYSIS OF GLUTAMINE SYNTHETASE GENES IN HIGHER PLANTS[1]

Gloria M. Coruzzi, Scott V. Tingey[2], Elsbeth L. Walker, Janice W. Edwards and Fong-Ying Tsai

Laboratory of Plant Molecular Biology, The Rockefeller University, New York, N.Y. 10021-6399

ABSTRACT We have examined the molecular basis for the differential expression of the distinct forms of glutamine synthetase (GS) in *Pisum sativum* and *Nicotiana plumbaginifolia*. Leaves of both species contain a chloroplast stromal GS polypeptide (GS_2) which is 44 kd in size. Cytosolic GS_1 in pea leaves has a subunit size of 38 kd which is identical to root GS. In the legume pea, a third form of GS (37 kd) is expressed most abundantly in nitrogen-fixing root nodules (GS_n). We have characterized cDNA clones encoding distinct GS polypeptides. The GS cDNA clones were used as molecular probes to characterize the different GS primary translation products, GS mRNAs, and GS genes in both *P. sativum* and *N. plumbaginifolia* (1,2). These results indicate that the different GS polypeptides are encoded by a family of homologous but distinct nuclear genes which are each differentially expressed *in vivo*.

INTRODUCTION

In higher plants, glutamine synthetase (GS, E.C. 6.3.1.2) is the major enzyme responsible for the assimilation of ammonia into organic form (3). GS functions in a number of plant organs (leaves, roots, and nodules) as well

[1] This work was supported by NIH Grant GM32877.
[2] Present address: E.I. DuPont de Nemours Co., Central Research and Development, Wilmington, Delaware 19898.

as within distinct subcellular compartments (chloroplast and cytosol)(4). Biochemical studies have shown that different GS isoforms could be distinguished by chromatography on DEAE-Sephacel (5). Based on enzyme activity, it has been shown that the different isoforms of GS are differentially distributed among various organs and organelles (4,5). Leaves contain two forms of GS; chloroplastic GS_2, and cytosolic GS_1 (4). It has also been shown that the level of chloroplastic GS_2 is increased in light vs. dark grown plants (6). Roots contain a single isoform of GS which is identical to cytosolic GS_1 in its elution profile on DEAE-Sephacel (7). In *Rhizobium* infected root nodules, the activity of plant GS has been shown to increase 14-fold compared to uninfected roots (8). This increase in plant GS activity in nodules correlates in some species (i.e. *Phaseolus vulgaris*) with the production of a novel isoform of GS (GS_{n1})(9).

We have begun to examine the genes encoding GS and their products in higher plants (1,2). An analysis of the plant genes encoding GS will answer the following questions: 1) Are the distinct isoforms of GS encoded by separate genes and if so how are they related? 2) Is chloroplastic GS_2 encoded by the nuclear or chloroplast genome and does the light-induction of GS_2 reflect light-induced gene expression? 3) Does the differential distribution of different GS isoforms in leaves, roots, and nodules reflect organ-specific gene expression? 4) Have the cis-acting DNA elements of the GS gene family evolved to promote differential expression of individual members in response to diverse factors such as light or *Rhizobium* infection?

RESULTS

GS Subunit Polypeptides

The GS holoenzyme in higher plants is composed of eight identical subunits (3). We have characterized the size and distribution of the GS subunit polypeptides which are present in various organs and organelles of a legume (*P. sativum*)(1) and a non-legume (*N. plumbaginifolia*)(2). GS polypeptides were detected with antibody to GS of *Nicotiana tabacum* (10) by Western blot analysis of total soluble protein (1,2) (Table 1).

In leaves, a 44 kd GS polypeptide is localized to the chloroplast stroma in both species (1,2). Pea leaves also contain cytosolic forms of GS (38 & 37 kd)(1). It had been shown previously that cytosolic GS_1 activity was not detected in leaves of members of the Solanaceae (11). Similarly, in N. plumbaginifolia, steady state levels of the 38 kd cytosolic GS_1 polypeptide are not detected in mature leaves (2).

TABLE 1. Sizes of GS subunit polypeptides.

	LEAVES		ROOTS	NODULES	
	chloroplast /	cytosol			REF
P. sativum	44*	38,37	38	44, 38, 37*	(1)
N. plumbaginifolia	44	n.d.	38	-----	(2)

Numbers indicate the size of the GS polypeptide in kilodaltons. Asterisk (*) indicates the predominant GS polypeptide when multiple forms are present in an organ. n.d. = not detected.

We have shown by lectin stain that the 44 kd chloroplastic GS_2 polypeptide contains no significant glycosylation which would affect its migration on SDS-PAGE (1,2). Therefore, the difference in migration of the chloroplastic and cytosolic GS polypeptides on SDS-PAGE is a true reflection of the distinct sizes of the polypeptides.

Roots of P. sativum and N. plumbaginifolia both contain a GS polypeptide which is 38 kd in size (1,2). In peas, this polypeptide comigrates on two-dimensional gels with the 38 kd cytosolic GS_1 found in leaves (1). Therefore, the 38 kd GS polypeptide in roots most likely corresponds to cytosolic GS_1 in leaves.

In the legume P. sativum, there is a third type of GS polypeptide which is 37 kd in size (1). On two-dimensional NEpHGE SDS-PAGE gels (12), this GS polypeptide appears as three charge variants (1). These 37 kd GS polypeptides are expressed most abundantly in nodules, we have therefore referred to them collectively as GS_n (1).

cDNA Clones Encoding GS

To determine whether the differential presence of the distinct GS polypeptides in leaves, roots, or nodules is the result of differential gene expression, cDNA clones encoding plant GS were isolated and characterized for use as molecular probes (1,2). We employed a cloned GS gene from Chinese Hamster (13) as a heterologous hybridization probe to isolate cDNA clones for GS from a cDNA library of *N. plumbaginifolia* (2). These clones were then used to isolate GS cDNA clones from pea cDNA libraries (1)(Figure 1).

In *P. sativum*, three distinct GS cDNAs have been characterized by nucleotide sequence analysis (pGS197, pGS134, pGS341)(1). pGS341 and pGS134 are 79% homologous at the DNA level and 89% homologous at the level of encoded amino acids (1). pGS341 is full length and encodes a 38 kd cytosolic GS polypeptide (1). The *N. plumbaginifolia* GS cDNAs also encode a 38 kd GS polypeptide that is 87% homologous to that deduced for cytosolic GS of pea (2).

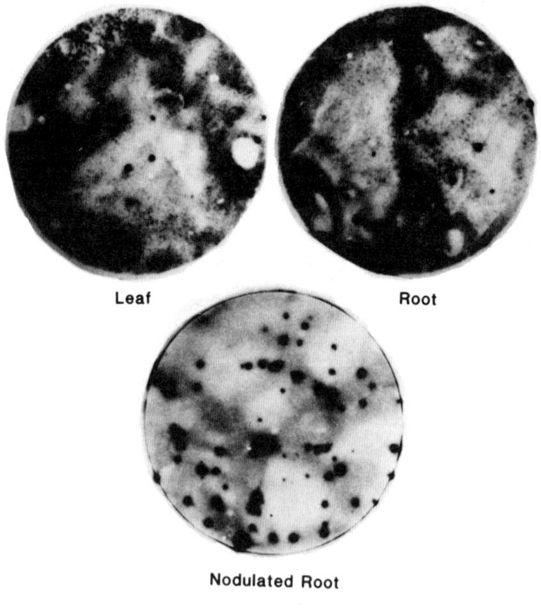

Figure 1. GS clones in cDNA libraries of *P. sativum*. Each plaque lift contains 25,000 recombinant λgt11 cDNA clones (1). GS cDNA clones are 10-fold more abundant in primary cDNA library of nodulated roots vs. leaves or roots.

The amino acid sequences deduced for plant cytosolic GS are 53% homologous to that deduced for animal GS (14). In pea, a cDNA clone has been characterized (pGS197) and proposed to encode chloroplastic GS (1). The encoded GS polypeptide is 76% homologous to that deduced for cytosolic GS and 48% homologous to animal GS (Figure 2).

```
           1                                               50
                    *  **  * **** ***   * * ***        **
1 MATSASSHLNKNIKQMYLCLPQGEKVQAMYIWVDGTGEGLRCKTRTLDCEPKCVEEL
2       MSSLSDLINFNLSDSTEKIIAEYIWVGGSGIDIRSKARTLPGPVSDPAKL
3 50aa..TINRVEDLLNLDITPFTDSIIAGYIWIGGTGIDVRSKSRTISKPVSHPSEV

                                                 100
* ** ***** *  *  *   * * * * *****  * ** * *       *  ** **
PEWNFDGSSTFQSEGSNSDMYLSPVAMFRDPFRRDPNKLVFCEVFKYNRKPAETNLRH
PKWNYDGSSTNQAPGKDSEVILYPQAIFKDPFRRGNNILVICDVYTPAGEPLPTNKRY
PKWNYDGSSTGQAPGEDSEVILYPQAIFKDPFRGGNNILVVCDAYTPAGEPIPTNKRH

                                          150
   *     *     ** * ******  **    * ***  * ********** *****
SCKRIM--DMVSNQHPWFGMEQEYTLMGTDGH-PFGWPSNGFPGPQGPYYCGVGADKA
NAAKIFSHPDVAAEVPWYGIEQEYTLLQKDINWPLGWPIGGYPGKQGPYYCGIGADKA
RAAEIFSNPKVEAEIPWYGIEQEYTLLQTNVKWPLGWPVGGYPGPQGPYYCAAGADKS

                            200
****** *** ******  * *** **** ***** **  **  *** * ** ** *
YGRDIVEAHYRACLYAGVKITGTNAEVMPAQWEFQIGPCEGIRMGDHLWVARFILHRV
YGRDIVDAHYKACLFAGINISGINGEVMPGQWEFQVGPSVGISAGDEIWAARYILERI
FGRDISDAHYKACIYAGINISGTNGEVMPGQWEWQVGPSVGIEAGDHIWASRYILERI

                      250
 *  **  ********* ************* ***  *   *  ****** **  **
CEDFGVIATFDPKPIPGNWNGAGCHTNFSTKAMREENGLKHIEEAIEKLSKRHRYHIR
TEIAGVVVSFDPKPIPGDWNGAGAHANFSTKSMRENGGYEVIKKAIEKLGLRHKEHIA
TEQAGVVLTLDPKPIEGDWNGAGCHTNYSTKSMREDGGFEVIKKAILNLSLRHKIHIE

               300
**   * * *** ***  ***** *****  ***  *      *********  * **
AY+LDNARGLTGFHETSNINDFSAGVANRSASIRIPRTVGQEKKGYFEDRRPSANCDP
AYGEGNERRLTGKHETADINVFSWGVANRGSSIRVGRDTEKDGKGYFEDRRPASNMDP
AYGEGNERRLTGKHETASINDFSWGVANRGCSIRVGRDTEKNGKGYLEDRRPASNMDP

      350
 **  *  * ** *
FAVTEAIVRTCLLNETGDEPFQYKN
YVVTSMIAETTILWKKP
YVVTALLAESTLLWE-PTLEAEALAAQKIALKV
```

Figure 2. Amino Acid sequences deduced for GS cDNAs of: 1. Hamster (14); 2. pea cytosolic (pGS341)(1); 3. pea chloroplastic (pGS197)(1). Amino acids are indicated in one letter code and numbers refer to pGS341. * = homology between animal and plant GS, + = amino acid insertion.

GS Primary Translation Products

GS mRNA hybrid-selected from poly(A)RNA of leaves, roots, or nodules encodes three distinct GS primary translation products (49, 38, and 37 kd) (1,2). In *P. sativum* and *N. plumbaginifolia*, the predominant GS primary translation product encoded by leaf poly(A)RNA is a 49 kd GS polypeptide which is proposed to be a precursor to the 44 kd chloroplast stromal GS_2 that contains a 5 kd chloroplast transit peptide (1,2). *In vitro* chloroplast uptake experiments have confirmed this hypothesis (manuscript in preparation).

Root GS mRNA encodes a 38 kd GS primary translation product in both species (1,2). In pea, the GS mRNA in nodules encodes 38 and 37 kd GS primary translation products (1). These results indicate that distinct GS mRNA species are differentially expressed in leaves, roots, and nodules.

In Vivo Expression of GS mRNA

In peas, the GS cDNAs encoding the cytosolic forms of GS hybridize to a 1.4 kb poly(A) RNA species (1). In nodules, the steady state level of this GS mRNA is induced 10-fold compared uninfected roots or leaves (1). The induction of GS mRNA in pea nodules is reflected in the screen of the primary cDNA libraries made from these organs (Fig. 1). In *N. plumbaginifolia* the GS cDNA also hybridizes to a 1.4 kb mRNA whose levels are 5-fold more abundant in roots than in leaves (2). In contrast, the pea cDNA clone proposed to encode chloroplastic GS_2 (pGS197), detects a larger GS mRNA species (1.5kb) which is expressed specifically in leaf poly(A)RNA (1). Recent data indicates that the steady state levels of this GS_2 mRNA are modulated by light (manuscript in preparation).

Nuclear Genes Encoding GS

Southern blot analysis, cDNA sequence analysis, and genomic clone analysis indicates that the distinct GS polypeptides are encoded by a small multigene family in both *P. sativum* (1) and *N. plumbaginifolia* (2). In *P. sativum* there are at least 3 nuclear genes for GS (1) and in *N. plumbaginifolia* at least 2 nuclear genes for GS (2). These genes are homologous but encode distinct GS polypeptides.

DISCUSSION

We have examined the molecular basis for the differential expression of the GS isoforms in higher plants. Leaves contain several types of GS polypeptides which are distinct in size and subcellular location (4). The 44 kd GS_2 polypeptide is localized to the chloroplast stroma (1,2). The leaf cytosolic GS_1 polypeptide (38 kd) is also present in roots (1,2). In the legume P. sativum, there is a third distinct GS_n polypeptide (37 kd) which is expressed most abundantly but not exclusively in root nodules (1). In contrast, studies on GS of P. vulgaris, indicate that a GS_{n1} polypeptide is uniquely expressed in nodule tissue (15). These results have recently been corroborated at the mRNA level (16,17). The difference in expression of the GS_n polypeptides of P. sativum and P. vulgaris may reflect a difference in gene expression between two species which possess distinct types of nodules (indeterminate vs. determinate, respectively)(18).

GS cDNA clones from N. plumbaginifolia and P. sativum were used to study the different GS amino acid sequences, GS primary translation products, GS mRNA species, and GS genes. The deduced amino acid sequences for the cytosolic forms of GS are highly homologous within members of the pea gene family (89%)(1), as well as with that deduced for cytosolic GS of N. plumbaginifolia (87%)(2), P. vulgaris (85%)(12), and M. sativa (87%)(19). The deduced amino acid sequence of plant cytosolic GS is 53% homologous to that deduced for animal GS (14). Recently, the nucleotide sequence of the GSI and GSII genes of Bradyrhizobium japonicum has shown that GSII is 46% homologous to higher plant cytosolic GS and shares only limited homology with the bacterial-like GSI gene (20). This evidence suggests that due to the intimate association of legumes and Rhizobium in the nodule environment, there has been a transfer of genetic information between the host and endosymbiont (20).

In plants, GS plays a key role in the reassimilation of ammonia lost during photorespiration (21). While there is controversy regarding which form of GS is responsible for this function, analysis of photorespiratory mutants suggests that the chloroplast form of GS may be involved in reassimilating photorespiratory ammonia in leaves (22). We have identified a pea GS cDNA clone as a candidate to encode the chloroplast form of GS in higher plants (1). More recently we have shown that chloroplastic GS_2 is synthesized as a 49 kd primary translation product (1,2) which is cleaved to the mature 44 kd chloroplast stromal GS (manuscript in preparation). The mRNA corresponding to chloroplastic GS_2 is expressed specifically in leaves (1) and its levels are modulated by light (manuscript in preparation). In contrast, the levels of mRNA for a cytosolic GS are low in leaves and roots and are induced 10-fold in pea root nodules (1). The induction of GS mRNA in pea nodules correspond well with the level of induction reported for a nodule specific GS clone of *P. vulgaris* (16).

Southern blot analysis of nuclear DNA as well as genomic clone analysis has revealed that the genes for higher plant GS are homologous but encode distinct GS polypeptides (1,2). Unlike other plant gene families studied to date (i.e. rbcS, Cab) (22-26), the individual members of the GS gene family encode unique gene products which are targeted to separate subcellular locations (chloroplast and cytosol). Another distinguishing feature of the GS gene family is that each member is differentially expressed in various plant organs (leaves, roots, or nodules) and individual members respond to different external factors (i.e. light, *Rhizobium* infection). Our continuing studies are directed at characterizing the family of GS genes and to define cis-acting DNA elements which may be required for their differential expression *in vivo*.

ACKNOWLEDGMENTS

We would like to thank Drs. B. Hirel and M. Lara for GS antibodies, and P. Sanders for the animal GS clone. We thank Dr. Nam-Hai Chua for helpful discussions.

REFERENCES

1. Tingey SV, Walker EL, Coruzzi GM (1987). Glutamine synthetase genes of pea encode distinct polypeptides which are differentially expressed in leaves, roots, and nodules. EMBO J 6: In Press.
2. Tingey SV, Coruzzi GM (1987). Glutamine synthetase of Nicotiana plumbaginifolia: cloning and in vivo expression. Plant Physiol: In Press.
3. Stewart GR, Mann AF, Fentem PA (1980). Enzymes of glutamate formation: glutamate dehydrogenase, glutamine synthetase, and glutamate synthase. Miflin B (ed). The Biochemistry of Plants, Acad Press, N.Y. Vol 5, p. 272.

4. Hirel B, Gadal P (1981). Glutamine synthetase isoforms in pea leaves: intracellular localization. Z Pflanzen 102:315.
5. McNally S, Hirel B (1983). Glutamine synthetase isoforms in higher plants. Physiol Veg 21(4):761.
6. Hirel B, Vidal J, Gadal P (1982). Evidence for a cytosolic dependent light induction of chloroplastic glutamine synthetase during greening of etiolated rice leaves. Planta 155:17.
7. Hirel B, Gadal P (1980). Glutamine synthetase in rice: a comparative study of the enzymes from roots and leaves. Plant Physiol 66:619.
8. Robertson JG, Farnden KJF, Warburton MP, Banks JM (1975). Induction of glutamine synthetase during nodule development in lupin. Aust J Plant Physiol. 2:265.
9. Lara M, Cullimore J, Lea PJ, Miflin BJ, Johnston AWB, Lamb JW (1983). Appearance of a novel form of plant glutamine synthetase during nodule development in Phaseolus vulgaris L. Planta 157:254.
10. Hirel B, Weatherly C, Cretin C, Bergounioux C, Gadal P (1984). Multiple subunit composition of chloroplastic glutamine synthetase of Nicotiana tabacum L. Plant Physiol 74:448.
11. McNally SF, Hirel B, Gadal P, Mann FA, Stewart GR (1983). Glutamine synthetase of higher plants. Plant Physiol 72:22.
12. O'Farrell PZ, Goodman HM, O'Farrell PH (1977). High resolution two-dimensional electrophoresis of basic as well as acidic proteins. Cell 12:1133.
13. Sanders PG, Wilson RH (1984). Amplification and cloning of the Chinese hamster glutamine synthetase gene. EMBO J 3:65.
14. Hayward BG, Hussain A, Wilson RH, Lyons A, Woodcock V, McIntosh B, Harris TJR (1986). The cloning and nucleotide sequence of cDNA for an amplified glutamine synthetase gene form the Chinese hamster. Nuc Acids Resh 14:999.
15. Lara M, Porta H, Padillla J, Folch J, Sanchez F (1984). Heterogeneity of glutamine synthetase polypeptides in Phaseolus vulgaris L. Plant Physiol 76:1019.
16. Cullimore J, Gebhardt C, Saarelainen R, Miflin BJ, Idler KB, Barker RF (1984). Glutamine synthetase of Phaseolus vulgaris L.: Organ-specific expression of a multigene family. J Mol Appl Genet 2:589.

17. Gebhardt C, Oliver JE, Forde BG, Saarelainen R, Miflin BJ (1986). Primary structure and differential expression of glutamine synthetase genes in nodules, roots, and leaves of Phaseolus vulgaris. EMBO J 5:1429.
18. Robertson JG, Farnden KJF (1980). Ultrastructure and metabolism of the developing legume root nodule. In Miflin BJ (ed) Acad Press, New York, "The Biochemistry of Plants", Vol 5:65.
19. Tischer E, DasSarma S, Goodman H (1986). Nucleotide sequence of an alfalfa glutamine synthetase gene. Mol Gen Genet 203:221.
20. Carlson TA, Chelm BK (1986). Apparent eukaryotic origin of glutamine synthetase II from the bacterium Bradyrhizobium japonicum. Nature 322:568.
21. Wallsgrove RM, Keys AJ, Lea PJ, Miflin BJ (1983). Photosynthesis, photorespiration and nitrogen metabolism. Plant Cell & Environ 6:301.
22. Somerville CR, Ogren WL (1982). Genetic modification of photorespiration. TIBS 7:171.
23. Coruzzi GM, Broglie R, Edwards C, Chua NH (1984). Tissue-specific and light-dependent expression of a nuclear gene encoding the small subunit of ribulose-1,5-bisphosphate carboxylase. EMBO J 3:1671.
24. Fluhr R, Moses P, Morelli G, Coruzzi G, Chua NH (1986). Expression dynamics of the pea rbcS multigene family and organ distribution of the transcripts. EMBO J 5:2063.
25. Dunsmuir P, Smith SM, Bedbrook J (1983). The major chlorophyll a/b binding protein of petunia is composed of several polypeptides encoded by a number of distinct nuclear genes. J Mol & App Genet 2:285.
26. Pichersky E, Bernatzky R, Tanksley SD, Breidenbach RB, Kausch AP, Cashmore A (1985). Molecular characterization of two clusters of genes encoding chlorophyll a/b-binding proteins in Lycopersicon esculentum (tomato) Gene 40:247.

Plant Gene Systems and Their Biology, pages 227–242

BIOCHEMISTRY AND MOLECULAR BIOLOGY OF STARCH SYNTHESIS[1]

Matthew Morell, Mark Bloom, Ray Larsen*, Thomas W. Okita* and Jack Preiss

Department of Biochemistry, Michigan State University, East Lansing, Michigan 48824 and *Institute of Biological Chemistry, Washington State University, Pullman, Washington 99164

ABSTRACT Regulation of starch synthesis in chlorophyllous and non-chlorophyllous tissues of plants, is exerted at the level of ADPglucose pyrophosphorylase, the enzyme catalyzing synthesis of ADPglucose, the glucosyl donor of starch synthase. 3-P-glycerate (3PGA) activates and orthophosphate (P_i) inhibits ADPglucose synthesis. 3PGA at high concentrations reverses or overcomes the inhibition caused by P_i. Pyridoxal-P (PLP) has been shown to activate spinach leaf ADPglucose pyrophosphorylase activity about 5- to 6-fold. It is not as effective as 3PGA since maximal stimulation by 3PGA is 25-fold. PLP can also inhibit the activation caused by 3PGA and similar to 3PGA it can overcome the inhibition by P_i. These results suggest that PLP binds at or close to the activator site. Reduction of the PLP onto the spinach leaf ADPglucose pyrophosphorylase with $NaBH_4$ gives an enzyme with 5 to 6-fold higher activity in the absence of activator than the untreated enzyme when about 0.5 mole of [^{3}H]-PLP is bound per mole of enzyme. This covalently modified enzyme is highly resistant to phosphate inhibition in contrast to the untreated enzyme which is quite sensitive to P_i inhibition (K_i=42 μM). The amino acid sequence at the

[1]This work was supported by research grants, USPHS AI22835, NSF DMB 85-10088, DMB-86-10319 and by a McKnight Foundation Training Grant.

PLP binding site of the 51 Kd subunit of the spinach leaf ADPG pyrophosphorylase has been determined. Using antibody prepared against spinach leaf ADPG pyrophosphorylase and λGT11, a cDNA clone of rice endosperm ADPglucose pyrophosphorylase has been isolated. It is an Eco RI fragment of 1700 Kb giving an open reading frame of 354 amino acids. This amino acid sequence corresponds to about 23% identity to the *Escherichia coli* ADPglucose pyrophosphorylase. There is identity in 5 of the 6 amino acids at a site reputed to be part of the binding site for ADPglucose in the *E. coli* enzyme. The amino sequence of the 3PGA (PLP) binding site appears to be situated near the carboxyl terminus of the subunit.

INTRODUCTION

ADPglucose pyrophosphorylase (ATP:α-glucose-1-P adenyl transferase, EC 2.2.7.27) is thought to play a key role in the regulation of the synthesis of starch in plants (1,2,3) and of glycogen in bacteria (4). The enzyme catalyzes the synthesis of ADPglucose and pyrophosphate from glucose-1-P and ATP and both plant and bacterial ADPglucose pyrophosphorylases are subject to allosteric regulation. The kinetic and physical properties of ADPglucose pyrophosphorylases from a number of plant and bacterial systems have recently been reviewed (4,5). Plant ADPglucose pyrophosphorylases, from both photosynthetic and non-photosynthetic tissues, are subject to allosteric activation and inhibition by 3-phosphoglycerate (3PGA) and inorganic phosphate (P_i), respectively (1-3,5-8). Evidence accumulated from intact leaf systems supports the postulation that the ratio of [3PGA] to [P_i] regulates the activation of ADPglucose pyrophosphorylase *in vivo* (9,10). Alteration of the levels of 3PGA and P_i will therefore determine the rate of synthesis of ADP-glucose, the glucosyl donor of the plant starch synthases.

Thus information on the structure-function relationship of the allosteric effector sites of the plant ADPglucose pyrophosphorylase is of interest and could be important in attempting to manipulate and modify various plant systems for increased economic and/or quality potential. It is quite possible a direction for improved nutritional quality of some plants (e.g. maize) may be to use the "freed energy" and carbon from reduced starch formation and direct them to formation and storage of quality protein. Improved control of starch biosynthesis and starch structure combined with control of source-sink relationship and sucrose translocation would enable one to utilize starch in more efficient ways in the growth and development of the plant.

This report is concerned with reviewing the current information on the structure of the plant ADPglucose pyrophosphorylase, the nature of the allosteric effector sites and the current attempts to clone the ADPglucose pyrophosphorylase gene.

RESULTS

Purification of Spinach Leaf ADPglucose Pyrophosphorylase.

Copeland (8) first reported a procedure for the purification of spinach leaf ADPglucose pyrophosphorylase to homogeneity, however, the amounts of enzyme obtained were insufficient for structural studies. A revised procedure for the purification of spinach leaf ADPglucose pyrophosphorylase has been developed (unpublished). The advantages of the procedure are that between 20 and 30 mg of enzyme, of greater than 90% purity, can be prepared from 10 Kg of spinach leaves with an overall yield of 30%. The initial steps in the purification were carried out essentially as described by Copeland (8), however, after DEAE-cellulose chromatography rapid and facile purification of the enzyme in high yield is achieved by anion-exchange chromatography on a Mono Q column using an FPLC system, and hydrophobic chromatography on ethyl-agarose.

Gel Electrophoresis.

A previous investigation (8) showed that native gel electrophoresis of a purified spinach leaf ADPglucose pyrophosphorylase preparation gave a single band of protein co-migrating with enzyme activity in 5 and 7.5% gels. In SDS-PAGE this enzyme preparation yielded 2 protein bands, staining with approximately equal intensity with Coomassie blue, of 54 and 51 Kd (erroneously reported to be 48 and

44 Kd, respectively (8). These results have been confirmed for spinach leaf ADPglucose pyrophosphorylase purified using the new procedure. The single protein band present following native gel electrophoresis was excised, equilibrated with SDS (11) and run in a 10% SDS-PAGE system. Both the 54 and 51 kd proteins were resolved and stained with about equal intensity with Coomassie blue (data not shown).

Antibodies raised previously against purified spinach leaf preparations (6) have been used to probe Western blots of spinach leaf extracts (12), however interpretation of the blots of crude extracts is difficult because the 54 kd protein comigrates with the large subunit of ribulose bisphosphate carboxylase (RUBISCO), and RUBISCO reacts with the anti-spinach leaf ADPglucose pyrophosphorylase serum (13,14). Two dimensional gel electrophoresis according to the method of O'Farrell (15) has been used to resolve the 54 kD protein from the large subunit of RUBISCO (Fig. 1). The SDS-PAGE second dimension was Western blotted and reacted with anti-spinach leaf ADPglucose pyrophosphorylase antisera (Figure 1). Both

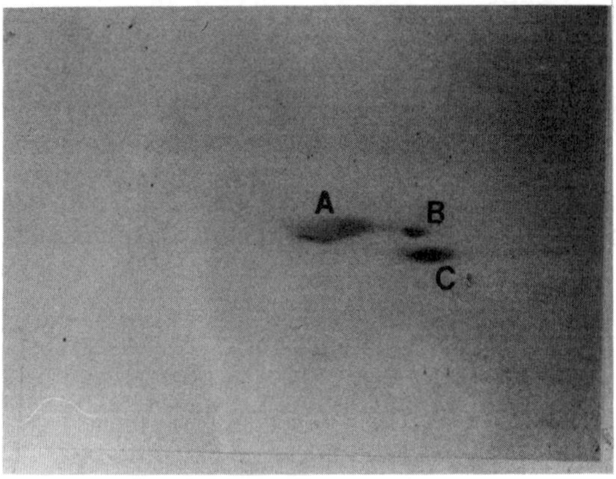

FIGURE 1. Western blot of a spinach leaf crude extract subjected to 2-dimensional electrophoresis. Isoelectrofocusing of the gel, pH 10 to 3, was done horizontally from left to right while SDS gel electrophoresis was done vertically. The gel was electroblotted to nitrocellulose paper and reacted with affinity purified antibody to spinach leaf ADPglucose pyrophosphorylase. A, RUBISCO. B, 54 kd subunit. C, 51 kd protein.

the 54 and 51 kD proteins are present in crude spinach leaf extracts and this indicates that the 51 kD protein does not arise as a result of proteolysis of the 54 kD protein during purification. If post-translational modification or proteolysis has occurred to generate the 51 kD protein from the 54 kD protein, the process must have occurred *in vivo*.

Separation of the 54 and 51 kd Proteins.

In order to investigate the structural properties of the 54 and 51 kd proteins in more detail, separation of the proteins was necessary. All attempts to separate the proteins in non-denaturing systems were unsuccessful. Isoelectric focusing of the proteins in urea indicated that the denatured 51 kD protein has a lower pI than the 54 kD protein. This difference in properties allows base-line separation of the two proteins by anion-exchange chromatography on a Mono Q column in 8 M urea (Fig. 2). The 54 kd polypeptide, having a higher pI under these conditions, elutes first.

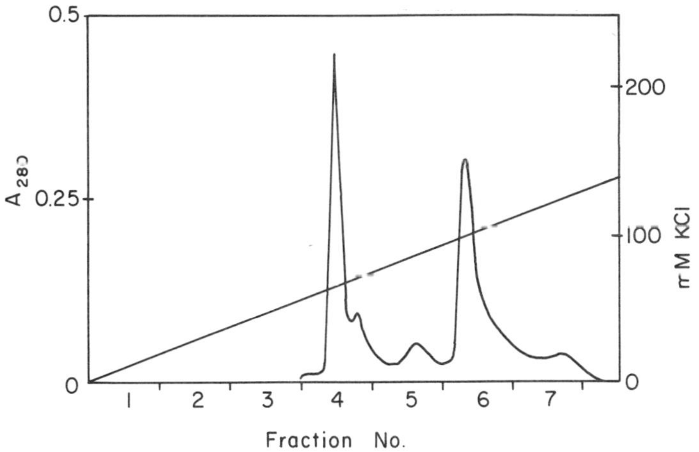

FIGURE 2. Separation of the 54 and 51 kd proteins by anion exchange chromatography in 8 M urea. The 54 kilodalton protein was the first peak eluted.

N-Terminal Sequence Analysis.

N-terminal sequence analysis of the two subunits shows that they are different (Fig. 3). The N-terminal sequences do not show homology with E. coli ADPglucose pyrophosphorylase (16) and the sequence data also eliminates the possibility that the 54 kD protein represents contamination by the large subunit of RUBISCO (17).

54 kilo- dalton fraction	^{1}Ser-Val-Thr-Ala-Asp-Asn-Ala-Ser-Glu-Thr^{10}- Lys-Val-Arg-Glu-Ile-Gly-Gln-Glu-Lys-( )^{20} (Ser)-(Ser)-
51 kilo- dalton	^{1}Val-Ser-Asp-Ser-Gln-Asn-Ser-Gln-(Asp)-(Gly)^{10} (Leu)-(Asp)-(Pro)-(Glu)-

FIGURE 3. N-terminal amino acid sequences of the spinach leaf ADPglucose pyrophosphorylase subunits. The amino acids placed in parenthesis indicate that identification is only tentative. Residue 20 of the 54 kilodalton subunit was not identified.

Peptide Mapping.

The possibility that the two proteins are sequentially related and have arisen through a gene duplication event or from post-translational modification was investigated by tryptic peptide mapping. Reverse phase HPLC of tryptic digests show that the 54 kd and 51 kd proteins gave very different peptide patterns (Figs. 4A and 4B, respectively). There were few major peptides that eluted with similar retention times in both digests. Thus, the peptide maps generated by digestion of the 54 and 51 kd proteins by trypsin show that there are major differences in structure between the two proteins. This discounts the possibility that the 51 kd protein is an in vivo degradation product of the 54 kd protein, and the two proteins are therefore the products of different genes.

Activation of ADPglucose Pyrophosphorylase by Pyridoxal Phosphate.

One possible way of manipulating ADPglucose pyrophosphorylase activity and therefore starch synthesis would be by modification of the activator and inhibitor binding sites. Thus, efforts were initiated to determine the nature of these effector binding sites.

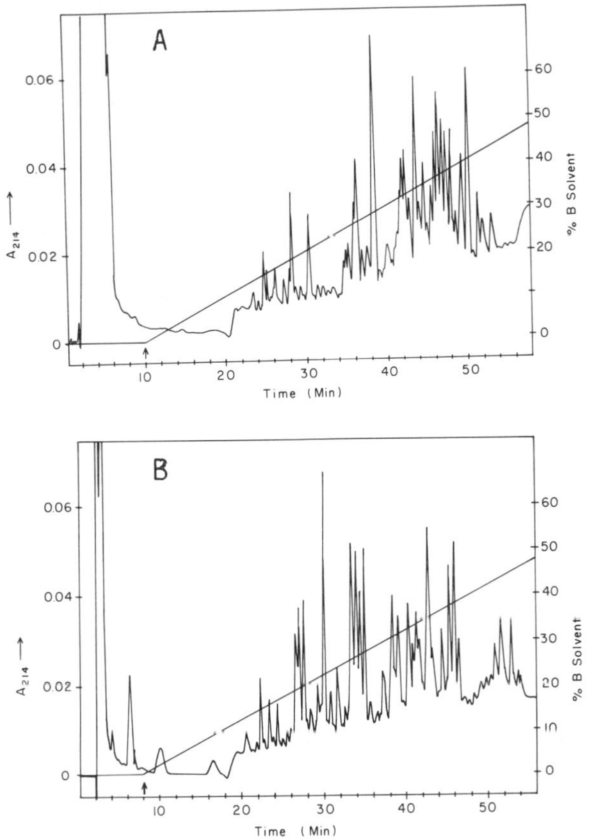

FIGURE 4. HPLC of the tryptic peptides of the 54 and 51 kilodalton subunits. Tryptic digests (2 nmol) were applied to a C18 reverse phase column. A 0-60% gradient of solvent A (0.1%, TFA) and solvent B (0.075% TFA in acetonitrile) was run in 60 min at a flow rate of 1 ml/min. A, 54 kd subunit. B, 51 kd subunit.

Previous results have shown that pyridoxal-P (PLP) was an site-specific probe for the chloroplast P_i/triose-P/3PGA translocator (18,19) as well as for the *E. coli* ADPglucose pyrophosphorylase (20,21). We therefore attempted to see if pyridoxal-P would be a site-affinity label for the spinach leaf ADPglucose pyrophosphorylase. Recent results have shown that PLP appears to be an effective activator of the spinach leaf ADPglucose pyrophosphorylase. Fifty percent of maximal stimulation requires 6 to 10 µM PLP while 15 to 30 µM 3PGA is required for half maximal activation (13,14). However, 3PGA stimulates enzyme activity 25-fold while maximal stimulation by PLP is only 6-fold. At concentrations over 50 µM PLP inhibition of ADPglucose synthesis is observed. Pyridoxal-P as does 3PGA (1,2) can also reverse P_i inhibition (14). PLP, due to its lower stimulation, inhibits 3PGA activation (14). In the presence of 1 mM 3PGA, 50% inhibition occurs with 0.39 mM PLP and in the presence of 0.5 mM 3PGA there is 50% inhibition with 0.25 mM PLP. The above results suggest that PLP binds at or near the 3-PGA activator site. PLP can be irreversibly covalently bound to the protein by reduction with $NaBH_4$. Thus, it was of interest to determine the properties of a reduced phosphopyridoxylated enzyme. If PLP binds to the activator site and places the enzyme in an active conformation, the modified enzyme is more active in the absence of activator. Thus, the activity would be less dependent on the activator for activity and the ratio of activity in the presence of 3PGA to activity in the absence of 3PGA would decrease.

Table 1 shows the relationship between incorporation of [^{3}H]-PLP in enzyme and the change in ratio of +3PGA/-3PGA activity. When reductive phosphopydoxylation is carried out with 10 µM PLP, there is a dramatic decrease in the +3PGA/-3PGA activity ratio from 25 to 5.4. About 1.3 mole of [^{3}H]-PLP per mole of tetrameric enzyme is incorporated. Further incorporation is seen at 50 µM [^{3}H]-PLP. The activity ratio of +3PGA/-3PGA is decreased to about 2 and about 2.6 mole of PLP is incorporated per mole of native enzyme.

The incorporation of [^{3}H]-PLP is inhibited by the allosteric effectors 3PGA and P_i (Table 1). Concomitant with this is the retainment of the high activity ratio of +3PGA/-3PGA. α-Glucose-6-P or 2-phosphoglycerate (2PGA), metabolites which are not activators of the spinach leaf

ADPglucose pyrophosphorylase do not inhibit the incorporation of [^{3}H]-PLP into enzyme. The substrates, ADPglucose and glucose-1-P have minimal effects on PLP incorporation. ATP has a small protective effect at 10 µM PLP. Pyrophosphate (PP_i) also protects significantly, 69% and 43%, at 10 µM and 50 µM PLP, respectively. Although PP_i is a substrate, it may also be considered as an analogue of the allosteric inhibitor, P_i. These results are consistent with the notion that [^{3}H]-PLP is binding at or near the activator site.

The ^{3}H-labeled PLP binds to both of the subunits of the spinach leaf ADPG pyrophosphorylase (Table 1). There is about 1.5 to 2-times more incorporation of [^{3}H]-PLP into the higher molecular weight peptide. The significance of this is not known at present, since the relative amounts of each subunit peptide present in the enzyme have not been conclusively established. At any rate, the incorporation of the labeled PLP into the two subunits of ADPG

TABLE 1
REDUCTIVE PHOSPHOPYRIDOXYLATION OF SPINACH LEAF ADPGLUCOSE PYROPHOSPHORYLASE IN THE PRESENCE OF SUBSTRATES AND EFFECTORS

Substrate or Effector	10 µM PLP				50 µM PLP			
	+3PGA/ -3PGA	mol PLP/ mol enzyme	% protection	mol PLP per 54 Kd/ mol PLP per 51 Kd	+3PGA/ -3PGA	mol PLP mol enzyme	% protection	mol PLP per 54 Kd/ mol PLP per 51 Kd
None, no BaH_4	25	--	--	--	23	--	--	--
None	5.4	1.27	0	0.85	2	2.6	0	1.6
1 mM ADPglu 6 mM $MgCl_2$	7.4	1.14	10	1.05	3.3	2.8	0	1.5
1 mM ATP 6 mM $MgCl_2$	8.2	0.87	32	1.21	3.4	2.7	0	1.3
1 mM G1P	9	1.27	0	0.84	3.3	2.6	0	1.4
1 mM PPi	15	0.4	69	0.64	7.5	1.5	43	1.1
1 mM 3PGA	18	0.11	91	1.97	15.0	0.74	72	1.8
1 mM Pi	32	0.32	75	4.8	22	1.1	59	2.2
1 mM 2PGA	8	1.05	17	0.8	4	2.4	10	1.5
1 mM G6P	4.5	1.28	0	1.18	3.7	2.5	6	1.5

TABLE 1. Reductive phosphopyridoxylation was carried out as described (13,14,20,21, Fig. 5) and with the various substrate or effectors added as indicated.

synthetase has enabled us to determine the sequence of the activator binding site of the 51 kd molecular weight subunit.

Amino Acid Sequence at the PLP Binding Site of the 51 kd Molecular Weight Subunit.

The native ADPglucose pyrophosphorylase was reacted with 50 µM PLP in the presence of the $NaBH_4$. The phosphopyridoxylated protein was then separated into its respected subunits as described in Fig. 2. Both labeled subunits were then carboxymethylated with iodoacetic acid and then proteolyzed with trypsin. Fig. 5 shows the HPLC pattern of the tryptic digest of the labeled 51 kd subunit. One major radioactive peak is observed which was further purified. The isolated peptide was then subjected to amino acid analysis and peptide sequencing. Fig. 6 shows the sequence obtained and compares it with the allosteric

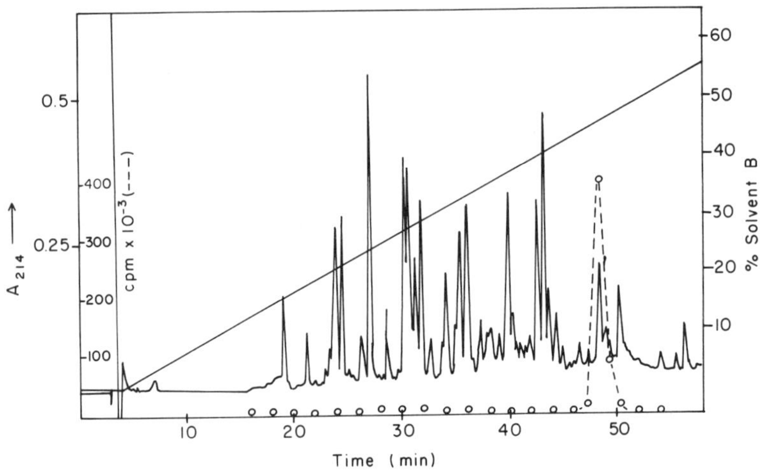

FIGURE 5. Elution profile of the tryptic peptide digest of the 51 kd protein after reductive phosphopyridoxylation of the native enzyme. Reductive phosphopyridoxylation of the enzyme was carried out in the presence of 50 µM [^{3}H]pyridoxal-P in the absence of protecting ligands. Chromatographic conditions were identical to those described in Fig. 4.

activator site found for the bacterial (*Escherichia coli*) ADPglucose pyrophosphorylase. The major activator for the *E. coli* enzyme is fructose 1,6-P_2 and thus different than the activator seen for the plant enzymes. As would be expected the amino acid sequences corresponding to the two different activator binding sites are not similar. The sequence seen for the bacterial enzyme contains more basic amino acid residues than the spinach leaf enzyme sequence.

It was previously shown that after phosphopyridoxylation, the enzyme becomes quite insensitive to P_i inhibition (13,14). Whereas the untreated enzyme was 50% inhibited by about 60 μM P_i, only 16% inhibition at 1.0 mM P_i

E. coli	^{1}-Arg-Leu-Lys-Asp-Leu-Thr-Asn-Lys-Arg-Ala
Spinach leaf	-Ser-Gly-Ile-Val-Thr-Val-Ile- - -Asp-Ala
E. coli	^{11}-Lys-Pro-Ala-Val
Spinach leaf	Leu-Ile-Pro-(Ser)

FIGURE 6. Comparison of the amino acid sequences of the *E. coli* activator binding site with the spinach leaf activator binding site. Residue 8 of the spinach leaf enzyme has not been definitively determined as lysine.

was noted for the enzyme chemically modified with 1 mM PLP. When chemically modified with 50 μM PLP, the enzyme shows 42% inhibition at 1 mM P_i. Both P_i and 3-PGA, when present during the reductive phosphopyridoxylation, prevent to a great extent the desensitization of the enzyme to P_i inhibition (13,14). These results are also in keeping with the view that PLP binds to the activator site and places the enzyme in a conformation more resistant to P_i inhibition.

Preparation of an ADPGlucose Pyrophosphorylase Complementary DNA Clone from Rice Endosperm Messenger RNA.

Recently, an ADPG pyrophosphorylase complementary DNA (cDNA) was isolated from rice endosperm tissue using the λgt11 technique as described by Huynh et al. (22) and using the antispinach leaf ADPG pyrophosphorylase (6). Recent sequencing of the cDNA fragment, which is about 1.5

kbp, was done based on the method of Henikoff (23). The deduced amino acid sequence derived from the nucleotide sequence has been determined (14). If one compares it with the known amino acid sequence of the *E. coli* ADPG pyrophosphorylase (16), about a 23% identity is found (Fig. 7). Of interest is the identity seen at amino acid residues 93 to 98. It was shown with *E. coli* enzyme that lysine residue 97 was chemically modified by PLP and that the residue is protected from chemical modification by the substrate ADPG plus Mg^{2+} (20,21). This would suggest that the lysine residue 97 is involved in binding of the substrate.

Of interest is the finding that the putative activator binding site for the spinach leaf enzyme is seen starting with amino acid 298 and ending at amino acid 308. Whereas the deduced amino acid sequence for the rice endosperm ADPglucose pyrophosphorylase at amino acid residues 309-311 is Leu-Leu-Ala, for the spinach leaf enzyme these amino acids are Ile-Pro-Ser.

It should be noted that the 3PGA binding site is near the C-terminal end according to the deduced amino acid sequence. This is different from what has been observed with respect to where the activator site is situated on the *E. coli* ADPGlucose pyrophosphorylase. The data indicate that a lysine residue, the 38th amino acid from the N-terminus of the *E. coli* enzyme is involved in binding of the activator, fructose 1,6-P_2 (20,21). It seems that in evolution of the ADPglucose pyrophosphorylase rearrangement of the structural gene may have occurred with modification of the specificity of the regulatory binding site.

SUMMARY

Recent advances have been made with respect to obtaining more information on the structure of the ADPglucose pyrophosphorylase, the nature and location of the allosteric site and the amino acid sequence of enzyme. Progress has also been made in isolating a cDNA clone of the ADPglucose pyrophosphorylase. It is hoped that in the future that a complete cDNA clone is obtained to enable us to get high expression of enzyme activity. This would allow us to study in greater detail the regulatory phenomena involved in the allosteric control of ADPglucose synthesis and the recently discovered covalent modification of the enzyme (phosphorylation) (14,24). This may also allow one in the future facile manipulation of the higher plant ADPglucose pyrophosphorylase gene expression as well as manipulation of the activity of the protein product.

```
    1                                                                            20
a.  Val Glu Val Leu Ala Ala Gln Gln Ser Pro Asp Asn Pro Asn Trp Phe Gln Gly Thr Ala
b.  Val Asp Leu Leu Pro Ala Gln Gln Arg Met Lys Gly Glu Asn Trp Tyr Arg Gly Thr Ala

    21                                                                           40
a.  Asp Ala Val Arg Gln Tyr Leu Trp Leu Phe Glu Glu His Asn Val Met Glu Phe Leu Ile
b.  Asp Ala Val Thr Gln Asn Leu Asp Ile Ile Arg Arg Tyr Lys Ala Glu Tyr Val Val Ile

    41                                                                           60
a.  Leu Ala Gly Asp His Leu Tyr Arg Met Asp Tyr Ser Lys Phe Ile Gln Ala His Arg Glu
b.  Leu Ala Gly Asp His Ile Tyr Lys Gln Asp Tyr Ser Arg Met Leu Ile Asp His Val Glu

    61                                                                           80
a.  Thr Asp Ser Asp Ile Thr Val Ala Ala Leu Pro Met Asp Glu Lys Arg Ala Thr Ala Phe
b.  Lys Gly Val Arg Cys Thr Val Val Cys Met Pro Val Pro Ile Glu Glu Ala Ser Ala Phe

    81                                                                ◆         100
a.  Gly Leu Met Lys Ile Asp Glu Glu Gly Arg Ile Val Glu Phe Ala Glu Lys Pro Lys Gly
b.  Gly Val Met Ala Val Asp Glu Asn Asp Lys Thr Ile Glu Phe Val Glu Lys Pro Ala Asn

    101                                                                         120
a.  Glu Gln Leu Lys Ala Met Met Val Asp Thr Thr Ile Leu Gly Leu Asp Asp Val Arg Ala
b.  Pro Pro Ser Met Pro Asn Asp Pro Ser Lys Ser Leu Ala Ser Met Gly Ile Tyr Val Phe

    121                                                                         140
a.  Lys Glu Met Pro Tyr Ile Ala Ser Met Gly Ile Tyr Val Ile Ser Lys Asn Val Met Leu
b.  Asp Ala Asp Tyr Leu Tyr Glu Leu Leu Glu Glu Asp Asp Arg Asp Glu Asn Ser Ser His

    141                                                                         160
a.  Gln Leu Leu Arg Glu Gln Phe Pro Gly Ala Asn Asp Phe Gly Ser Glu Val Ile Pro Gly
b.  Asp Phe Gly Lys Asp Leu Ile Pro Lys Ile Thr Glu Ala Gly Leu Ala Tyr Ala His Pro

    161                                                                         180
a.  Ala Thr Asn Ile Gly Met Arg Val Gln Ala Tyr Leu Tyr Asp Gly Tyr Trp Glu Asp Ile
b.  Phe Pro --- Leu Ser Cys Val Gln Ser Asp Pro Asp Ala Glu Pro Tyr Trp Arg Asp Val

    181                                                                         200
a.  Gly Thr Ile Glu Ala Phe Tyr Asn Ala Asn Leu Gly Ile Thr Lys Lys Pro Val Pro Asp
b.  Gly Thr Leu Glu Ala Tyr Trp Lys Ala Asn Leu Asp --- Leu Ala Ser Val Val Pro Lys

    201                                                                         220
a.  Phe Ser Phe Tyr Asp Arg Ser Ala Pro Ile Tyr Thr Gln Pro Arg His Leu Pro Pro Ser
b.  Leu Asp Met Tyr Asp Arg Asn Trp Pro Ile Arg Thr Tyr Asn Glu Ser Leu Pro Pro Ala

    221                                                                         240
a.  Lys --- Val Leu Asp --- --- --- --- --- Ala Asp Val Thr Asp Ser Val Ile Gly Glu
b.  Lys Phe Val Gln Asp Arg Ser Gly Ser His Gly Met Thr Leu Asn Ser Leu Val Ser Gly

    241                                                                         260
a.  Gly Cys Val Ile Lys Asn Cys Lys Ile His His Ser Val Val Gly Leu Arg Ser Cys Ile
b.  Gly Cys Val Ile Ser Gly Ser Val Val Val Gln Ser Val Leu Phe Ser Arg Val Arg Val

    261                                                                         280
a.  Ser Glu Gly Ala Ile Ile Asp Lys Asn Ala Arg Ile Gly Asp Asn Val Lys Ile Ile Asn
b.  Asn Ser Phe Cys Asn Ile Asp Ser Ala Val Leu Leu Pro Glu Val Trp Val Gly Arg Ser

    281                                                                         300
a.  Val Asp Asn Val Gln Glu Ala Ala Arg Glu Thr Asp Gly Tyr Phe Ile Lys Ser Gly Ile
b.  Cys Arg Leu Arg Arg Cys Val Ile Asp Arg Ala Cys Val Ile Pro Glu Gly Met Val Ile

    301                                                                         319
a.  Val Thr Val Ile Lys Asp Ala Leu Leu Leu Ala GLu Gln Leu Tyr Glu Val Ala Ala STOP
b.  Gly Glu Asn Ala Glu Glu Asp Ala Arg Arg Phe Tyr Arg Ser Glu Glu Gly Ile Val Leu

    321                                                     335
b.  Val Thr Arg Glu Met Leu Arg Lys Leu Gly His Lys Gln Glu Arg STOP
```

FIGURE 7. Comparison of the primary sequence of the rice endosperm (lane A) and *E. coli* (lane B) ADPglucose pyrophosphorylase. Only the last 332 amino acids of the *E. coli* enzyme are shown. The homologies are indicated by the enclosed boxes. The sequencing procedures have been described (14). The black diamond indicates the lysine residue that may be involved in substrate binding and the underlined sequence is the putative activator binding site (see Fig. 6).

ACKNOWLEDGEMENTS

We thank Dr. Y.M. Lee of the Biochemistry Department at Michigan State University for his determination of the amino acid sequences of the spinach leaf ADPglucose pyrophosphorylase subunits and the amino acid composition analyses.

REFERENCES

1. Ghosh HP, Preiss J (1966). Adenosine diphosphate glucose pyrophosphorylase. A regulatory enzyme in the biosynthesis of starch in spinach chloroplasts. J Biol Chem 241:4491-4504.
2. Preiss J, Ghosh HP, Wittkop J (1967). Regulation of the biosynthesis of starch in spinach leaf chloroplasts. In Goodwin TW (ed): "Biochemistry of Chloroplasts," Vol 2, Academic Press, Inc, New York, pp 131-153.
3. Sanwal GG, Greenberg E, Hardie J, Cameron EC, Preiss J (1968). Regulation of starch synthesis in plant leaves: Activation and inhibition of ADPglucose pyrophosphorylase. Plant Physiol 43:417-427.
4. Preiss J (1984). Bacterial glycogen synthesis and its regulation. Ann Rev Microbiol 38:419-458.
5. Preiss J (1982). Regulation of the biosynthesis and degradation of starch. Ann Rev Plant Physiol 33:431-454.
6. Plaxton WC, Preiss J (1987). Purification and properties of non-proteolytic degraded ADPglucose pyrophosphorylase from maize endosperm. Plant Physiol, in press.
7. Sowokinos JR, Preiss J (1982). Pyrophosphorylases in Solanum tuberosum III. Purification, physical, and catalytic properties of ADPglucose pyrophosphorylase in potatoes. Plant Physiol 69:1459-
8. Copeland L, Preiss J (1981). Purification of spinach leaf ADPG pyrophosphorylase. Plant Physiol 68:996-1001.
9. Santarius KA, Heber U (1965). Changes in the intracellular levels of ATP, ADP, AMP and P_i and regulatory function of the adenylate system in leaf cells during photosynthesis. Biochim Biophys Acta 102:39-54.

10. Heldt HW, Chon CJ, Maronde D, Stankovic, ZS, Walker DA, Kraminer A, Kirk MR, Heber U (1977). Role of orthophosphate and other factors in the regulation of starch formation in leaves and isolated chloroplasts. Plant Physiol 59:1146-1155.

11. Cleveland DW, Fischer SG, Kirschner MW, Laemmli UK (1977). Peptide mapping by limited proteolysis in sodium dodecyl sulfate and analysis by gel electrophoresis. J Biol Chem 252:1102-1106.

12. Krishnan HB, Reeves CD, Okita TW (1986). ADPglucose pyrophosphorylase is encoded by different mRNA transcripts in leaf and endosperm of cereals. Plant Physiol 81:642-645.

13. Preiss J, Morell M, Bloom M, Knowles V, Lin T-P (1987). Starch synthesis and its regulation. In Biggins J (ed): "Proceedings of the VII International Congress on Photosynthesis," Martinus Nijhoff Publishers, Amsterdam, in press.

14. Preiss J, Bloom M, Morell M, Knowles VL, Plaxton WC, Okita TW, Larsen R, Harmon AC, Putnam-Evans C (1987). Regulation of starch synthesis: enzymological and genetic studies. In Bruening G, Kosuge T, Harada J, Hollaender A (eds): "Tailoring Genes for Crop Improvement: An Agricultural Prospective," New York: Plenum Press, pp 133-152.

15. O'Farrell PH (1975). High resolution two-dimensional electrophoresis of proteins. J Biol Chem 250:4007-4021.

16. Baecker PA, Furlong CE, Preiss J (1983). Biosynthesis of bacterial glycogen. Primary structure of *Escherichia coli* ADPglucose synthetase as deduced from the nucleotide sequence of the *glg*C gene. J Biol Chem 258:5084-5088.

17. Zurawski G, Perrot B, Bottomley W, Whitfeld, PR (1981. The structure for the gene for the large subunit of ribulose 1,5-bisphosphate carboxylase from spinach chloroplast DNA. Nucl Acids Res 9:3251-3270.

18. Fliegge R, Flügge, Werden K, Heldt HW (1978) Specific transport of inorganic phosphate, 3-phosphoglycerate and triose phosphates across the inner membrane of the envelope in spinach chloroplasts. Biochim Biophys Acta 502:232-247.

19. Flügge U-I, Heldt, HW (1984). The phosphate-triose phosphate-phosphoglycerate translocater of the chloroplast. Trends Biochem Sci 9:530-533.

20. Parsons TF, Preiss J (1978). Biosynthesis of bacterial glycogen. Incorporation of pyridoxal-phosphate into the allosteric activator site and an ADPglucose-protected pyridoxal-P binding site of *Escherichia coli* B ADPglucose synthase. J Biol Chem 253:6197-6202.

21. Parsons TF, Preiss J (1978). Biosynthesis of bacterial glycogen. Isolation and characterization of the pyridoxal-P allosteric activator site and the ADPglucose-protected pyridoxal-P binding site of *Escherichia coli* B ADPglucose synthase. J Biol Chem 253:7638-7645.

22. Huynh TV, Young RA, Davis RW (1984). Constructing and screening cDNA libraries in λgt10 and λgt11. In Glover D (ed): "Cloning Techniques: A Practical Approach," Oxford: IRL Press, pp 49-78.

23. Henikoff S (1984). Unidirectional digestion with exonuclease III creates targeted breakpoints for DNA sequencing. Gene 28:351-359.

24. Harmon AC, Putnam-Evans C, Cormier MJ (1987). Calcium-dependent but calmodulin-independent protein kinase from soybean. Plant Physiol, in press.

Plant Gene Systems and Their Biology, pages 243–250

THE ALTERNATIVE OXIDASE OF *SAUROMATUM GUTTATUM*[1]

Lee McIntosh, Roxy L. Nickels and Thomas E. Elthon

M.S.U./D.O.E. Plant Research Laboratory, Biochemistry Department, Michigan State University, East Lansing, Michigan 48824

ABSTRACT The presence of an alternative, cyanide-resistant, pathway of respiration in plants was first discovered over 50 years ago. It has been intensively investigated at the physiological level, but with relatively little biochemical characterization. We have solubilized the alternative oxidase from the inner mitochondrial membrane of the aroid *Sauromatum guttatum* Schott. The inflorescence of *Sauromatum* shows both developmental and tissue-specific control of the alternative pathway activity. A partially purified preparation of the alternative oxidase has been employed to raise antibodies against this complex. These polyclonal antibodies will immunoprecipitate the enzyme activity from a solubilized extract and have now been employed to localize candidates for the structural polypeptides of the alternative oxidase.

INTRODUCTION

Mitochondrial respiration in most species of higher plants is, at one level, more complex than that present in mammalian systems. One major contribution to this complexity comes from the presence in plant systems of an alternative, cyanide-resistant, respiratory pathway (1). The alternative pathway has also been studied in some fungi

[1]This work was supported by a National Science Foundation Postdoctoral Fellowship DMB-8508782 (T.E. Elthon) and Department of Energy Contract DE-AC02-76ERO-1338.

such as Neurospora (2). In the alternative pathway electrons are still donated to oxygen to form water, however, an alternative terminal oxidase is involved which is insensitive to cyanide and specifically inhibited by hydroxamic acids (3). Another unique feature of this pathway is that it is not linked to ATP production (4).

The functional significance of the alternative pathway is only agreed upon for plants displaying thermogenic, CN-resistant respiration in their inflorescences, such as Sauromatum guttatum (1,4). Heat produced by electron flow through the alternative pathway in these inflorescences is used to volatilize insect attractants, thus facilitating pollination.

The question remains: what is the function of the alternative pathway in plant tissues other than thermogenic inflorescences? One hypothesis concerning the role of the alternative pathway proposes that it serves as a method to rid the cell of excess or "luxury" carbohydrates (5). This "overflow" hypothesis predicts that the alternative oxidase would be active only when the cytochrome pathway is saturated or limited: an idea well supported by published reports for many plant species (for review see 6). At many points during development ATP may be in excess, whereas carbon skeletons for metabolic processes could be limiting. Therefore, the alternative pathway might allow the TCA cycle to continue to function producing carbon skeletons for biosynthetic purposes.

Relatively little has been reported about the alternative oxidase at the biochemical or protein levels. It was initially solubilized using mitochondria from the thermogenic spadix (the terminal portion of the inflorescence) of Arum maculatum (7). More recent efforts at solubilizing a stable, partially purified alternative oxidase have been reported by Elthon and McIntosh (8) for S. guttatum, and Bonner et al. (9) for A. maculatum. In order to obtain information about the protein constituents of the alternative oxidase we have applied biochemical and immunological tools to the examination of the alternative pathway solubilized from S. guttatum.

MATERIALS AND METHODS

Plant Material. S. guttatum plants were maintained in a glasshouse as previously described (8). Plants were grown from corms and remained in the vegetative phase for approximately 6-8 months. Following the vegetative phase, corms were removed from the soil, and remained visibly dormant until floral development ensued (approximately 4-8 weeks). Corms were also received as generous gifts from Dr. Bastiaan J.D. Meeuse, Botany Department, University of Washington, Seattle, Washington. Isolation of mitochondria, assay of the alternative oxidase activity, and solubilization were performed following published procedures (8).

CM-Sepharose CL-6B Fractionation. A 2 ml column was used for the initial fractionation of the alternative oxidase solubilized in BIGCHAP (N,N-bis-3-D-glucoamidopropyl deoxycholamide) as described (8). The column was equilibrated in BIGCHAP buffer (30 mM MOPS, 1 mM EDTA, and 0.5% BIGCHAP at pH 6.8). Two ml (2.0 mg protein) of solubilized alternative oxidase was applied to the column. The column was washed with BIGCHAP buffer, and the alternative oxidase eluted with BIGCHAP buffer plus 100 mM NaCl.

Antibody Production. Antibodies were produced in eight-week old Balb/c female mice. Proteins were injected in PBS (10 mM NaH_2PO_4, 150 mM NaCl, pH 7.2) and Freund's incomplete (1:3 ratio).

Gel Electrophoresis and Immunoblots (Western blots). SDS-PAGE and Western blots were performed as previously described (8).

RESULTS

Developmental Control of Alternative Oxidase Activity. In the inflorescence of *S. guttatum* the alternative pathway is expressed in cortex tissue within various regions of the inflorescence (8, Table 1). The alternative oxidase is also developmentally controlled. The upper portion of the *S. guttatum* inflorescence, the appendix, produces heat on the single day of anthesis (D-0). Five days before the appendix becomes thermogenic (D-5), the alternative pathway is present in relatively low amounts when compared to the cytochrome chain. During the next four days, the alternative pathway capacity increases dramatically. This event, along with an increased flow of substrates through glycolysis (4), is responsible for the production of heat during flowering. Three days after flowering (D+3), the the alternative pathway has returned to its previous low level. It should be noted that other parts of the inflorescence, such as the female floral region, express little or no alternative oxidase activity.

TABLE 1

ALTERNATIVE OXIDASE ACTIVITY IN *S. GUTTATUM*

Mitochondrial source	Solubilized alt. oxidase
	natoms O/min/mg protein
Appendix:	
D-5	11.0
D-1	72.2
D-0	85.8
Female Floral Region	0

Antibody Production. Polyclonal antibodies were raised against the solubilized and partially purified CM-Sepharose fraction of the alternative oxidase. The plant material employed for this purification was D-1 appendix tissue which has the alternative pathway in abundance. As can be seen from Figure 1, there are still a number of proteins present in the CM fraction. However, there is a considerable decrease in proteins present when compared to total mitochondrial proteins. The antibodies produced to this fraction were employed to immunoprecipitate solubilized alternative oxidase activity from solution (Table 2). Of the proteins present in the CM-Sepharose fraction, only 13 are antigenic.

Alternative Pathway Components. Immunoprecipitation of the alternative pathway with CM-Sepharose derived antibodies indicated that some, if not all, of the components are reactive to these antibodies. With this in mind, we have begun to employ them to identify the polypeptide constituents of the alternative oxidase. Mitochondria were isolated from D-5 and D-1 appendix tissue; uninduced and induced tissue, respectively. The mitochondria were

TABLE 2

IMMUNOPRECIPITATION WITH CM-SEPHAROSE DERIVED ANTIBODIES

Type of serum	Bound activity natoms O/min/mg protein
Pre-immune	1.6
CM-Sepharose	11.3

solubilized (materials and methods), activities measured, and proteins separated on SDS polyacrylamide gels. These gels were then immuno-blotted and probed with the CM-antibodies. As shown in Figure 1, a number of proteins react in the D-5 lane, the preparation containing little activity. However, in the lane containing D-1 proteins from a preparation high in activity, there are three additional proteins present. This experiment has been repeated for other tissues in the inflorescence, such as the female floral region which do not express the alternative pathway. In these experiments the three additional proteins mentioned are not present (data not shown).

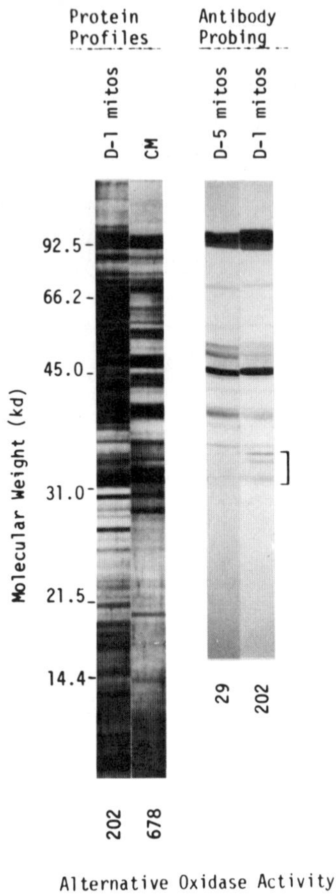

FIGURE 1. Stained gels and immumoblots of solubilized mitochondrial proteins from S. guttatum appendix tissue. The first lane shows a protein profile of total mitochondrial proteins from D-1 tissue. The CM lane contains those polypeptides present in the CM-Sepharose preparation which was injected for antibody production. The D-5 lane contains D-5 mitochondrial proteins from a solubilized preparation which has been probed with CM-sepharose derived antibodies. The D-1 lane contains D-1 mitochondrial proteins blotted with CM-antibodies. This lane shows three different polypeptides (when compared to the D-5 lane) that cluster around 35 kilodaltons. The alternative pathway activity, expressed as natoms O/min/mg protein, is shown for each preparation at the bottom of the lane.

DISCUSSION

The alternative pathway of respiration in higher plants has been investigated at the physiological level for decades. Biochemical analysis, coupled with molecular biological techniques, should now enable isolation of the protein components of the alternative oxidase. Our recent work has allowed us to (1) produce antibodies which will immunoprecipitate the alternative pathway activity and (2) to implicate at least three polypeptides as components of the alternative oxidase. Further work is needed to demonstrate the exact identity of these proteins. Specifically, antibodies to individual polypeptides which will react on Western blots and also immunoprecipitate the activity are desired.

Identification of the structural components of the alternative oxidase will also allow the genetics of this respiratory system to be studied in higher plants. The best genetic data for the alternative pathway comes from systems other than plants. In *Neurospora*, inhibitors of cytoplasmic protein synthesis and nuclear mRNA synthesis prevent appearance of the alternative pathway, while chloramphenicol (an inhibitor of mitochondrial protein synthesis) leads to an increase in activity (2). Genetic analysis of the alternative pathway in *Neurospora* has shown that it is encoded in the nucleus (10). Development of the alternative pathway in plants such as *S. guttatum* is dependent on cytoplasmic protein synthesis indicating a nuclear-encoded message for at least one component (11). Questions concerning the function and developmental regulation of the alternative pathway can only be answered through complete biochemical characterization and, ultimately, genetic manipulation of the alternative oxidase.

ACKNOWLEDGEMENTS

We wish to acknowledge gifts of plant material, helpful scientific discussions and general encouragement from Bastiaan J.D. Meeuse.

REFERENCES

1. Lance CM, Chauveau M, Dizengremel P (1985). The cyanide-resistant pathway of plant mitochondria. In Douce R, Day DA (eds): "Encyclopedia of Plant Physiology: Higher Plant Cell Respiration," Berlin: Springer-Verlag, p 202.

2. Edwards DL, Unger BW (1978). Cyanide- and hydroxamate-resistant respiration in Neurospora crassa. J. Bact. 133:1130.
3. Schonbaum GR, Bonner WD, Storey BT, Bahr JT (1971). Specific inhibition of the cyanide-insensitive respiratory pathway in plant mitochondria by hydroxamic acids. Plant Physiol. 47:124.
4. Meeuse BJD (1975). Thermogenic respiration in aroids. Ann. Rev. Plant Physiol. 26:117.
5. Lambers H (1982). Cyanide-resistant respiration: a non-phosphorylating electron transport pathway acting as an energy overflow. Physiol. Plant. 55:478.
6. Lambers H (1985). Respiration in intact plants and tissues: its regulation and dependence on environmental factors, metabolism and invaded organisms. In Douce R, Day DA (eds): "Encyclopedia of Plant Physiology: Higher Plant Cell Respiration," Berlin: Springer-Verlag, p 418.
7. Rich P (1978). Quinol oxidation in Arum maculatum mitochondria and its application to the assay, solubilisation and partial purification of the alternative oxidase. FEBS Lett. 96:252.
8. Elthon TE, McIntosh L (1986). Characterization and solubilization of the alternative oxidase of Sauromatum guttatum mitochondria. Plant Physiol. 82:1.
9. Bonner WD, Clarke SD, Rich PR (1986). Partial purification and characterization of the quinol oxidase activity of Arum maculatum mitochondria. Plant Physiol. 80:838.
10. Bertrand H, Argan CA, Szakacs NA (1983). Genetic control of the biogenesis of cyanide insensitive respiration in Neurospora crassa. In Schweyen RJ, Wolf K, Kaudewitz F (eds): "Mitochondria 1983," Berlin: Walter de Gruyter & Co., p 495.
11. McIntosh L, Meeuse BJD (1978). Control of the development of cyanide-resistant respiration in Sauromatum guttatum (Araceae). In Ducet G, Lance C (eds): "Plant Mitochondria," Amsterdam: Elsevier/North Holland Biomedical Press, p339.

Plant Gene Systems and Their Biology, pages 251–259

BIOCHEMICAL AND MOLECULAR CHARACTERIZATION OF THE UBIQUITIN-DEPENDENT PROTEOLYTIC PATHWAY IN HIGHER PLANTS[1]

Richard D. Vierstra, Thomas J. Burke, Janelle Cortner, Peggy H. Hatfield, Merten Jabben, and John Shanklin

Department of Horticulture, 1575 Linden Drive, University of Wisconsin-Madison, Madison, Wisconsin 53706

ABSTRACT Ubiquitin is a highly-conserved, 76 amino acid protein which plays an integral role in ATP-dependent protein degradation in mammalian cells through its covalent ligation to other intracellular proteins. We have discovered that higher plants also have ubiquitin and have identified recently several enzymatic activities required for both formation and degradation of ubiquitin-protein conjugates *in vitro*. The conjugating enzymes have been detected in all plant species examined, require ATP, and will use both endogenous and exogenous proteins as substrates. One function of the ubiquitin system in plants has been recently identified from studies of the Pfr-specific degradation of the photomorphogenic pigment, phytochrome. Photoconversion of the stable Pr form to Pfr induces the rapid formation of ubiquitin-phytochrome conjugates which appear to serve as intermediates of phytochrome catabolism. Studies on the molecular organization and regulation of the ubiquitin have been initiated recently in *Arabidopsis* with the sucessful cloning and characterization of ubiquitin gene(s). DNA sequence analysis indicates that the *Arabidopsis* gene(s), like those from mammals and yeast, code for a polyubiquitin precursor and that the derived amino acid sequence is identical to that determined for the protein from several other plant species.

[1]This work supported by grants from NSF (DMB-8409210), USDA-CRGO (85-CRCR-1-1547) and the Research Division of the College of Agriculture and Life Science (Hatch 2858).

INTRODUCTION

Intracellular protein degradation is an important process in the regulation of plant metabolism and development and, ultimately, in the expression of recombinant DNA gene products in transgenic plants. To date, both the pathways responsible for protein catabolism and the mechanisms that govern selectivity remain obscure in plants.

A few years ago, an important proteolytic pathway was discovered in mammalian cells involving the small (76 amino acids) protein, ubiquitin (1,2). Ubiquitin, as the name implies, has been found in all eukaryotes and is one of the most conserved proteins yet identified, being invariant in all animal species from insects, fish and birds to humans. In this proteolytic pathway, first identified in reticulocytes, one or more ubiquitin molecules are covalently ligated to proteins destined for catabolism in an ATP-dependent process (Fig. 1). Ligation is accomplished by a multi-enzyme system and occurs via a novel peptide bond between the carboxy-terminus of ubiquitin and free amino groups (N-terminal and lysyl) on the protein. The conjugated proteins are then degraded by protease(s) which recognize such modified proteins with the concomitant release of intact ubiquitin. As a result of this cycle, ubiquitin serves as a resuable recognition signal for selective proteolysis. In terms of function, this pathway appears to be the major route for degrading short-lived proteins which includes many important regulatory proteins. It also assumes the role as the cell's "housekeeper", removing abnormal proteins produced by biosynthetic errors or as a result of denaturation that would otherwise accumulate.

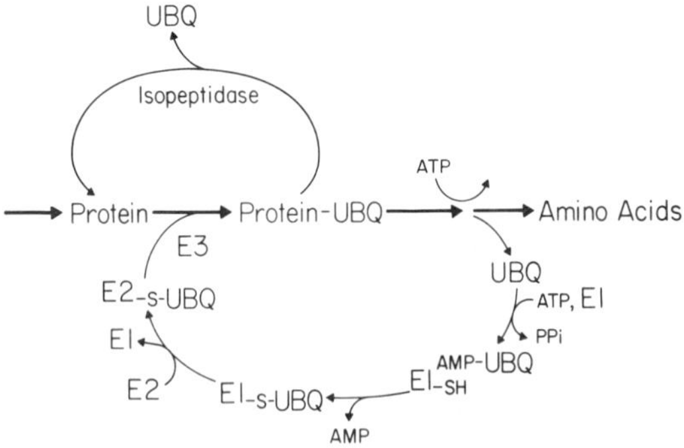

FIGURE 1. Proposed pathway for ubiquitin(UBQ)-dependent proteolysis [from (1)].

Until recently, little was known concerning how plants degrade intracellular proteins (3). Although a large number of proteolytic activities has been identified, the specific role that each protease plays still remains a mystery. This report describes our recent discovery that a ubiquitin-dependent proteolytic pathway exists in plants that represents a major avenue for protein degradation.

RESULTS AND DISCUSSION

From the immunological studies of Goldstein et al. (4) suggesting that plants contain ubiquitin, we began to investigate whether plants do indeed have ubiquitin and the associated proteolytic pathway. Immunoblot analysis with anti-human-ubiquitin antibodies revealed the presence of ubiquitin in crude extracts from several plant species with an apparent molecular mass identical to animal ubiquitin [8500 kDa (5)]. The protein was present in all plant species and tissues examined suggesting an important role in plant cell physiology. Subsequent characterization of ubiquitin purified from oats indicated that it is remarkably similar to its animal counterpart (5); differing in amino acid sequence by only three residues out of 76 (6) and having a near identical tertiary structure as determined by X-ray crystallography (7).

In addition to ubiquitin, plants also have the enzymes necessary for ligating ubiquitin to other proteins. As with the animal enzymes, they require ATP, will conjugate both endogenous and exogenous proteins, and are inhibited by thiol-reactive agents (8). The sensitivity to thiol-reactive agents suggests that the mechanism of ubiquitin ligation involves thiolester intermediates similar to those identified in mammalian cells (1). We have also detected activities that degrade ubiquitin conjugates in vitro. Whether these proteases are specific for ubiquitin conjugates is not yet known but the fact that they are inhibited by hemin, a potent and specific inhibitor of ubiquitin-conjugate specific protease(s) from reticulocytes, would suggest that they are.

One of the best plant sources for the conjugating enzymes we have identified is wheatgerm. From this source, a stable extract can be obtained that will effectively conjugate ubiquitin to either endogenous proteins or purified proteins such as lysozyme (Fig. 2). Importantly, it is free of ubiquitin-inactivating protease(s) that cleave ubiquitin at the carboxy-terminal glycine dipeptide thus rendering ubiquitin inactive with respect to ligation (5). When lysozyme (14.3 kDa) is used as a substrate, conjugated forms with molecular masses ranging from 21 kDa up to greater than 100 kDa can be obtained. Lysozyme contains only seven possible ubiquitin-conjugation sites; six lysines and the amino terminus. As a result, any lysozyme-ubiquitin conjugates with molecular masses >74 kDa should represent conjugates where some ubiquitin moieties are linked indirectly to lysozyme through polyubiquitin structures.

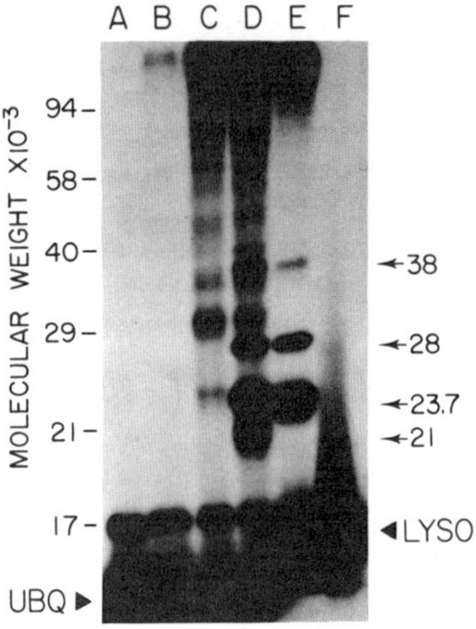

FIGURE 2. Demonstration of ATP-dependent ubiquitin-conjugating activity in crude extracts from wheatgerm. ^{125}I-ubiquitin was added to a wheatgerm extract and incubated for 90 min at 30°C either without ATP (Lane B), with ATP (Lane C), or with ATP and unlabelled lysozyme (Lane D). Lane E contains an extract incubated for 90 min at 30°C with ATP, unlabelled ubiquitin and ^{125}I-lysozyme Lanes A and F contain ^{125}I-ubiquitin and -lysozyme, respectively. Reactions were terminated by boiling in SDS and 2-mercaptoethanol and aliquots subjected to SDS-PAGE and autoradiography. Arrows to the right indicate the apparent molecular masses of several lysozyme-ubiquitin conjugates formed *in vitro*.

The physiological relevance of a ubiquitin-dependent proteolytic pathway in the physiology of plants is not yet understood. The sheer number of diverse conjugates present *in vivo* (5) would suggest that ubiquitin is not conjugated to only a few select proteins. Whether these proteins represent conjugates awaiting degradation is not known but studies with a specific protein, phytochrome, would suggest that they are. This photomorphogenic photoreceptor is rapidly degraded after photoconversion (t/2 1-2h) from the extremely stable Pr form (t/2 >100h) to the Pfr form. We have recently discovered that soon after Pfr formation, multiple forms of ubiquitin-phytochrome conjugates appear _in vivo_ (9). Kinetic analyses indicate that these conjugates represent intermediates of degradation and are themselves rapidly degraded once produced (t/2 30min). Ubiquitin-phytochrome conjugates are associated primarily with aggregates of phytochrome [previously identified as "pelletable" or "sequestered" (10)] that are formed rapidly (t/2 2sec) after photoconversion to Pfr. Taken together, these data suggest that this rapid aggregation is the first step of phytochrome catabolism, followed by formation of ubiquitin-phytochrome conjugates on these aggregates and finally, degradation of the conjugates.

In an effort to determine how ubiquitin-dependent proteolysis is regulated in plants during development and in response to stress, a molecular approach is being used to study the expression of ubiquitin gene(s). We chose _Arabidopsis thaliana_ for such a study because of its small genome, its short life cycle which should facilitate developmental studies spanning the life of the plant, and its promising molecular genetics. Immunoblot analyses of crude extracts from etiolated seedlings have demonstrated that _Arabidopsis_ contains free ubiquitin and a substantial quantity of ubiquitin conjugates.

Putative ubiquitin genes were isolated from an _Arabidopsis_ lambda gt11 genomic library (provided by E. Tobin and B Kohom) using anti-oat ubiquitin antibodies. From approximately 150,000 plaques, four positive plaques were identified. Secondary screens with both anti-oat and anti-human ubiquitin antibodies identified subsequently two recombinant phage coding for a protein immunoreactive to both antibody types (Fig. 3). The presence of ubiquitin coding sequences in these phage were confirmed by hybridization of the DNA to cDNA and genomic clones of ubiquitin from chicken and yeast, respectively (provided by M. Schlesinger and A. Varshavsky). A 1.3 kbp DNA fragment containing almost the entire ubiquitin coding region was identified, subcloned into a Gemini 4 plasmid vector, and used for subsequent analyses.

Partial DNA sequencing of the positive clones revealed that their inserts do indeed contain ubiquitin coding sequences (Fig. 4). It also showed that the ubiquitin

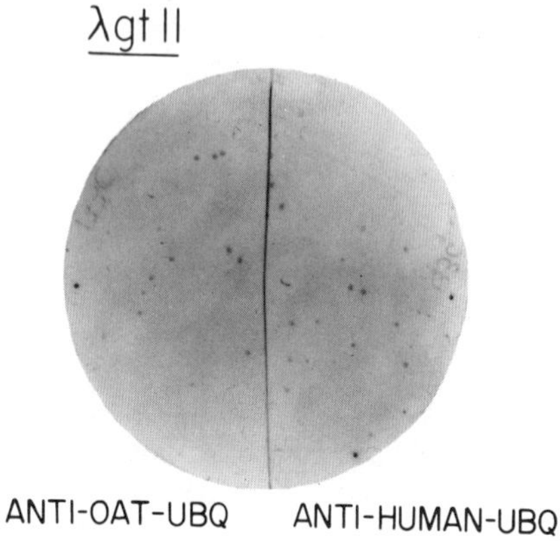

FIGURE 3. Secondary immunoscreen of an *Arabidopsis thaliana* lambda gt11 genomic library with rabbit antibodies against oat and human ubiquitin. Primary screening of the genomic library was accomplished with anti-oat ubiquitin antibodies. The presence of anti-ubiquitin antibodies was detected using alkaline phosphatase-conjugated goat anti-rabbit IgG in conjunction with nitroblue tetrazolium and 5-chloro-4-bromo-3-indolyl phosphate.

gene(s) in *Arabidopsis*, like those from mammals, yeast and barley (12,13) encode a unique polyubiquitin precursor protein. The coding frames for the ubiquitin monomers are linked it a head to tail fashion, C-terminal glycine of one repeat linked to N-terminal methionine of the next. Subsequent restriction mapping with Sac I, which cuts once within the sequence for each ubiquitin monomer, indicated that one of the *Arabidopsis* genes codes for a precursor protein containing **five** ubiquitin repeats. Since such polyproteins have never been detected *in vivo* in any eukaryote, it suggests that the precursor is rapidly cleaved into monomers soon after translation of the mRNA.

The derived amino acid sequence of *Arabidopsis* ubiquitin when compared with the amino acid sequence of ubiquitin from other plant and animal species analyzed to date illustrates the magnitude of sequence conservation. The sequence is totally conserved in all the higher plants investigated, oats (6), barley (13) and *Arabidopsis*, differs by only two residues between plants and yeast (12). It differs by only three residues between plants and all animal forms (11).

```
GC TCA GAC ACA ATC GAT AAT GTC AAG GCA AAG ATT CAA GAC AAG GA- GGA ATT CCA
...Ser Asp Thr Ile Asp Asn Val Lys Ala Lys Ile Gln Asp Lys Glu Gly Ile Pro
    20                             30

   CCA GAT CAG CAG AGA CTC ATA TTT GCT GGA AAA CAG CTT GAA GAT GGA CGC ACA
   Pro Asp Gln Gln Arg Leu Ile Phe Ala Gly Lys Gln Leu Glu Asp Gly Arg Thr
           40                                      50

   CTC GCT GAT TAC AAC ATC CAA AAG GAG TCT ACC CTC CAT CTT GTG CTT CGT CTC
   Leu Ala Asp Tyr Asn Ile Gln Lys Glu Ser Thr Leu His Leu Val Leu Arg Leu
                   60                                      70
                  |
   CGT GGT GGT ↓ ATG CAG ATA TTT GTG AAG ACC CTT ACC GGA AAA ACC ATT ACT CTT
   Arg Gly Gly Met Gln Ile Phe Val Lys Thr Leu Thr Gly Lys Thr Ile Thr Leu
           76  1                                   10
               Sac I
   GAG GTT GAG AGC TCA GAC ACA ATT GAT AAT GTC
   Glu Val Glu Ser Ser Asp Thr Ile Asp Asn Val...
                   20
```

FIGURE 4. Partial nucleotide and derived amino acid sequence of a ubiquitin genomic clone from *Arabidopsis*. Direct double-stranded sequencing was performed of the 1.3 kbp subclone. Arrow indicates the junction between the individual ubiquitin repeats.

Interestingly, the locations of all the amino acid substitutions in the tertiary structure of the molecule are clustered and spatially removed from the carboxy-terminal conjugation site.

In addition to the structural complexity within the individual genes of *Arabidopsis*, the organization and expression of the genes appears also to be complex. Genomic Southern blot analysis demonstrates that ubiquitin is encoded by a multigene family comprised of at least four members. Likewise, Northern blot analysis of both total and poly $(A)^+$ RNA isolated from green tissue indicates four abundant transcripts accumulate; ranging in size from 850 to 1900 bases. The size of these mRNA is consistent with the possibility that they encode polyubiquitin presursors containing up to 8, 6, 5, and 3 ubiquitin monomers.

CONCLUSIONS

It is now evident that higher plants contain ubiqutiin and many of the enzymes necessary for ubiquitin-dependent proteolysis. The fact that plants also contain a multitude of ubiquitin conjugates and that one particular protein, phytochrome, appears to be degraded via a ubiquitinated intermediate suggests that this pathway is important to the physiology of plants. Further characterization of this pathway at the molecular level should help elucidate the specific periods in the plant's life when ubiquitin plays an important role.

	1				5					10					15					20
Arabidopsis:	Met-	Gln-	Ile-	Phe-	Val-	Lys-	Thr-	Leu-	Thr-	Gly-	Lys-	Thr-	Ile-	Thr-	Leu-	Glu-	Val-	Glu-	Ser-	Ser-
Oat:	-	-	-	-	-	-	-	-	-	-	-	-	-	-	-	-	-	-	-	-
Barley:	-	-	-	-	-	-	-	-	-	-	-	-	-	-	-	-	-	-	-	-
Yeast:	-	-	-	-	-	-	-	-	-	-	-	-	-	-	-	-	-	-	-	-
Animal:	-	-	-	-	-	-	-	-	-	-	-	-	-	-	-	-	-	-	Pro	-

	21				25					30					35					40
Arabidopsis:	Asp-	Thr-	Ile-	Asp-	Asn-	Val-	Lys-	Ala-	Lys-	Ile-	Gln-	Asp-	Lys-	Glu-	Gly-	Ile-	Pro-	Pro-	Asp-	Gln-
Oat:	-	-	-	-	-	-	-	-	-	-	-	-	-	-	-	-	-	-	-	-
Barley:	-	-	-	-	-	-	-	-	-	-	-	-	-	-	-	-	-	-	-	-
Yeast:	-	-	-	-	-	-	-	Ser	-	-	-	-	-	-	-	-	-	-	-	-
Animal:	-	-	-	Glu	-	-	-	Ala	-	-	-	-	-	-	-	-	-	-	-	-

	41				45					50					55					60
Arabidopsis:	Gln-	Arg-	Leu-	Ile-	Phe-	Ala-	Gly-	Lys-	Gln-	Leu-	Glu-	Asp-	Gly-	Arg-	Thr-	Leu-	Ala-	Asp-	Tyr-	Asn-
Oat:	-	-	-	-	-	-	-	-	-	-	-	-	-	-	-	-	-	-	-	-
Barley:	-	-	-	-	-	-	-	-	-	-	-	-	-	-	-	-	-	-	-	-
Yeast:	-	-	-	-	-	-	-	-	-	-	-	-	-	-	-	-	Ser	-	-	-
Animal:	-	-	-	-	-	-	-	-	-	-	-	-	-	-	-	-	Ser	-	-	-

	61				65					70						76
Arabidopsis:	Ile-	Gln-	Lys-	Glu-	Ser-	Thr-	Leu-	His-	Leu-	Val-	Leu-	Arg-	Leu-	Arg-	Gly-	Gly
Oat:	-	-	-	-	-	-	-	-	-	-	-	-	-	-	-	-
Barley:	-	-	-	-	-	-	-	-	-	-	-	-	-	-	-	-
Yeast:	-	-	-	-	-	-	-	-	-	-	-	-	-	-	-	-
Animal:	-	-	-	-	-	-	-	-	-	-	-	-	-	-	-	-

FIGURE 5. Amino acid sequence comparison of ubiquitin determined for several plants, animals and yeast. Sequence data for oats was obtained from (6); barley (13); yeast (12) and animals (11).

REFERENCES

1. Hershko A, Ciechanover A (1982). Mechanisms of intracellular protein breakdown. Annu Rev Biochem 51:335-364.
2. Finley D, Varshavsky A (1985). The ubiquitin system: functions and mechanisms. Trend Biochem Sci 117:343-347.
3. Vierstra R (1987). Ubiquitin, a key component in the degradation of plant proteins. Physiol Plant (in press).

4. Goldstein G, Scheid J, Hammerling O, Boyse E, Schlesinger D, Niall HD (1975). Isolation of a polypeptide that has lymphocyte-differentiating properties and is probably represented universally in living cells. Proc Natl Acad Sci USA 72:11-15.
5. Vierstra RD, Langan SM, Haas, AL (1985). Purification and initial characterization of ubiquitin from the higher plant, Avena sativa. J Biol Chem 260:12015-12021.
6. Vierstra RD, Langan SM, Schaller GE (1986). Complete amino acid sequence of ubiquitin from the higher plant, Avena sativa. Biochemistry 25:3105-3108.
7. Vijay-Kumar S, Buggs CE, Wilkinson KD, Hatfield PM, Vierstra RD, Cook WJ (1987). Comparison of the three-dimensional structures of yeast and oat ubiquitin with human ubiquitin. J Biol Chem (submitted).
8. Vierstra R (1987). Demonstration of ATP-dependent, ubiquitin-conjugating activities in higher plants. Plant Physiol (in press).
9. Shanklin J, Jabben M, Vierstra RD (1987). Red light-induced formation of ubiquitin-phytochrome conjugates: identification of possible intermediates of phytochrome degradation. Proc Natl Acad Sci USA 84:359-363.
10. McCurdy DW, Pratt LH (1986) Kinetics of intracellualr redistribution of phytochrome in Avena coleoptiles after photoconversion to the active, far-red-absorbing form. Planta 167:330-336.
11. Schlesinger DH, Goldstein G (1975). Molecular conservation of 74 amino acid sequence of ubiquitin between cattle and man. Nature 255:423-424.
12. Ozkaynak E, Finley D, Varshavsky A (1984). The yeast ubiquitin gene: head to tail repeats encoding a polyubiquitin precursor protein. Nature 312:663-666.
13. Gausing K, Barkardottir R (1986). Structure and expression of ubiquitin genes in higher plants. Eur J Biochem 158:57-62.

V. TRANSGENIC EXPRESSION AND PROMOTER ANALYSIS

Plant Gene Systems and Their Biology, pages 263–277

DEVELOPMENTAL AND ENVIRONMENTAL REGULATION OF THE MAIZE ALCOHOL DEHYDROGENASE 1 (Adh1) GENE: PROMOTER-ENHANCER INTERACTIONS

Jim Peacock, David Wolstenholme[1], John Walker, Karam Singh, Danny Llewellyn, Jeff Ellis and Liz Dennis

Division of Plant Industry, CSIRO, Canberra, A.C.T., Australia

ABSTRACT The alcohol dehydrogenase genes of maize (Adh1 and Adh2) show tissue specific and environmental regulation of their expression. We describe experiments on the molecular basis of anaerobic induction of these genes in maize protoplasts and transgenic tobacco plants. Mutation of the Adh1 promoter linked to the bacterial cat gene has defined a 40 bp region which controls anaerobic expression. This promoter region, the anaerobic regulatory element (ARE) functions in maize and tobacco indicating a conservation of the regulatory mechanism in these distantly related plant species. A model invoking positive regulatory factors is postulated for the anaerobic regulation of the Adh1 gene.

INTRODUCTION

Differential gene expression is important to plants for their development and for their ability to survive changes in their environment. Few genes, if any, are likely to be truly constitutive and be expressed in every cell of a plant. Many genes are expressed only in a particular range of cells and tissues and at particular times in the life cycle of a plant. Some genes are responsive to certain environmental conditions and are induced in cells in which the gene would otherwise not be

[1]On leave from Department of Biology, University of Utah, Salt Lake City, U.S.A.

expressed. The mechanisms controlling these different patterns of gene regulation are now beginning to be understood in a few plant gene systems. As expected from earlier analyses in animal systems, there are specific DNA sequences, usually upstream of the transcribed region of the gene, which are important for the regulation of gene activity(1). In animal systems, at least some of these short DNA sequences interact with a protein which binds to the region(2). These trans-acting proteins, either directly or through interactions with other proteins, affect the frequency of transcription of specific genes.

The presence of trans-acting proteins implies that other gene loci, which may or may not be linked to the gene in question, are important in the control of a gene's expression, but one single regulatory protein could conceivably be involved in the co-ordinate expression of a number of different genes. This could apply to genes that are all active in a particular tissue at a specific developmental time or genes which are all switched on by a particular environmental stress condition.

Already there are some data which suggest that the cis-positioned DNA control sequences 5' of genes are conserved through plant evolution(3), and this, in turn, implies that genes involved in coding for regulatory proteins will also be conserved units of genetic information. If this is so it is likely to be of some practical significance in that genes used in trans-specific and trans-generic transfers could be associated with controls so that they would have their normal expression patterns.

Some gene regulation events, such as the environmental responses, are reversible, but others involved in developmental and cellular differentiation may not be. However, cells from a range of tissues, in many plant species, maintain the potential for dedifferentiation and redifferentiation and hence must be capable of altering developmental patterns of gene activation and inactivation. In plants we have no information about the activity of particular genes at the earliest stages of embryonic development. Are all genes active or are all genes inactive in the original zygotic cell? Are genes progressively and specifically turned on or shut-down at particular times and places in the succeeding cascade of cell divisions? In a few gene systems there is now enough molecular characterisation data to begin to answer these questions.

The alcohol dehydrogenase (Adh) 1 and 2 genes in maize present us with opportunities to investigate a number of the fundamental principles of gene regulation in plants. These genes display tissue-specific and developmentally regulated patterns of expression and are inducible by one or more specific environmental conditions.

TISSUE-SPECIFIC EXPRESSION OF THE ALCOHOL DEHYDROGENASE GENE

Freeling(4) documented a striking example of tissue-specific expression of the maize Adh1 gene in his demonstration that the gene is expressed after meiosis in the male gametophyte. This expression is under the control of the haploid genome, as shown by a 1:1 segregation of pollen grains staining for ADH enzyme activity in a plant heterozygous for functional and non-functional (mutant) alleles of Adh1. If the gene were under control of the diploid genome, ADH activity would be detectable in all pollen grains of the heterozygous plant. Stinson and Mascarenhas(5) have shown that ADH activity can be first detected shortly after meiosis in the single nucleate male gametophyte and that it steadily increases up to the stage of the second mitotic division of the gametophyte. There is no information as to whether the genes are expressed in the premeiotic diploid cells but Stinson and Mascarenhas report that there is no enzyme activity in meiotic cells.

We have recently investigated, by histochemical methods, the tissue-specific expression of the Adh1 gene in the mature dry kernel and in germinating kernels. The aleurone is the single outermost layer of cells of the endosperm, and is derived from the triploid cell which is one of the two fertilization products. In dry kernels of the Berkeley inbred lines carrying the Fast (F) or Slow (S) alleles of the Adh1 gene (BKS and BKF (6)), and in another series of lines developed by Nuffer ((7) and having a different genetic background) there is ADH activity in the aleurone (Fig.1). The cell layers outside the aleurone, the pericarp, are derived from maternal tissue and do not have any functional ADH enzyme. The starchy endosperm cell layers beneath the aleurone layer also contain no active enzyme.

Both the developing pericarp and the developing endosperm contain active ADH. Loss of ADH from the pericarp of dry kernels therefore occurs at a time when this tissue's primary remaining function is one of

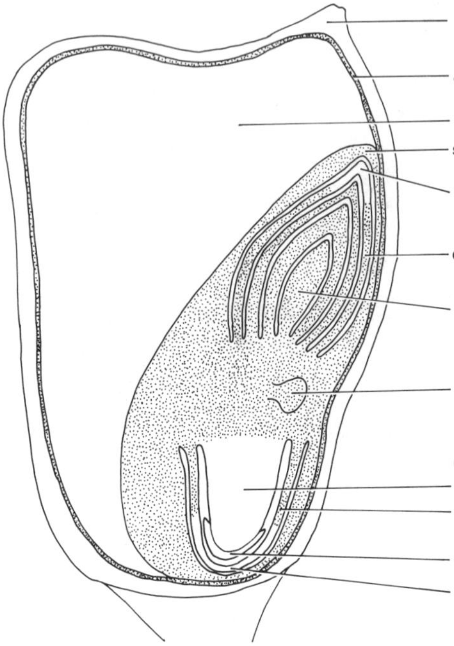

FIGURE 1. Distribution of ADH enzyme activity in thin sections of developing kernels of a particular generation of inbred Berkeley maize, homozygous for either Adh1-1S (BKS) or Adh1-1F (BKF). Tissues expressing ADH (+ve) are stippled.

protection. Loss of ADH from the endosperm correlates with the completion of storage of materials that will be drawn on by the future developing embryo.

In both the Berkeley and the Nuffer lines, all cells of the dormant embryo stain ADH positive. It appears that this staining and that of the aleurone results mainly from ADH1 as the equivalent cells of kernels of an Adh1-null mutant (gamma-25; 8) do not stain.

The distribution of ADH1 is essentially the same in tissues of kernels that have germinated for up to 72 hours as it is in dry kernels, and again these observations hold true for both the Berkeley and the Nuffer lines. However, in a specific generation of BKS and BKF kernels that were the product of selfing, we have observed a curious, aberrant distribution of ADH1 in germinating embryos. In dry BKS and BKF kernels of this particular generation the distribution of ADH1 is as described above. However, following induction (by soaking) of germination, ADH1 activity becomes progressively lost from the primary root, from the tip of the coleorhiza and from the tip of the coleoptile (Fig.1). There is a sharp demarcation along the files of cells in the major axis of the developing plant; cells in the shoot region contain ADH1 activity and cells in the root region do not contain ADH1 activity. This is also a region of contrasting cell morphologies; the shoot region cells are larger in volume, more elongate, and highly vacuolate: the root cells are smaller and have dense cytoplasm. A similar morphological differentiation is shown between the cells in the tip of the coleorhiza versus the more distal cells in which the Adh1 gene is expressed. No differentiation of morphology is obvious in those cells of the coleoptile which do, or which do not, contain ADH1 activity. The lack of ADH1 activity in primary seminal root does not extend to the lateral seminal root initials present in the embryo, which do show ADH1 activity.

Although up to 70% and 60%, respectively, of these BKF and BKS kernels germinate into plants, none of the BKF seedlings, and a maximum of 50% of the BKS seedlings that we examined had an extended primary seminal root. The BKS primary roots that do extend are short and their distal end and root cap lack ADH1 activity. As in unextended primary roots, there is a distinct line of demarcation between cells that do, and do not, contain ADH1 activity, and again this is correlated with distinctly different cell morphologies. As this generation of both BKS and BKF kernels resulted from selfing, and the kernels were stored together for about five years, the extent to which genetic and environmental factors contributed to the unusual ADH1 distributions, and to failure of primary root development is presently unclear. Also, it is not possible from the available data to decide whether the loss of ADH1 activity at the noted sites during germination is a special case of differential gene expression, or a general consequence of differential cell death.

ENVIRONMENTAL INDUCTION OF EXPRESSION OF THE ALCOHOL DEHYDROGENASE GENE

Expression of the Adh1 gene of maize in the seedling root is induced by anaerobic conditions(9). Using a transient expression system we have demonstrated that the

5' region of the Adh1-S gene, from base pairs -1094 to +106, will promote the expression of a chloramphenicol acetyl transferase (cat) coding region(10). Expression of this AdhCAT chimeric gene, introduced into maize protoplasts by electroporation, is stimulated by a reduction in the concentration of oxygen as determined by analysis of CAT enzyme activity or by S1 nuclease mapping of the transcript. The expression of the introduced AdhCAT gene parallels the expression of the endogenous Adh1 gene under the different oxygen concentrations(10).

Identification of the cis acting DNA signals required for anaerobic expression in maize protoplasts has been determined by a functional analysis of a set of in vitro mutagenised AdhCAT chimeric gene constructions(11). The promoter region of the AdhCAT was modified by 5' deletions to define the 5' end of the regulatory element. The 3' limit of necessary sequences was determined by transferring the promoter to a truncated 35S promoter, and by linker-scanning constructions which generated clustered point mutations. We have shown that there is a 40 bp anaerobic regulatory element (ARE) within the Adh1 promoter, at position -140 to -100. The ARE is necessary for anaerobically regulated expression of the gene in maize cells, and attached to an unrelated promoter, confers anaerobically regulated expression on the heterologous gene. The linker-scanner mutations indicate a substructure to the ARE, identifying two critical sequence regions.

Anaerobically Induced Expression of the Adh1 Gene Suggests a Model of Positive Regulation.

Our results suggest that the ARE acts in conjunction with a trans-acting factor to activate gene expression during anaerobiosis. Present experimental data suggesting positive regulation for the Adh1 gene are: 1) deletions or clustered point mutations either have no effect or abolish expression, i.e. no mutations lead to a high level of constitutive expression; 2) the ARE can be transferred to a heterologous promoter, in this case a deleted cauliflower mosaic virus 35S gene promoter that retains its TATA box and cap site but lacks other required upstream sequences, and the ARE confers both expression and regulation on this heterologous gene.

We propose two models by which a trans-acting factor in conjugation with the ARE activates gene expression during anaerobiosis (Fig.2). In Model I, the binding of the factor is the determining step and occurs only under anaerobic conditions. In Model II, the factor is bound to the ARE in all oxygen concentrations and is activated by anaerobiosis, perhaps through a conformational change, so that it promotes transcription. For the sake of simplicity we have assumed the presence of another generalised cellular binding factor that recognises the TATA box and that some interaction (contact) between this factor and upstream factors is important for the formation of an active transcription complex.

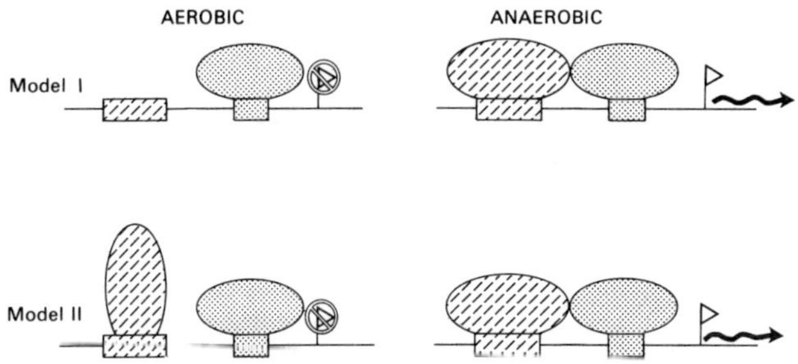

FIGURE 2. Anaerobic regulation of the Adh1 gene in maize cells: interactions between specific DNA sequences and trans-acting DNA binding proteins. The two alternative models are explained in the text. Hatched areas represent the maize ARE sequence and its protein factor; stippled areas are the TATA box and its corresponding factor. The flag designates the start point for transcription and the wavy arrow the level of transcription. Inhibition of transcription is indicated by a barred circle over the flag. Model 1 requires the anaerobically induced binding of a factor to the ARE in such a way that in conjunction with the TATA factor, transcription is activated. In Model II the factor is bound under aerobic conditions but is in such a conformation that interaction cannot occur and transcription is not possible. When an anaerobically-induced conformational change occurs, the ARE factor interacts with the TATA factor and transcription is activated.

Data From an Adh1-35S Hybrid Promoter Suggests that Model II is Correct.

The Adh1 ARE has been linked to a truncated 35S promoter deleted to either -90 or -45. The Adh1-ARE/-45 35S construct produced an anaerobically regulated promoter (Fig.3) while the Adh1-ARE/-90 35S construct gene showed constitutive high level expression. A chimeric construct which has identical spacing between the ARE and the 35S TATA box as the Adh1-ARE/-90 35S construct, but which has extra Adh DNA sequences in place of 35S sequences, (Adh1-ARE-89/-45 35S), is still anaerobically regulated. This suggests that the constitutive effect depends on another factor binding to 35S DNA in the -90 to -45 region.

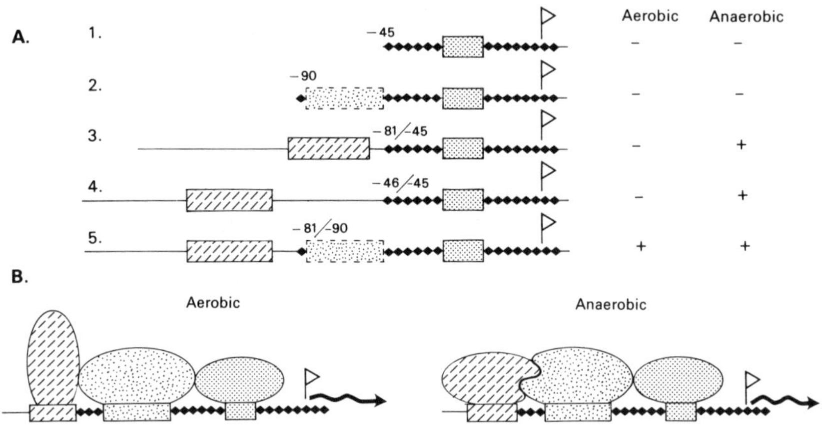

FIGURE 3. Adh1ARE interactions with a 35S truncated promoter. A: Hybrid promoter constructions between maize Adh1 and various deleted 35S promoter segments. 35S DNA is indicated by diamonds, Adh1 DNA by a continuous line. Numbers indicate deletion endpoints relative to the start of transcription at +1. Random stippled areas are the putative binding site within 35S DNA for an as yet undefined protein factor. Other symbols are explained in Fig.2. Results at the right indicate expression in maize protoplasts incubated under aerobic or anaerobic conditions. B: Proposed model of protein factor binding and interaction under aerobic and anaerobic conditions.

This -90/-45 35S region is not an effective promoter on its own, but the 35S gene does require these sequences for high level expression. Since the Adh1-ARE/-90 35S construct operates under aerobic, as well as anaerobic, conditions we conclude that the ARE factor must be bound under aerobic regimes, and can, even in its aerobic conformation, interact effectively with the putative 35S factor and the TATA factor. These data are inconsistent with Model I and add support to Model II (see Fig.2).

The OCS Enhancer Increases the Level of Adh1 Expression in Maize Protoplasts. Regulation is Dependent on its Relative Position.

We have defined a sequence from the octopine synthase gene(12) which increases the level of expression of the maize Adh1CAT construct by a factor of at least five when

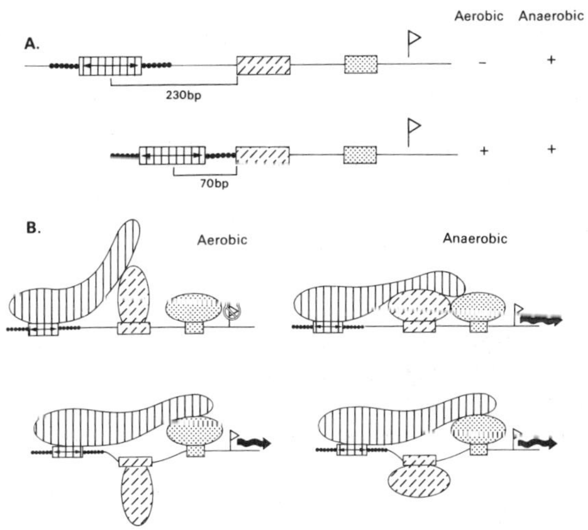

FIGURE 4. Interactions of the OCS enhancer with the maize ARE: effect of distance on anaerobic regulation. A: OCS/AdhARE hybrid promoter constructions. Beads represent OCS DNA and the boxed double-ended arrow the OCS enhancer palindrome. Distances between the palindrome and ARE are indicated. Other symbols and expression in maize protoplasts are as previously described. B: Proposed model of protein factor binding and interactions. The OCS protein is shown with vertical stripes and binds to the palindrome sequence. Other symbols are given in Fig.2.

it is introduced into maize protoplasts. Deletion analysis indicated a 16 bp palindrome sequence was responsible for the enhancing activity; testing of a synthesised 16 bp sequence confirmed its role(13).

When the OCS enhancer is placed directly adjacent to the ARE, the OCS-AdhCAT construct shows constitutive expression in maize protoplasts and CAT expression is not increased by anaerobic stress. The factor that presumably binds to the OCS region may hinder the binding of the ARE factor. However when the OCS enhancer is moved ∿ 150 bp further upstream, anaerobic regulation is restored. This result again suggests that a factor is bound to the ARE under aerobic conditions and that this ARE factor prevents the interaction of the OCS factor with the TATA box factor. Under anaerobic conditions the ARE factor no longer prevents the OCS/TATA interaction, and may itself be involved in the interaction.

The Maize *Adh1-1S* Promoter is Weakly Expressed in Transgenic Tobacco Unless the OCS Enhancer is Present.

Two *Adh/CAT* fusion constructs have been tested for expression in transformed *Nicotiana tabacum* plants(12). The first construct has *Adh* sequences from -1094 to +106 and the second from -140 to +106. Neither construct promotes expression of the CAT enzyme significantly over background. However expression of the longer promoter construct was detectable in anaerobically induced tobacco tissue by S1 analysis. Although the expression was low, the 5' end of the transcript of the hybrid gene corresponded to the normal cap site of the *Adh* transcript in maize. The promoter therefore functions in transgenic tobacco to give a low level, but nevertheless accurate, anaerobic transcription initiation.

Addition of upstream elements from the octopine synthase gene to the *AdhCAT* construct at -140 increases the level of expression of CAT a further five to ten fold. Expression is induced by anaerobiosis and the transcripts are initiated at the same site as they are in maize. This result is different to that seen in maize protoplasts where the OCS enhancer in the same position results in constitutive, not inducible, expression of the *Adh1* promoter. This difference could relate to differential opportunities for *in situ* molecular interactions between the OCS factor, and the maize and tobacco ARE factors with the TATA factor. Deletion analysis of the maize promoter

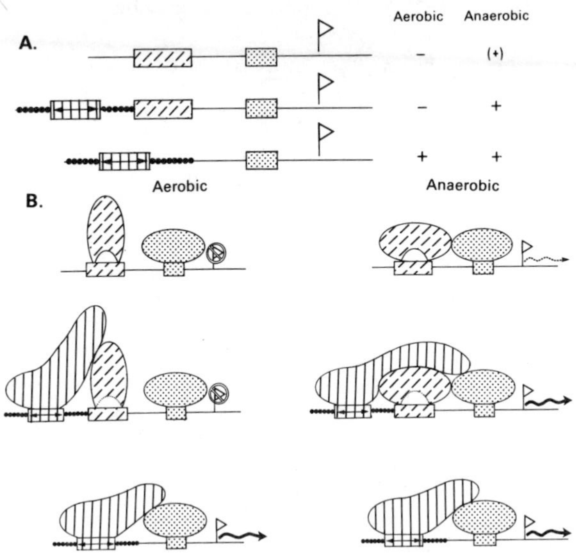

FIGURE 5. Expression of the maize Adh promoter in transgenic tobacco plants. A: Adh and OCS/Adh hybrid promoter constructions introduced into stably transformed tobacco plants. Expression under aerobic and anaerobic conditions is indicated. Symbols are explained in Figs.2 and 4. Weak expression from the maize promoter is indicated by the small plus sign in parentheses.
B: Proposed model to account for the different pattern of expression. The protein factors are endogenous tobacco binding proteins instead of maize proteins shown in previous figures. The limited contact between the tobacco ARE factors and the maize ARE sequences indicates a limited level of conservation of the system as discussed in the text.

assayed in transgenic tobacco shows that the same regions are important for anaerobic regulation in tobacco and that factors responsible for anaerobic induction must be conserved between the two distantly related genera.

These results can be explained in terms of Model II with the assumption that the tobacco ARE factor binds weakly unless stabilised by a bound OCS factor (Fig.5). The tobacco ARE binding factor can prevent the OCS factor from interacting with the TATA factor unless conditions are anaerobic when the ARE factor's conformation supports a tripartite, active transcriptional complex. These results are not specifically dependent upon the sequences of the OCS enhancer since the same results are obtained when the enhancer region from the *35S* promoter is substituted for the OCS enhancer(12).

CONCLUSIONS

Based on our data, we have formulated a model of positive regulation for the induction of the *Adh1* gene of maize under anaerobic conditions. We have presented the simplest model consistent with our data - undoubtedly the situation is more complex. The model makes a number of predictions that can be tested experimentally. For example, we predict that there is a factor bound to the ARE even under aerobic conditions. *In vivo* footprinting should confirm whether this prediction holds. The model also predicts that some clustered point mutations (linker-scanning mutations) in the region between -90 and -45 of the CaMV *35S* promoter should convert constitutive expression of the *Adh*-ARE/*35S* promoter to inducible expression, due to inhibition of binding of the *35S* -90/-45 factor.

An anaerobic response is known to occur in many species of plants. Our data show that the factor regulating the *Adh1* gene is sufficiently conserved between tobacco and maize to give anaerobically regulated expression of the maize promoter in tobacco, although, in tobacco, participation of an upstream enhancer complex is required for high level expression. However not all binding factors are conserved. We have isolated an anerobically regulated *Adh* promoter from *Pisum sativum*(14) which functions in tobacco but not in maize protoplasts. The nopaline synthase promoter (*nos*) also functions poorly in maize(10,15). In this case it may be the interaction of the maize TATA binding protein with the *nos* TATA box that is at fault. The OCS enhancer does not interact with either the *Pisum Adh* or *nos* promoters to promote high level expression in maize.

Since the anaerobic response in maize involves the co-ordinate induction of approximately 20 polypeptides(6) it might be expected that a common trans-acting factor would interact with all these promoters. Certainly it would be more economical in terms of the numbers of genes encoding regulatory products if batteries of genes were responsive to the same factor. The maize *Adh2* gene is another well characterised gene which is induced by anaerobic conditions(3). The nucleotide sequence in the promoter region does show islands of conservation with Adh1, including the two ARE regions of the Adh1 promoter we have identified by *in vitro* expression studies as being critical for anaerobic induction(11). Conservation of sequence suggests conservation of the factors binding to the sequences. Other anaerobically induced genes that have been examined, maize aldolase(16) and sucrose synthase(17), do not present such simple comparisons. The promoter region of aldolase shows a lesser level of homology to the Adh1 ARE and sucrose synthase shows conservation of part of the ARE sequence (a stretch of 13 bases), but the sequence segment is located in the first intron.

Adh2, like Adh1, is anaerobically induced, but as it is not active in the same tissues in which we see developmentally programmed expression of the Adh1 gene, it is clear that Adh2 is not recognised by the same factors for tissue specific expression as Adh1. The signals for anaerobic and tissue specific expression must involve different stretches of DNA, and different or additional trans-acting factors. We hope to be able to localise the regulatory sequences controlling tissue-specific expression.

Adh1 gene expression can also be turned off from a developmentally programmed active state. In the presence of light, ADH1 activity is lost from young shoots and in general any green tissue does not contain ADH activity. In mature maize leaves Adh1 cannot be induced even by anaerobic conditions(18); there appears to be an irreversible switch-off of the gene during plant development. This does not necessarily imply a modification of the Adh1 locus itself, but could reflect the loss of expression of specific trans-acting factors. The isolation and characterisation of the trans-acting factors associated with different regions of the Adh1 promoter, should contribute to a more complete understanding of the developmental and environmental regulation of this gene system.

ACKNOWLEDGEMENT

Agrigenetics Research Corporation provided support for this research.

REFERENCES

1. Gluzman Y (1985). Eukaryotic transcription. The role of cis- and trans-acting elements in initiation. Cold Spring Harbor: Cold Spring Harbor Laboratory.
2. Briggs MR, Kadonaga JT, Bell SP, Tjian R (1986). Purification and biochemical characterization of the promoter-specific transcription factor, Sp1. Science 234:47.
3. Dennis ES, Sachs MM, Gerlach WL, Finnegan EJ, Peacock WJ (1985). Molecular analysis of the alcohol dehydrogenase 2 (Adh2) gene of maize. Nucleic Acids Res 13:3, 727.
4. Freeling M (1976). Intragenic recombination in maize: pollen analysis methods and the effect of parental Adh1+ isoalleles. Genetics 83:701.
5. Stinson J, Mascarenhas JP (1985). Onset of alcohol dehydrogenase synthesis during microsporogenesis in maize. Plant Physiol 77:222.
6. Sachs MM, Freeling M, Okimoto R (1980). The anaerobic proteins of maize. Cell 20:761.
7. Nuffer MG (1953). Similarity of Mp and Ac mutator systems. Maize Genet Coop Newsletter 27:68.
8. Freeling M, Bennett DC (1985). Maize Adh1. Ann Rev Genet 19:297.
9. Hageman RH, Flesher D (1960). The effect of an anerobic environment on the activity of alcohol dehydrogenase and other enzymes of corn seedlings. Arch Biochem Biophys 87:203.
10. Howard EA, Walker JC, Dennis ES, Peacock WJ (1987). Regulated expression of an alcohol dehydrogenase 1 chimeric gene introduced into maize protoplasts. Planta 170:535.
11. Walker JC, Howard EA, Dennis ES, Peacock WJ (1987). DNA sequences required for anaerobic expression of the maize alcohol dehydrogenase 1 gene. Proc Natl Acad Sci (in press).
12. Ellis JG, Llewellyn DJ, Dennis ES, Peacock WJ (1987). Maize Adh1 promoter sequences control anaerobic regulation: addition of upstream promoter elements from constitutive genes is necessary for expression in tobacco. EMBO J 6:11.
13. Ellis JG, Llewellyn DJ, Walker JC, Dennis ES, Peacock WJ (1987). The OCS element : A 16 bp palindrome essential for activity of the octopine synthase enhancer (submitted).

14. Llewellyn DJ, Finnegan EJ, Ellis JG, Dennis ES, Peacock WJ (1987). Structure and expression of an alcohol dehydrogenase 1 gene from *Pisum sativum* (cv. Greenfeast). J Mol Biol (in press).
15. Fromm M, Taylor LP, Walbot V (1985). Expression of genes transferred into monocot and dicot plant cells by electroporation. Proc Natl Acad Sci USA 82:5822.
16. Dennis ES, Gerlach WL, Walker JC, Peacock WJ (1987). An anaerobically regulated aldolase gene of maize. (in press).
17. Werr W, Frommer WB and Starlinger P (1985). The structure of the sucrose synthase gene on chromosome 9 of *Zea mays* L. EMBO J 4:1373.
18. Okimoto R, Sachs M, Porter E, Freeling M (1980). Patterns of polypeptide synthesis in various maize organs under anaerobiosis. Planta 150:89.

Plant Gene Systems and Their Biology, pages 279–288

CIS-REGULATORY ELEMENTS IN HEAT SHOCK AND T-DNA PROMOTERS[1]

William B. Gurley[2], Wesley B. Bruce[2], Eva Czarnecka[2], Ram Bandyopadhyay[2], Ronald T. Nagao[3], and Joe L. Key[3]

[2]Department of Microbiology and Cell Science
1059 McCarty Hall, University of Florida
Gainesville, Florida 32611

[3]Botany Department, University of Georgia
Athens, Georgia 30602

ABSTRACT Promoters of the environmentally inducible heat shock (HS) gene (Gmhsp17.5-E) of soybean and two constitutive genes of T-right (Ti plasmid) show a tripartite organization consisting of TATA, TATA-proximal, and -distal domains. For the HS gene, the TATA-proximal domain (-155 to -40) contains three heat shock consensus elements (HSEs) and is the major determinant of thermal induction. The Gmhsp17.5-E TATA-distal region also contains up to eight HSEs, but they appear to play a minor role in thermal induction. The 780 and agropine synthase (AGS) genes of T-DNA have TATA-proximal and -distal domains with unique properties. The 780 activator (TATA-distal element) is enhancer-like as judged by its bidirectionality and flexibility in location. The 780 TATA-proximal element is not essential for full activity, whereas the AGS TATA- proximal element seems to be absolutely required.

INTRODUCTION

The genes under study have promoters that are representative of two general types: those that are environmentally inducible, such as the heat shock (HS)

[1] This work was supported by contracts to W.B.G. and J.L.K. from Agrigenetics Research Associates.

promoter from soybean, and two T-DNA promoters that are constitutive in their expression. The T-DNA promoters from the _780_ and _AGS_ genes are located on T-right. Our overall approach was to identify functional domains of these promoters by assessing the effects of mutations on _in vivo_ transcriptional activity in sunflower tumors. The method of gene transfer into plants utilized Ti plasmid-based vectors harbored in _Agrobacterium tumefaciens_.

METHODS

Vectors used to incorporate mutated promoters into the genome of sunflower tumor cells contained a homologous reference gene as an internal standard so that transcripts derived from the altered promoters could be precisely quantified. Relative transcription levels (RTLs) were determined by a comparison of test and reference gene transcript abundance. By deleting a small portion of the reference gene leader sequence, RNA derived from the reference gene could be distinguished from the longer test gene RNA by S1 nuclease hybrid protection mapping. The use of the _Agrobacterium tumefaciens_/T-DNA system provided the advantages of a low copy number vector and stable insertion into the plant genome. The variability of expression often encountered due to random integration was minimized to 10% (+/-) of wild type activity by pooling RNA from 200 to 300 tumors for each promoter mutation. Since most tumors are the result of multiple transformation events, it is estimated that each batch of pooled tumors represented at least 1,000 integrations. The use of the reference gene vector and large numbers of tumors insures not only reproducibility sufficient to resolve promoter fine structure, but allows one to directly compare results obtained from tumor batches inoculated months (or years) apart. The results from S1 nuclease analysis of tumor RNA from the -476 5'-deletion construct of the _780_ promoter are presented in Table 1 as a demonstration of reproducibility of this method. Each test:reference gene ratio was obtained by measuring radioactivity in test and reference transcript bands from separate S1 nuclease mapping gels.

TABLE 1
REPRODUCIBILITY OF TRANSCRIPT QUANTIFICATION

Date of RNA isolation	Ratio test:reference	Average
10-04-85	8.62	8.95
	9.44	
	8.81	
01-16-86	10.43	
	10.42	
	9.35	
	10.02	
	9.52	9.95
08-31-86	8.50	
	10.30	
	11.40	10.06

RESULTS

Heat Shock Promoter Structure

One HS promoter we are characterizing is from the soybean Gmhsp17.5-E gene which encodes a low molecular weight HSP of 17.5 kDa (1,2,3). Seven 5'-deletions and one fill-in insertion at -94 have been analyzed (Fig. 1) for thermoinducible expression in sunflower tumors using the reference gene vector. When sequences 5' to position -591 were deleted, a 25% drop in transcriptional activity was observed. Although this region contains two HS consensus elements (HSEs), the most significant loss of activity occurred when an AT-rich region with no HSEs (-767 to -591) was removed. Surprisingly, removal of sequences between -591 and -259 had little effect on promoter activity even though this region contains two clusters of three HSEs each.

A dramatic loss of promoter strength did occur, however, upon removal of sequences between -259 and -179. The only obvious consensus sequence located within these 80 bp is a secondary TATA and an overlapping upstream dyad (5'-TATAAAGAATTTC-3') which is also found in several Drosophila HS genes (4). This upstream dyad has not been shown to be involved in thermal induction in animal systems, but its conservation in soybean HS genes, along with the results of our deletion analysis, suggest that it may play a role in the transcriptional activation of HS promoters. Deletion into the TATA-proximal cluster of HSEs (position -100 to -40) resulted in a further reduction in activity.

DNAse I footprinting studies (Czarnecka, et al., unpublished) with Gmhsp17.5-E indicate that the TATA-proximal domain of the promoter (-40 to -156) is bound in a DNA protein complex *in vitro* by nuclear factors from soybean plumules. This protected region not only contains three HSEs, but other sequences as well which do not conform to known consensus motifs. The potential complexity of the TATA-proximal domain suggests that more than one type of transcription factor may interact in this area to promote thermoinducible expression.

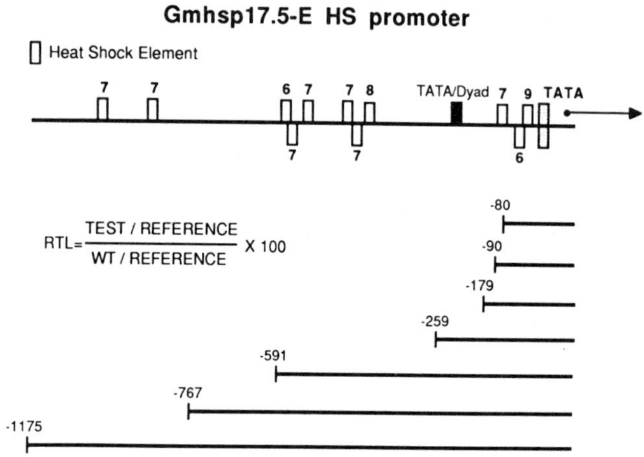

FIGURE 1. 5'-deletions of the soybean HS gene Gmhsp17.5-E in sunflower tumors. RTLs were determined by S1 nuclease mapping of poly(A) RNA using a homologous reference gene incorporated in the T-DNA-based vector as an internal standard.

Our overall view of the structure of the Gmhsp17.5-E promoter delegates the functions of thermoinducibility and the amplitude of the response to the TATA-proximal domain. The positioning of HSEs within this region is generally conserved among the plant HS genes sequenced to date (1,3,5,6). Other *cis*-acting elements also seem to be present within this domain, but their composition appears to be variable between the various HS genes of soybean. The upstream, or TATA-distal, domain seems to be multicomponent with a great deal of variation in its subelement composition. The role, if any, that the upstream cluster of HSEs plays is not yet clear. Single base mutations are being introduced into the TATA-proximal HSEs to evaluate the contribution of these sequences to thermal induction and *in vitro* binding affinities of HS transcription factors.

780 Gene Promoter Structure

The *780* gene is located adjacent to the left border of T-right. Internal and 5'-deletions of the *780* promoter have been analyzed (7) in tumors and a model of its promoter organization has been proposed. The *780* gene promoter (Fig. 2) contains three domains: the activator (TATA-distal), upstream element (TATA-proximal) and TATA.

The activator has enhancer-like properties and stimulates transcription of the "core" promoter (TATA and TATA-proximal element) by 200-fold. It has been shown to function bidirectionally when positioned 5'- to the start of transcription, but has no activity in either orientation when located 3' to the gene. Flexibility in spacing of the activator was demonstrated by its activity (210%) when positioned 613 bp upstream of TATA by the insertion of a fragment of Phi X 174 RF DNA between the activator and TATA (Bruce et al., unpublished). A discrete subelement within the activator is located near its 5'- border since removal of 31 bp in this area reduced activity by approximately 45% (7). This subelement contains three direct repeats of 7 bp. Repeat "c" (5'-TTGAAAA-3") is found at two other positions within the 5'-flanking region, and is also present upstream of seven out of thirteen known genes of octopine type T-DNA. Removal of a second activator component located between

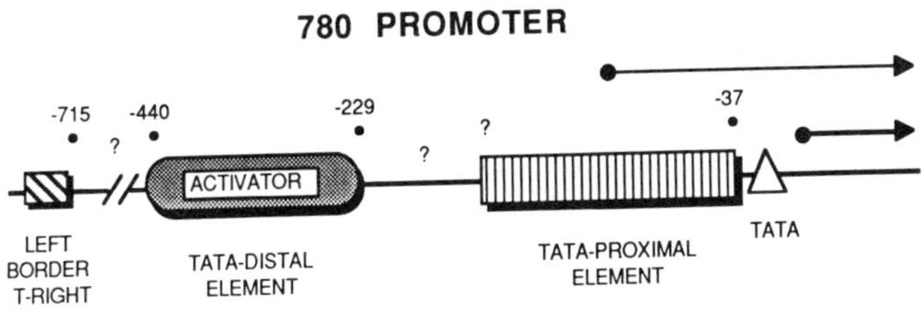

FIGURE 2. Model of the *780* gene promoter determined by analysis (7) of 5' and internal deletion mutants.

-396 and -229 resulted in loss of activity from an RTL of 50 at -396 to 1 at -229.

The upstream element corresponds to the CCAAT box region and seems to contribute positively to the function of the wild type gene. Although it can be removed entirely without decreasing promoter activity, small scale disruptions have a strong negative influence on activity.

DNAse I footprinting and gel retardation studies using the _780_ activator indicate that nuclear factors are present in untransformed soybean plumules that are able to bind specifically _in vitro_. The formation of a low mobility protein:DNA complex was competed by unlabeled probe and was not inhibited by the addition of heterologous DNA (data not shown). The DNAse I protection pattern appears to be complex suggesting the involvement of more than one _trans_-acting factor.

AGS Promoter Structure

The _AGS_ gene encodes a 1550 base transcript and is located adjacent to the right border of T-right. We have analyzed 13 5'-deletions and 17 internal deletions using a chimeric _AGS_-5'/_780_-3' gene in a reference gene vector (Bandhyopadhyay et al., unpublished). Our results indicate that the overall organization of the _AGS_ promoter is similar to that of _780_ in that two major domains are located upstream of TATA. The TATA-proximal domain (-164 to TATA) of the _AGS_ promoter appears to be multicomponent and essential for promoter activity unlike the analogous region of the _780_ gene. The region between -165 and -206 seems to

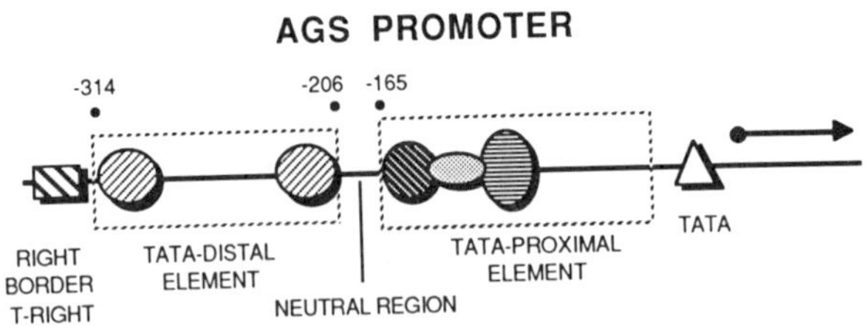

FIGURE 3. Model of the _AGS_ promoter determined by analysis of 5' and internal deletion mutants using a reference gene vector (Bandyopadhyay et al., unpublished). The test and reference genes contained the 5'-flanking sequences of _AGS_ joined to the coding region of the _780_ gene.

be neutral and can be replaced by a synthetic oligonucleotide with no effect on transcriptional activity.

The TATA-distal domain is located immediately adjacent to the right border of T-right (-314) and extends 3' to the neutral region at -206. This element can be positioned 41 bp closer to TATA and still function, but it does not appear to possess the strong enhancer-like properties of the 780 activator. The AGS TATA-distal element does not activate a -37 5'-deletion (TATA only) of the 780 gene and therefore seems to require other upstream elements in addition to TATA.

DISCUSSION

Mutational studies on plant promoters have shown a high degree of similarity in the overall organization of plant and animal regulatory sequences. In general, the TATA, TATA-proximal element, and TATA-distal element (enhancer-like) organization of plant promoters seem to closely parallel that of animal genes. From the limited number of plant genes examined, there seem to be several differences in detail that are unique to plant promoters. For example, the TATA-proximal element in animal genes usually extends 5' to around -110, whereas in several plant genes (maize ADH-1 [8], AGS, 780, and soybean Gmhsp17.5-E) this element seems to include additional upstream sequences, out to position -140 to -164.

Enhancers are a class of promoter regulatory elements found in animal genes that are not always present, or required, for transcriptional activity. Enhancers are located outside of the TATA and TATA-proximal regions and can function to activate transcription over large distances of up to several kilobases. The properties of mammalian enhancers have been categorized (9) as follows: 1) they have the ability to potentiate, or stimulate, transcription from heterologous promoter elements; 2) they can often function without other upstream elements or TATA; 3) they act bidirectionally; and 4) they are active when positioned both 5' and 3' to the gene and also when they are located in introns. Although bacterial regulatory sequences can act at long range (10), all those characterized (gal, ara, and deo) play a role only in the repression of transcription. Enhancer-like elements are also present in plant genes with the best characterized found in the rbc (11,12), cab (13), OCS (8), and 780 genes (7). Unlike mammalian enhancers, there have been no published reports of plant enhancer-like elements activating transcription when placed 3' to the gene.

The HS genes of soybean have regulatory sequences in the promoter that show a close homology to the HS consensus element (HSE) (5'-CT-GAA--TTC-AG-3') of Drosophila which has been shown to be essential for thermoinducibility of transcription (14,15,16). As with animal HS genes the dominant regulatory elements seem to be located in the TATA-proximal domain. The Gmhsp17.5-E promoter has three regions of homology to the HSE located between -49 and -99. Removal of the distal HSE (7 out of 10 bp match) by 5'-deletion, or disruption by a 4bp insertion, drastically reduced thermoinducible activity suggesting that multiple HSEs within the TATA-proximal domain contribute strongly to induction. Deletion of HSEs located far upstream had little or no effect on transcription. Additional sequence motifs such as the SV40 enhancer core (11 out of 14 bp match) and a CCAAT box-like sequence are also found in the TATA-proximal domain but their role is still undetermined. The possibility of multiple classes of cis-acting elements within the TATA-proximal domain is not implausible since the analogous region of the human HSP70 promoter has been shown to contain binding sites for three types of trans-acting factors (17).

The soybean HS promoter also has sequences located upstream of the TATA-proximal domain which collectively contribute from 25 to 50% of the activity. This region has no recognizable consensus sequences other than a series of HSEs and the Drosophila HS dyad. A sequence comparison with the upstream regions of other HS genes in plants (1,3,5,6) suggests that the composition and spacing of TATA-distal elements are variable. At present it seems likely that this region should be viewed as a collection of individual elements rather than a single highly organized domain.

Unlike the HS genes which are induced to very high levels by environmental stress, the 780 and AGS genes of T-DNA are expressed at moderate levels with no apparent control by environmental or developmental parameters. Even though most T-DNA genes appear to be constitutive, the results of our studies indicate that T-DNA promoters have a complex structure, certainly possessing components other than simply TATA and CCAAT. The presence of TATA-distal elements (activators) that exert a positive effect on transcription in T-DNA promoters indicates that these promoters are truly eukaryotic in design and function. The overall tripartite (TATA, TATA- proximal element, and enhancer-like element) organization of the constitutive T-DNA promoters closely parallels that found in genes under very stringent developmental and environmental control suggesting that highly regulated transcriptional expression may be the result of varying degrees of specialization of the more fundamental components of promoter structure.

Just as mammalian viral genes have provided much basic information regarding eukaryotic promoter structure, the T-DNA genes are well suited to serve as models for promoter organization in higher plants. The benefits of wide host range derived by *Agrobacterium* have driven selection in T-DNA promoters for utilization of eukaryotic transcriptional control mechanisms that are highly conserved among a wide range of higher plant species. Although T-DNA genes may be bacterial in evolutionary origin, and certainly by lineage, they have achieved a close mimicry of plant promoters from a functional standpoint and in their interactions with *trans*-acting factors.

REFERENCES

1. Czarnecka E, Gurley WB, Nagao RT, Mosquera LA, Key JL (1985). DNA sequence and transcript mapping of a soybean gene encoding a small heat shock protein. Proc Natl Acad Sci, USA 82:3726.
2. Gurley WB, Czarnecka E, Nagao RT, Key JL (1986). Upstream sequences required for efficient expression of a soybean heat shock gene. Mol Cell Biol 6:559.
3. Nagao RT, Czarnecka E, Gurley WB, Schoffl F, Key JL (1985). Genes for low-molecular weight heat shock proteins of soybeans: sequence analysis of a multigene family. Mol Cell Biol 5:3417.
4. Southgate R, Ayme Y, Voellmy R (1983). Nucleotide sequence analysis of the *Drosophila* small heat shock gene cluster at locus 67B. J Mol Biol 165:35.
5. Schoffl F, Raschke E, Nagao RT (1984). The DNA sequence analysis of soybean heat-shock genes and identification of possible regulatory promoter elements. EMBO J 3:2491.
6. Rochester DE, Winter JA, Shah DM (1986). The structure and expression of maize genes encoding the major heat shock protein, hsp 70. EMBO J 5:451.

7. Bruce WB, Gurley WB (1987). Functional domains of T-DNA promoter active in crown gall tumors. Mol Cell Biol 7:59.
8. Ellis JG, Llewellyn DJ, Dennis ES, Peacock WJ (1987). Maize Adh-1 promoter sequences control anaerobic regulation:addition of upstream promoter elements from constitutive genes is necessary for expression in tobacco. EMBO J 6:11.
9. Chambon P, Dierich A, Gaub M-P, Jakowley S, Jongstra J, Krust A, LePence J-P, Oudet P, Reudelhuber T (1984). Promoter elements of genes coding for proteins and modulation of transcription by estrogens and progesterone. In Greep RO (eds): "Recent progress in hormone research," New York: Academic Press 40:1.
10. Dandanell G, Valentin-Hansen, P, Larsen JEL, Hammer K (1987). Long-range cooperativity between gene regulatory sequences in a prokaryote. Nature 325:823.
11. Timko M. Kausch A, Castresana C, Fassler J, Herrera-Estrella L, Van de Broek G, Van Montagu M, Schell J, Cashmore A (1985). Light regulation of plant gene expression by an upstream enhancer-like element. Nature 318:579.
12. Fluhr R, Kuhlemeier C, Nagy F, Chua N-H (1986). Organ-specific and light-induced expression of plant genes. Science 232-1106.
13. Simpson J, Schell J, Van Montagu M, Herrera-Estrella L (1986). Light-inducible and tissue-specific pea cab gene expression involves an upstream element combining enhancer- and silencer-like properties. Nature 323:551.
14. Pelham HRB (1982) A regulatory upstream promoter element in the Drosophila HSP70 heat-shock gene. Cell 30-517.
15. Mirault M-E, Southgate R, Delwart E (1982). Regulation of heat-shock genes: A DNA sequence upstream of Drosophila hsp70 genes is essential for their induction in monkey cells. EMBO J 1:1279.
16. Pelham HRB, Bienz M (1982). A synthetic heat-shock promoter element confers heat-inducibility on the herpes simplex virus thymidine kinase gene. EMBO J 2:1473.
17. Wu BJ, Williams GT, Morimoto RI (1987). Detection of three protein binding sites in the serum-regulated promoter of the human gene encoding the 70-kDa heat shock protein. Proc Natl Acad Sci, USA 84:2203.

Plant Gene Systems and Their Biology, pages 289–299

EXPRESSION OF PETUNIA *rbcS* GENE FUSIONS IN TRANSFORMED TOBACCO PLANTS

Caroline Dean, Mitchell Favreau, Stanley Tamaki, Jonathan Jones, Pamela Dunsmuir and John Bedbrook.

Advanced Genetic Sciences, Inc.
Oakland, California 94608

ABSTRACT We have confirmed the relative steady state mRNA levels of the eight petunia *rbcS* genes in leaf tissue using a primer extension analysis. We have then used the 5' and 3' fragments from the most strongly (SSU301) and the most weakly (SSU911) expressed *rbcS* genes to direct the expression of foreign coding regions when introduced into tobacco. These experiments suggest that sequences within the coding region or the introns of the SSU301 gene contribute to its high level of expression.

INTRODUCTION

The nuclear genes which encode the small subunit of ribulose bisphosphate carboxylase (RbcS) have been characterized from a variety of species (1-7). Multiple nuclear genes encode the RbcS polypeptide in all the plant species studied to date. In most species it is not clear whether all of the multiple genes are active or whether the multiple genes show different patterns of expression with respect to development, tissue specificity or light. We have characterized the organization and expression of the eight *rbcS* genes of *Petunia* (Mitchell) (8-10). The genomic linkage and structure of the petunia *rbcS* genes is summarized in Fig.1. We have measured the level of expression of all eight genes in petunia leaf tissue and have analyzed the expression of five of the genes in four other organs of petunia (9). Homology analysis to three different petunia *rbcS* cDNA clones (1) and nucleotide sequence analysis of the genes (10) shows that the eight petunia *rbcS* genes can be divided into three groups (71,117 and 51), which we now term subfamilies. Two of the subfamilies (71 and 117) contain single genes: SSU301 and SSU611 respectively. The 51 subfamily contains the remaining six genes: SSU491; SSU231; SSU112; SSU211; SSU911; SSU11A (8). The three subfamilies of genes potentially encode different mature RbcS polypeptides, with all of the amino acid differences being clustered at the C-terminus of the polypeptide (10).

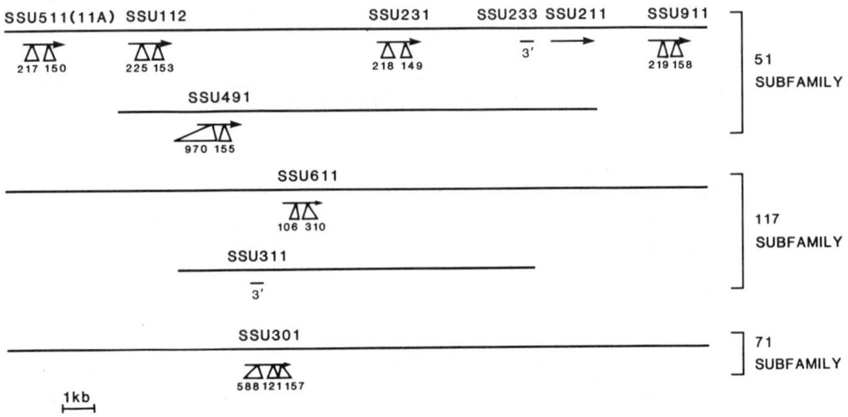

Figure 1. Schematic representation of the linkage and gene structure with respect to intron number and size, of the eight rbcS genes in petunia. The introns are drawn as triangles below the genes.

We present here, firstly, a primer extension analysis to measure the relative expression levels of the eight petunia rbcS genes in petunia leaf tissue. This analysis was carried out to resolve a confusion in the literature (9,11) as to which of the petunia rbcS genes is the most strongly expressed gene in leaf tissue. Secondly we present results of experiments using selected 5' and 3' fragments from two different petunia rbcS genes to drive the expression of foreign coding regions in transformed plants. These experiments were to test whether the levels of expression characteristic of these two rbcS genes could be reproduced if the gene flanking regions were directing the synthesis of foreign coding regions.

RESULTS

Relative expression levels of the rbcS genes in leaf tissue.

We have previously reported on the relative expression levels of each of the eight petunia rbcS genes in petunia leaf tissue (9). The steady state RNA levels of the genes were assayed by hybridizing Northern blots and slot blots with four gene specific probes cloned from the 3' untranslated regions of the genes. For the genes of the 51 subfamily which are too closely related to allow construction of gene specific probes, we used the frequency of cloning in a cDNA library to estimate their relative expression level (9). The combination of these two approaches allowed us to estimate the level of expression in leaf tissue for each of the eight petunia rbcS genes, these are: SSU301 47.3%; SSU611 23.2%; SSU491 7%; SSU112 5.4%; SSU911 >1%; SSU11A + SSU231 15.2%; SSU211 1.9%.

Tumer et al (11) have recently reported on the level of expression of two petunia rbcS genes, SSU8 which corresponds to SSU301 in our terminology, and SSU11A. In contrast to our results they find that the gene SSU301 accounts for only 5% of the total rbcS expression and SSU11A accounts for 40%. In order to resolve this discrepancy we reassayed the relative expression levels of the petunia rbcS genes using a primer extension analysis (Dean C, Favreau M, Dunsmuir P, Bedbrook J in preparation). The results of this analysis are shown in Fig.2.

Three oligonucleotides 301T, 611T and 911T were used in the analysis. The oligonucleotides 301T and 611T are specific for the genes SSU301 and SSU611 respectively. The oligonucleotide 911T anneals and extends from the six highly related genes of the 51 subfamily. It is possible, however, to discriminate between the primer extended fragments from the 51 subfamily genes because each of the genes has a 5' untranslated leader sequence of a distinct length. This results in multiple primer extended fragments, each of which can be identified as being synthesized from the individual 51 subfamily genes. In these experiments the amount of primer extended fragment for each gene is a direct measure of the steady state RNA in the petunia leaf RNA. The results shown in Fig.2 therefore confirm our previous data; the gene SSU301 shows the highest levels of steady state RNA of all the petunia rbcS genes. The gene SSU611 shows the next highest levels, followed by individual genes of the 51 subfamily (the gene SSU11A shows 5 fold lower levels than SSU301).

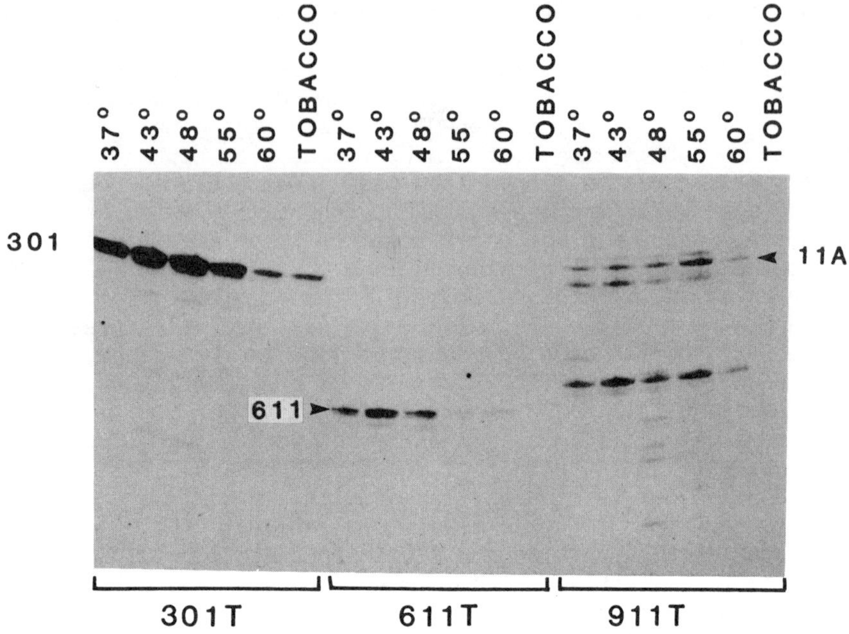

Figure 2. Primer extension analysis of the rbcS genes in petunia leaf tissue. The three oligonucleotides 301T, 611T and 911T were annealed to total petunia leaf RNA at several temperatures. The primer extended fragments corresponding to the genes SSU301, SSU611 and SSU11A are indicated. The tobacco RNA sample was from a plant carrying the petunia SSU301 gene.

Preparation of the SSU301 rbcS gene fusions.

We wanted to test whether the high level of expression (2% poly A RNA) of SSU301 could be reproduced if the 5' and 3' sequences of SSU301 were fused to foreign coding regions and the gene fusion were introduced into plants. We prepared an expression cassette containing the 5' and 3' flanking regions of the SSU301 gene. Our approach has been to create restriction endonuclease sites by oligonucleotide site-directed mutagenesis (12). We introduced restriction sites both at the ATG translation initiation codon and immediately following the translation stop codon. In parallel, the matching restriction sites were also introduced into the foreign coding regions. In this way we can create almost perfect translational fusions between the *rbcS* gene and the foreign coding region. Translational fusions have been shown to give higher levels of chimaeric RNA than the corresponding transcriptional fusion (13). We created NcoI restriction sites (CCATGG) at the ATG codon of the SSU301 gene and also at the ATG codons of three foreign genes, namely chloramphenicol acetyltransferase (CAT) (14), octopine synthase (OCS) (15) and the *Serratia marcescens* chitinase gene (*chiA*) (16).

In order to mimic as closely as possible the wild-type SSU301 gene, the 3' untranslated region of the SSU301 gene was included in the gene fusions. A BglII restriction site (AGATCT) was introduced into the SSU301 gene immediately downstream of the translation stop codon (TAA). Matching BglII restriction sites were also introduced into the CAT and OCS coding regions immediately downstream of the translation stop codons. The DNA fragment carrying the *chiA* coding sequence contained a BamHI restriction site 30bp 3' to the translation stop codon. This site was used to join the *chiA* coding region to the SSU301 3' tail via a BamHI/BglII fusion.

Analysis of the tobacco transformants carrying the SSU301 gene fusions.

The SSU301 CAT, OCS and *chiA* fusions were introduced into tobacco (*N. tabacum* cv. W38) via an *Agrobacterium tumefaciens* binary vector (17) using co-cultivation of tobacco protoplasts with *Agrobacterium* (18) or using the technique of leaf-disc transformation (19). The individual tobacco plants were assayed after they had been in the greenhouse 25 days, a time previously established to show the highest levels of *rbcS* steady state RNA (Dean C, Favreau M, Dunsmuir P, Bedbrook J in preparation). The steady state levels of the SSU301 CAT, OCS or *chiA* RNA were assayed using primer extension with oligonucleotides complementary to either the CAT, OCS or the *chiA* coding region. Approximately fifteen independent tobacco transformants were assayed for each of the constructions.

Figure 3 shows the primer extension analysis of 16 tobacco transformants carrying the SSU301-CAT fusion. The levels of the chimaeric SSU301-CAT RNA vary to a large degree between the individual transformants. This was also observed in the tobacco transformants carrying the SSU301-OCS and *chiA* constructions. This phenomenon which has been observed now in many studies has been termed between-transformant variability. The variability between independent transformants makes comparison of several constructions very difficult.

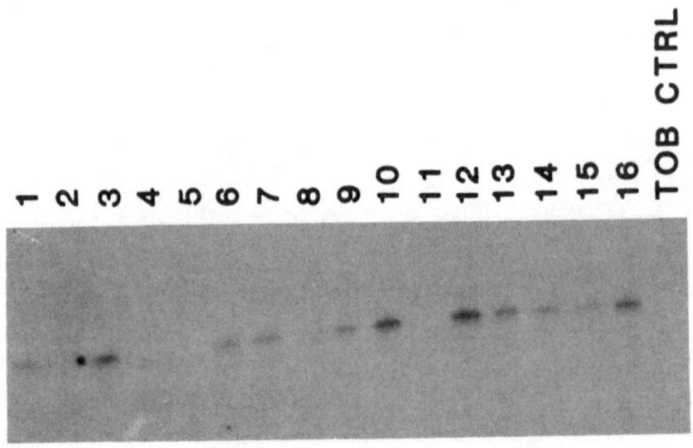

Figure 3. Primer extension analysis of tobacco transformants carrying the SSU301-CAT fusion. Sixteen transformants and an nontransformed tobacco control are shown.

To circumvent this problem we have repeated the primer extension analysis on pooled RNA samples from the SSU301-CAT, OCS and *chiA* transformants. Equal amounts of RNA isolated from each of the individual transformants were pooled in order to establish an average value for expression of each of the different fusions. Figure 4 shows the primer extension analysis on the pooled RNA samples. In these experiments the amount of primer extended fragment for each gene fusion is a direct measure of the steady state RNA present in the tobacco RNA. When we compare the steady state RNA levels of the different SSU301 fusions with the steady state RNA levels of an intact SSU301 gene which was introduced into tobacco plants we see striking differences. The three SSU301 gene fusions show between 15 (for *chiA*) and 500 (for OCS) fold lower levels of steady state RNA than the intact SSU301 gene. The RNA levels of the SSU301-chiA fusion are 4 fold higher than those of the SSU301-CAT fusion and the SSU301-CAT fusion shows 8 fold higher levels of RNA than the SSU301-OCS fusion.

The differences in the steady state RNA levels of the SSU301 fusions presumably reflect different stabilities of the chimaeric messages. The low levels of chimaeric SSU301-OCS RNA are in contrast to the high levels of chimaeric RNA observed when the OCS coding region and 3' OCS sequences were fused to 5' sequences from a chlorophyll a/b binding protein (*Cab*) gene (13). In other fusions, notably *chiA* fusions, the SSU301-*chiA* fusion gave 4 fold higher levels of RNA than a *Cab*-*chiA* fusion.

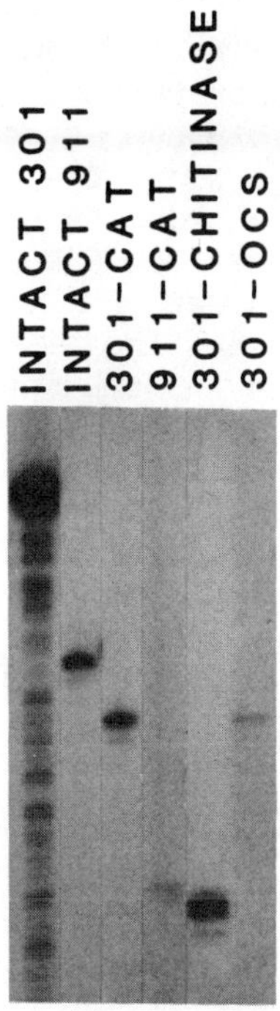

Figure 4. Primer extension analysis on pooled RNA samples from the tobacco transformants carrying petunia rbcS gene fusions.

We have tested whether the RNA levels of the SSU301-OCS fusion are increased when the SSU301 3' sequences are replaced by OCS 3' sequences. However, we saw no significant differences in the level of SSU301-OCS RNA between the two different 3' constructions, therefore we think that the low level of expression of the SSU301-OCS fusion cannot be attributed to the SSU301 3' sequences.

Analysis of SSU911 gene fusions.

We wanted to investigate why the levels of steady state RNA from the SSU301 gene fusions were so much lower than the steady state RNA levels of the intact SSU301 gene. In order to see if this was a general phenomenon for rbcS fusions we constructed a fusion between the weakly expressed petunia SSU911 rbcS gene and the CAT coding region. Using site-directed mutagenesis, we introduced an NcoI restriction site around the ATG codon of the SSU911 gene, and a BglII restriction site immediately downstream of the translation termination codon of the SSU911 gene. We constructed an SSU911-CAT fusion and then then introduced the SSU911-CAT fusion and in parallel an intact SSU911 gene, into tobacco plants. The tobacco plants were analyzed after they had been in the greenhouse 25 days and the steady state levels of RNA from the SSU911-CAT fusion and the intact SSU911 gene were assayed using a primer extension analysis.

Figure 5 shows the primer extension analysis of RNA from 10 tobacco plants carrying the SSU911-CAT fusion and 10 tobacco plants carrying the intact SSU911 gene. As in petunia the steady state RNA level of the SSU911 gene is low. In contrast to the SSU301 results however, the RNA level of the SSU911-CAT fusion is very similar to that of the intact SSU911 gene.

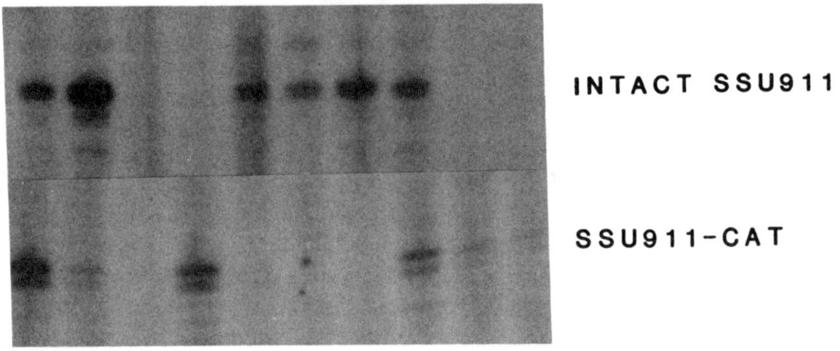

Figure 5. Primer extension analysis of tobacco transformants carrying the intact SSU911 gene and the SSU911-CAT fusion.

The primer extension analysis was repeated on pooled RNA samples to enable the comparison of the intact SSU911 with the SSU911-CAT fusion to be made without the complication of between-transformant variability. This is shown in Fig.4 alongside the SSU301 fusions. The SSU911-CAT fusion is expressed at a 4 fold lower level than the intact SSU911 gene. This compares with the 65 fold difference between RNA levels from the SSU301-CAT fusion and the intact SSU301 gene.

DISCUSSION

We have used selected 5' and 3' fragments from the petunia rbcS genes to drive the expression of foreign coding regions in transformed plants. We chose to use SSU301 which we have confirmed to be the most strongly expressed rbcS gene in leaf tissue, and SSU911, the most weakly expressed gene. We find very large differences in the steady state RNA levels of the three SSU301 gene fusions (CAT, OCS and chiA) in tobacco leaf cells. None of these fusions give levels of RNA that approach the level of RNA from the intact SSU301 gene. This does not appear to be the case for the rbcS gene SSU911. The SSU911-CAT fusion achieves RNA levels in the tobacco cells which are only 4 fold lower than the intact SSU911 gene. This suggests that sequences within the coding region or introns of the SSU301 gene contribute to the high level of expression of that gene. Recent experiments using reciprocal fusions between the two rbcS genes SSU301 and SSU911, where the sequences of the two genes were exchanged at the ATG codon, have shown that sequences both 5' and 3' to the ATG contribute to the high level of expression of the SSU301 gene (Dean C, Favreau M, Dunsmuir P, Bedbrook J in preparation). We have also introduced into tobacco an SSU301 gene that lacks introns. This experiment demonstrated that sequences within the introns were important for high level expression of this gene. Clearly we cannot achieve levels of RNA with the SSU301 gene fusions that match the RNA levels of the intact SSU301. However, the expression cassette (AGS007) containing the 5' and 3' sequences of the SSU301 gene can be used to obtain relatively high levels of expression of gene fusions (for example 0.1% of poly A RNA for the chiA fusion) in transformed tobacco plants.

REFERENCES

1. Dunsmuir P, Smith S, Bedbrook, J (1983). A number of different nuclear genes for the small subunit of RuBPCase are transcribed in petunia. Nuc Acid Res 11:4177.
2. Berry-Lowe S, McKnight T, Shah D, Meagher R (1982). The nucleotide sequence, expression and evolution of one member of a multigene family encoding the small subunit of ribulose bisphosphate carboxylase. J Mol Appl Genet 1:483.
3. Broglie R, Coruzzi G, Lamppa G, Keith B, Chua N-H (1983). Structural analysis of nuclear genes encoding for the precursor to the small subunit of wheat ribulose-1,

5-bisphosphate carboxylase. Biotechnology 1:55.
4. Cashmore A (1983). Nuclear genes encoding the small subunit of ribulose-1,5-bisphosphate carboxylase. In Kosuge T, Meredith CP, Hollander A (eds): "Genetic Engineering of Plants," Plenum Press New York, p 29.
5. Mazur B, Chui C-F (1985). Sequence of a genomic DNA clone for the small subunit of ribulose-bisphosphate carboxylase-oxygenase from tobacco. Nuc Acid Res 13:2373.
6. Pichersky E, Bernatzky R, Tanksley S, Cashmore A (1986). Evidence for selection as a mechanism in the concerted evolution of *Lycopersicon esculentum* (tomato) genes encoding the small subunit of ribulose-1, 5-bisphosphate carboxylase/oxzygenase. Proc Natl Acad Sci USA 83:3880.
7. Wimpee C, Stiekema W. Tobin E (1983). Sequence heterogeneity in the RuBP carboxylase small subunit gene family of *Lemna gibba*. In "Plant Molecular Biology" UCLA Symposium on Molecular and Cellular Biology New Series (ed) Goldberg RB. New York: Alan R. Liss, 12:391.
8. Dean C, van den Elzen P, Tamaki S, Dunsmuir P, Bedbrook J (1985). Linkage and homology analysis divides the eight genes for the small subunit of petunia ribulose-1, 5-bisphosphate carboxylase into three gene families. Proc Natl Acad Sci USA 82: 4964.
9. Dean C, van den Elzen P, Tamaki S, Dunsmuir P, Bedbrook J (1985). Differential expression of the eight genes of the petunia ribulose bisphosphate carboxylase small subunit multi-gene family. EMBO J 4: 3055.
10. Dean C, van den Elzen P, Tamaki S, Black M, Dunsmuir P, Bedbrook J (1987). Molecular characterization of the *rbcS* multi-gene family of *Petunia* (Mitchell). Mol Gen Genet in press.
11. Turner N, Clark W, Tabor G, Hironaka C, Fraley R, Shah D (1986). The genes encoding the small subunit of ribulose-1,5-bisphosphate carboxylase are expressed differentially in petunia leaves. Nuc Acid Res 14: 3325.
12. Kunkel T (1985). Rapid and efficient site-specific mutagenesis without phenotypic selection. Proc Natl Acad Sci USA 82: 488.
13. Jones J, Dunsmuir P, Bedbrook J (1985). High level expression of introduced chimaeric genes in regenerated transformed plants. EMBO J 4: 2411.
14. Alton N, Vapnek D (1979). Nucleotide sequence analysis of the chloramphenicol resistance transposen Tn9. Nature 282: 864.
15. De Greve H, Dhaese P, Seurinck J, Lemmers S, van Montagu M, Schell J (1983). Nucleotide sequence and transcript

map of the *Agrobacterium tumefaciens* Ti Plasmid-encoded octopine synthase gene. J Mol Appl Genet 1: 499.

16. Jones J, Grady K, Suslow T, Bedbrook J (1986). Isolation and characterization of genes encoding two chitinase enzymes from *Serratia marcescens*. EMBO J 5: 467.
17. van den Elzen P, Lee K, Townsend J, Bedbrook J (1985). Simple binary vectors for DNA transfer to plant cells. Plant Mol Biol 5: 149.
18. Marton L, Wullems G, Molendijk L, Schilperoot R (1979). In vitro transformation of cultured cells from *Nicotiana tabacum* by *Agrobacterium tumefaciens*. Nature 277: 129.
19. Horsch R, Fry J, Hoffman N, Eichholtz D, Rogers S, Fraley R (1985). A simple and general method for transferring genes into plants. Science 227: 1229.

Plant Gene Systems and Their Biology, pages 301–310

EFFECT OF 35S LEADER MODIFICATIONS ON PROMOTER ACTIVITY

D.A. Pierce, I.J. Mettler, A.R. Lachmansingh,
L.M. Pomeroy, E.A. Weck and D. Mascarenhas

Stauffer Chemical Co., 1200 S. 47th St.,
Richmond, CA 94804

ABSTRACT Several versions of the 35S promoter from CaMV have been attached to CAT or NEO marker genes and tested for promoter activity in maize and tobacco. Three versions were created by use of available restriction sites and differ only in the length of the leader region (3, 131 or 198 nucleotides) following the CAP site. These constructions showed striking differences in activity between maize and tobacco. The analysis was extended to include a series of Bal31 deletions which generated a wider range of leader lengths. The differences in expression shown by this series are not strictly a function of leader length. The activity of two of the leader constructions in growth-selected maize transformants showed differences in activity similar to those detected by transient gene expression.

INTRODUCTION

The Cauliflower Mosaic Virus (CaMV) has been widely used as a source for strong promoters in gene expression studies in plant cells (1,2). The work described here is an analysis of the activity of different length versions of the 35S promoter, which is known to be more active than the 19S promoter. Since the first open reading frame occurs over 500 bp downstream of the CAP site for the 35S RNA, we were interested in testing several different lengths of the leader for their effect on expression in both maize and tobacco. Previous work (3) has identified an enhancer region 110 bp upstream of the 35S promoter, but no study has been published on the leader region itself. Here we have made use of the transient assay system for chloramphenicol acetyltransferase (CAT) activity to monitor expression of 35S promoter constructions.

METHODS

Protoplasts were prepared from the maize cell culture line Black Mexican Sweet (BMS) or from leaves of the *Nicotiana tabacum* line SR1. Marker cassettes were introduced as supercoiled DNA into protoplasts by means of electroporation. Standard electroporation conditions are described in Fig. 1. For transient assays, acetylation of chloramphenicol (CAM) was determined by either of two methods, as indicated in Fig. 2. Marker cassettes carried the 3' region from the octopine synthase gene (OCS) of T-DNA and either the bacterial gene for chloramphenicol acetyltransferase (CAT) or the bacterial gene for neomycin phosphotransferase II (NEO). The CAT and NEO gene fragments contribute 15 and 30 bp, respectively, to the total length of untranslated leader in the final 35S promoter constructions. The false start ATG 16 bp upstream of the NPTII coding region in the original Tn5 version has been deleted. Tn9 served as the source of the CAT gene.

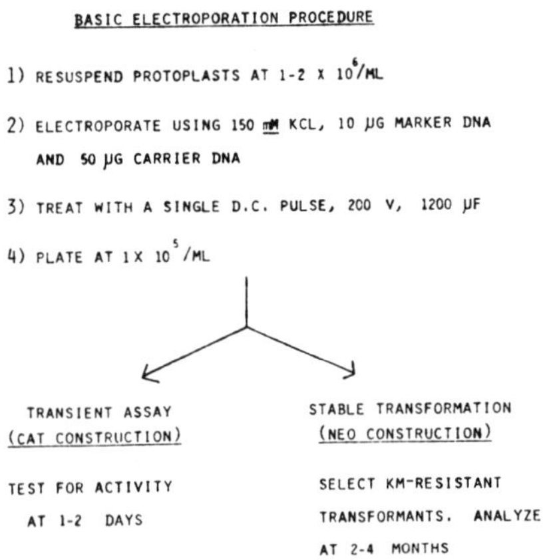

Figure 1. Basic electroporation procedure.

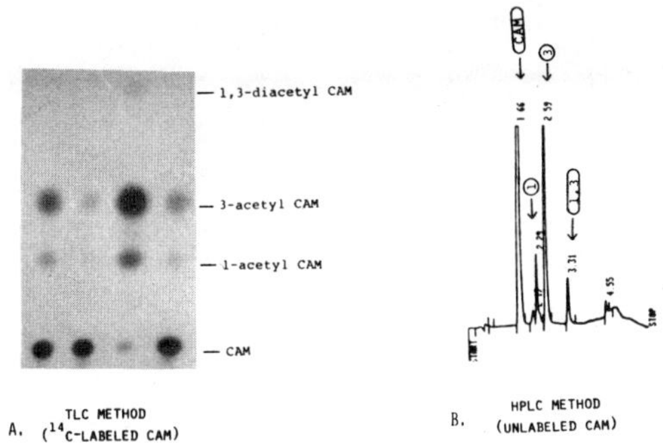

Figure 2. Methods for CAT assays. A) TLC method using ^{14}C-chloramphenicol (4): shown is an autoradiogram of labeled CAT assay products separated by thin-layer chromotography. Results are quantitated by scraping and counting regions corresponding to acetylated forms of chloramphenicol. B) HPLC method using unlabeled chloramphenicol (5): shown is a profile of unlabeled CAT assay products separated by HPLC on a C-18 column. Results are quantitated using a peak area integrator.

RESULTS

Electroporation and Assay Conditions

We have found that the activity of the marker cassette is increased when calf thymus DNA is included as carrier (Fig. 1). We also routinely include a 5 min. cooling period on ice between steps 3 and 4. The two assay methods we have used for CAT activity (Fig. 2) give very comparable results. The HPLC technique is linear over three orders of magnitude, and in our hands, is more reliable, less tedious and almost as sensitive as the TLC method.

35S Promoter Constructions

Fig. 3 shows the sequence in the region of the 35S promoter. The initial set of constructions made use of three available restriction sites (HphI, DdeI, RsaI) within the first 200 bp after the CAP site. They retain 3, 131 and 198 bp, respectively, of the original 35S leader. These promoter fragments were cloned into CAT-OCS cassettes and tested for

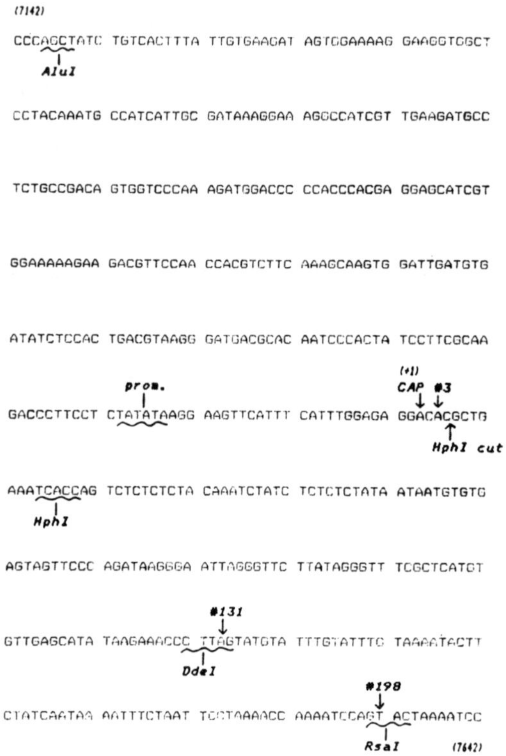

Figure 3. Sequence of CaMV in the region of the 35S promoter. The sequence shown includes nucleotides 7142-7642 of the CaMV genome (6). All subclones described in this paper were recloned from pDOB412, a caMV clone (Strasbourg strain) provided by K.E. Richards. Shown are the TATATA box, CAP site (7) and restriction sites used to generate the initial set of three leader lengths (HphI, DdeI, RsaI). T4 DNA polymerase was used to remove the one-base 3' overhang generated by HphI. DdeI and RsaI ends were treated with Klenow enzyme in the presence of all four deoxyribonucleotides. XbaI linkers were attached to each 3' end. The name of each subclone (e.g., #3) indicates the number of nucleotides of the original 35S leader present in that construction. The AluI site 288 bp upstream of the CAP site was used to provide the 5' end of the 35S promoter subclones.

activity. Results are shown in Table 1. It can be seen that the patterns are very different for BMS and tobacco and show a wide range of activity, especially for BMS. These results suggested that it might be useful to examine a more extensive range of leader lengths.

TABLE 1
EFFECT OF LEADER LENGTH ON PROMOTER ACTIVITY

Cut site	35S leader length (bp)	Total leader length (bp)[a]	Relative CAT activity[b] BMS	tobacco
HphI	3	41	17	100
DdeI	131	169	100	35
RsaI	198	236	7	61
no promoter	---	---	0.1	0.01

[a]Total leader length includes the indicated 35S region plus an additional 38 bp contributed by XbaI linkers and the CAT gene fragment.
[b]Relative CAT activity was determined by assigning a value of 100 to the most active construction in each cell line.

The scheme for using Bal31 nuclease to generate additional leader lengths is shown in Fig. 4. Six additional leader versions 13, 51, 72, 90, 105 and 157 bp in length were recovered. Their endpoints are shown in Fig. 5. Results on expression of the corresponding 35S CAT OCS cassettes are presented in Fig. 6. With the exception of the 198 bp leader which contains a functional polyadenylation signal, all lengths beyond 3 or 13 bp show higher activity in BMS. One clone in particular (131 bp) shows a very high level of activity in the maize line. The pattern in tobacco is extremely different and shows much less variation across leader length.

The effect of leader length is also seen in stable BMS transformants when the NEO gene is substituted for CAT. Analysis of expression at the protein level (Fig. 7A) shows a 10X higher average activity for the 131 bp leader compared to the 3 bp version. This is consistent with the 5-10X difference observed for transient CAT assays in BMS protoplasts (Table 1; Fig. 6). Preliminary results indicate that the RNA levels in these same transformed calli are also higher

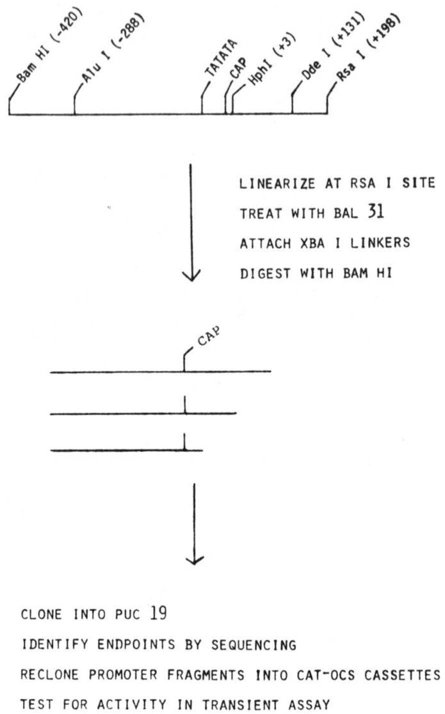

Figure 4. Scheme for Bal31 digestion. Linearized DNA was digested with Bal31 (0.1 u/ug) for 60, 90 or 120 sec. at 22°, phenol-extracted, and then treated with Klenow enzyme before attachment of linkers. Three size classes of Bal31-treated DNAs were recovered from a low-melting temperature agarose gel and recloned into pUC19. Dideoxysequencing on supercoiled DNA was performed using CsCl-purified or miniprep DNA. The endpoints of six clones used for further activity assays are indicated in Fig. 6. These six fragments were then cloned into CAT-OCS cassettes, using the upstream AluI site to generate the 5' end.

for the longer leader (Fig. 7B), but the average difference (3-4X) does not completely account for the 10X difference in NPTII activity.

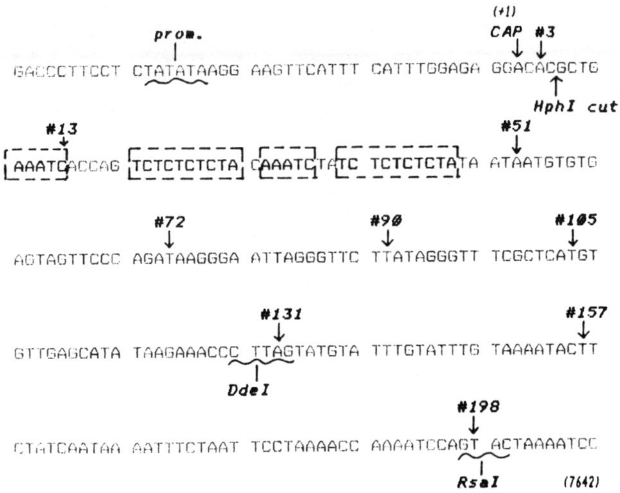

Figure 5. Position of Bal31 endpoints. The Bal31 series includes leader lengths of 13, 51, 72, 90, 105 and 157 bp in addition to the original lengths of 3, 131 and 198 bp. Tandem oligonucleotide repeats are indicated by boxed areas.

DISCUSSION

These results have shown that the activity of the 35S promoter varies with the length of the leader. The variation is more striking in BMS than in tobacco, but in neither system is the level of activity strictly proportional to leader length. At least for BMS, it is possible that there is a minimal leader length (longer than the leader in #3 or #13 cassettes) which is required for higher levels of expression. Examination of the CaMV sequence indicates that there is also an interesting region of tandem repeats within the first 47 bp of the leader (Fig. 5, boxed areas). Work is in progress to determine whether these repeats affect expression. At the present time, we do not know what additional factor(s) may be responsible for the particularly high activity of the 131 bp leader in BMS.

The effect of the leader is not unique to the CAT constructions or to the transient assay system in protoplasts, since similar results are seen in stable BMS transformants when the NEO gene is used. Although results on RNA analysis are preliminary, a comparison of NPTII and RNA levels sug-

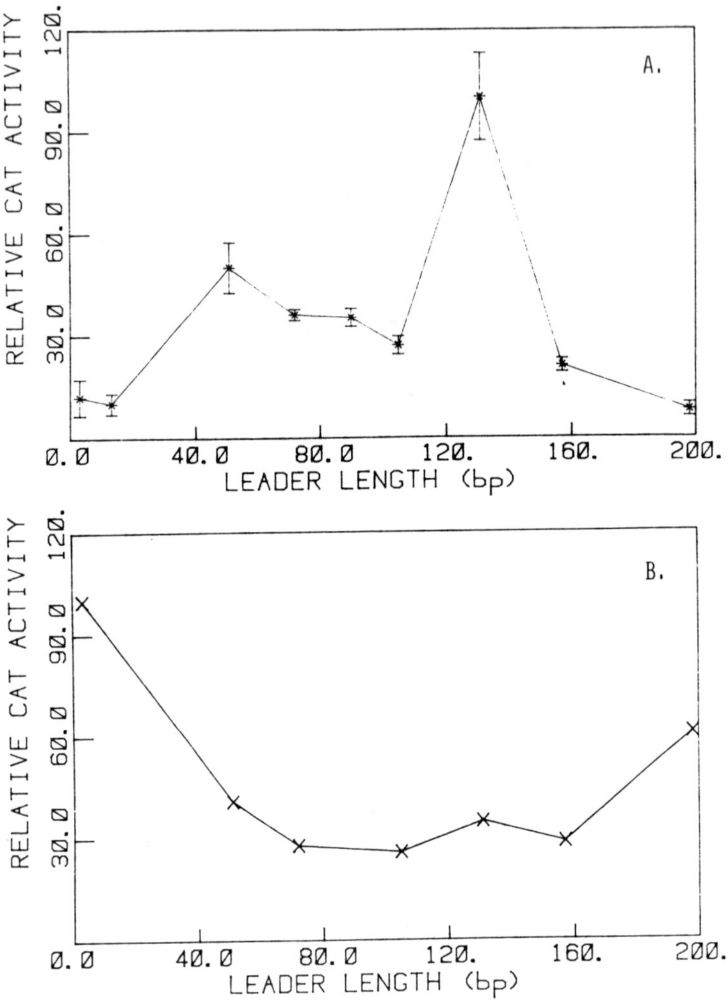

Figure 6. Activity of 35S leader series. A) Shown is the CAT activity profile of the 3, 13, 51, 72, 90, 105, 131, 157 and 198 bp leader versions in BMS. Standard errors for duplicate assays are included. Duplicates are independent electroporated protoplast samples using the same DNA construction. B) Shown is the CAT activity profile of the 3, 51, 72, 105, 131, 157 and 198 bp leader versions in tobacco. Leader length refers only to the nucleotides contributed from the 35S promoter fragment.

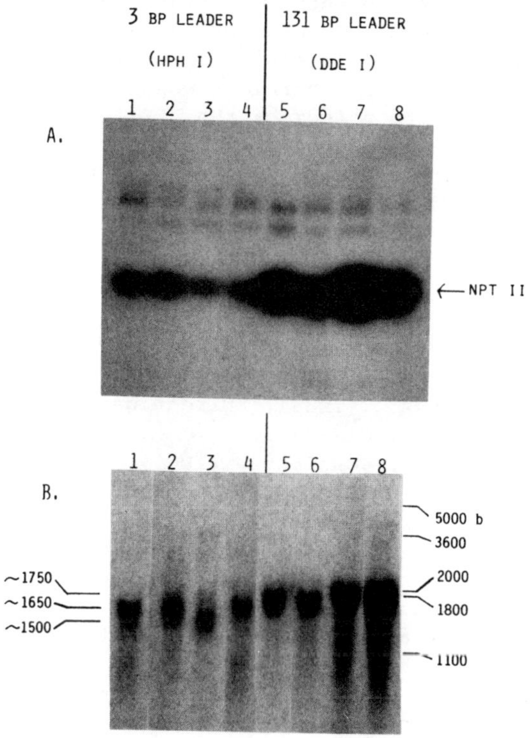

Figure 7. Analysis of activity in stable BMS transformants. The 3 and 131 bp leader versions were recloned into NEO-OCS cassettes and used for transformation of BMS protoplasts. Transformed calli were selected on 100 ug/ml kanamycin. Four independent calli from each leader construction were analyzed for expression (A) at the protein level by NPTII assays (8) and (B) at the RNA level by Northern analysis. For analysis of RNA, 30 ug of total BMS cell RNA were fractionated on each slot of a formaldehyde-agarose gel. The RNA pattern was transferred to a nitrocellulose filter and hybridized with a ^{32}P-labeled NEO probe.

gests that leader length may affect the efficiency of translation, at least in BMS.

The DNAs described here are chimaeric constructions

tested in two cell types which are not natural hosts for CaMV. Therefore, it is likely that these results are an indication of the requirements for high level expression of these 35S promoter cassettes in BMS and tobacco, rather than a reflection of the regulation of expression of CaMV itself in these cell types. Further work would need to be done to determine whether the effect of the 35S leader in BMS and tobacco reflects a more general difference between monocots and dicots. It would also be interesting to see whether variations in leader length can affect the activity of promoters from other sources.

REFERENCES

1. Paszkowski J, Shillito RD, Saul M, Mandak V, Hohn T, Hohn B, Potrykus I (1984). Direct gene transfer to plants. EMBO J 3:2717-2722.
2. Fromm M, Taylor LP, Walbot V (1985). Expression of genes electroporated into monocot and dicot plant cells. Proc Natl Acad Sci 82:5824-5828.
3. Odell JT, Nagy F, Chua N-H (1985). Identification of DNA sequences required for activity of the Cauliflower Mosaic Virus 35S promoter. Nat 313:810-812.
4. Gorman, C, Moffat L, Howard B (1982). Recombinant genomes which express chloramphenicol acetyl transferase in mammalian cells. Mol Cell Biol 2:1044-1051.
5. Young SL, Jackson AE, Puett D, Melner MH (1985). Detection of chloramphenicol acetyl transferase activity in transfected cells: a rapid and sensitive HPLC-based method. DNA 4:469-475.
6. Hohn T, Richards K, Lebeurier G (1982). Cauliflower Mosaic Virus on its way to becoming a useful plant vector. Current Topics in Microbiology and Immunology 96:193-236.
7. Guilley H, Dudley RK, Jonard G, Balazs E, Richards K (1982). Transcription of Cauliflower Mosaic Virus DNA: detection of promoter sequences and characterization of transcripts. Cell 30:763-773.
8. Reiss B, Sprengel R, Will H, Schaller H (1984). A new sensitive method for qualitative and quantitative assay of neomycin phosphotransferase in crude cell extracts. Gene 30:211-218.

Plant Gene Systems and Their Biology, pages 311–320

INTRODUCTION OF THE P-ELEMENT FROM *DROSOPHILA* INTO TOBACCO

J.M. Martinez-Zapater, R. Finkelstein and C.R. Somerville

MSU-DOE Plant Research Laboratory
Michigan State University
East Lansing, MI 48824

ABSTRACT In order to clone genes from higher plants by transposon tagging it would be advantageous to have a transposable element in which transposition is regulated by an external stimulus. A possible candidate for such an element is the P-element of *Drosophila*, one of the most thoroughly characterized eukaryotic transposable elements. We have employed a P-element derivative in which the transposase gene is under the control of the *Drosophila* 70 Kd heat shock protein (HSP70) promoter and examined the expression of this construct in transgenic tobacco plants. Our preliminary results indicate that either the transcription or the processing of the transcript is abnormal in tobacco and that the P-element does not excise at a high frequency, if at all.

INTRODUCTION

In those plants which are amenable to genetic studies various strategies have been proposed that would allow the

[1] This work was supported in part by a Fulbright fellowship from the Spanish Government to JMZ, a National Science Foundation postdoctoral fellowship to RF and grants from the NSF (#PCM8351595) and the U.S. Department of Energy (#AC02-76ER01338).

isolation of any gene which can be defined by a mutation. For instance, once a mutation has been characterized and mapped to a chromosome, the gene might be isolated by walking along the chromosome from a linked restriction fragment length polymorphism (RFLP). This will probably be feasible only in species such as Arabidopsis which has a small genome and litle repetitive DNA (1,2). To determine whether the gene corresponding to the mutant allele has been isolated it would be necessary to demonstrate complementation of the mutant phenotype by the cloned DNA sequence. Thus, a disadvantage of this approach is the necessity of many rounds of plant transformation and regeneration.

A potentially attractive alternative is to employ insertional mutagenesis, using either the T-DNA of the Ti plasmid or a transposable element to induce mutations. A fragment from the element can then be used as a molecular probe to isolate the affected gene from a genomic library of the mutant plants. However, because each transformation event results in only one or a few independent insertions of T-DNA into the plant genome, it would be necessary to have a very efficient system for regenerating large numbers of independent transformants in order to use the Ti plasmid in a directed search for mutants. Also, the presence of somaclonal variation in regenerated plants may confuse the assignment of the cause of mutant phenotypes to T-DNA insertion. Thus, depending on the frequency of transposition, the use of an endogenous element may be a more efficient method of insertional mutagenesis. Indeed, only transposable elements have been effectively used to clone plant genes on the basis of phenotype (3,4,5,6).

The choice of a suitable plant transposon is relatively limited. Maize contains several genetically well characterized transposable element systems, one of which (Ac) has been demonstrated to be active in tobacco (7). Although Ac may prove to be very useful as a broad host range plant transposon, several features of this and other known plant transposable elements must be improved to make them efficient mutagens. Of particular concern are the frequency of transposition, stability of the transposon induced mutations, randomness of the transposition target and transposon copy number. Although transposition randomness probably can not be improved at present, the other features may, in principle, be modified by altering well characterized transposable elements and using them in an appropriate plant host.

Only a few eukaryotic transposable elements are sufficiently well-characterized structurally and functionally to allow their modification. We chose the P-element of Drosophila for the following reasons. First, the element is autonomous in Drosophila, but we can not exclude the possibility that species-specific cofactors are required. Second, although transposition is postranscriptionally regulated in Drosophila at the level of intron splicing (8), deletion of the germ-line specific intron relegates control to the transposase promoter. Transposition frequency can be increased by placing expression of the transposase gene under control of a heat-shock inducible promoter (9). The system has been manipulated succesfully in Drosophila to use as a transposon tagging system and for transformation of germ-line cells with DNA flanked by P-element borders (9,10). Third, recent results indicate that transposition is non-replicative and that excision of the element is highly correlated with its insertion in a different part of the genome (9). This limits the number of copies that can be acumulated per genome.

RESULTS

The Drosophila P-element contains four exons which encode the transposase (11, 12). Transposition is normally restricted to germ-line cells because, although the first and second introns are correctly processed in all cells, the third intron is removed only in germ-line cells (8). In order to reduce potential problems with incorrect splicing of the P-element transcript by plant cells we used a modified P-element in which the third intron had been removed by in vitro mutagenesis (8). Although an element lacking all three introns has been constructed, it does not transpose in Drosophila for unknown reasons (Laski, pers. com.). In order to control expression of the P-element in plants, the P-element promoter (8) was replaced by the promoter from the HSP70 gene of Drosophila. When the P-transposase gene is placed under the control of this promoter in Drosophila, transposition increases ten-fold after heat-shock induction (9). This promoter is also heat-shock inducible in Nicotiana (13).

Although replacement of the P-element promoter by the heat-shock inducible promoter should prevent transposase

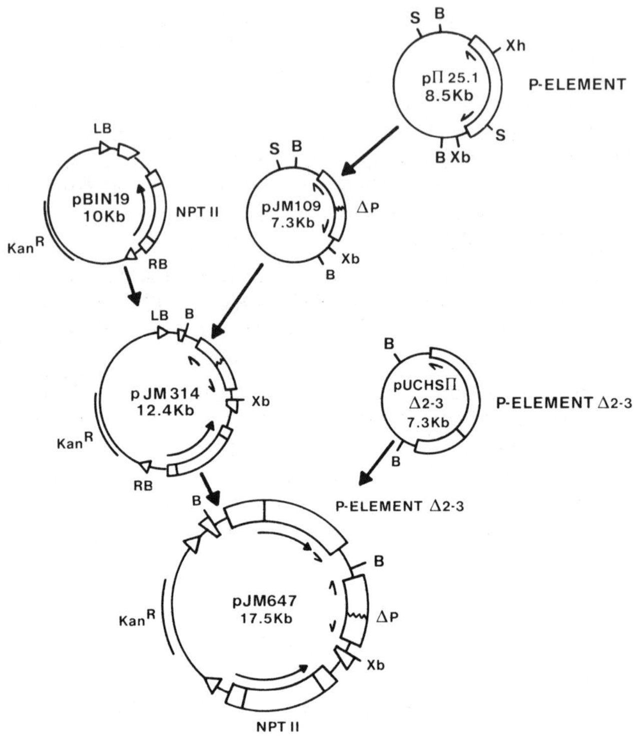

Polylinker: EcoR I, Sst I, Kpn I, Sma I, Bam HI, Xba I, Sal I, Hind III

B = Bam HI
S = Sal I
Xb = Xba I
Xh = Xho I

Figure 1. Construction of a binary Agrobacterium vector carrying a heat shock inducible P-element. The defective P-element was created by deletion of the internal XhoI-SalI restriction fragment of the plasmid pπ25.1 (11). A BamHI-XbaI fragment containing this defective P-element was ligated into the polylinker of the binary vector pBIN19 (15). Finally a BamHI fragment from pUChsπΔ2-3 (8) containing the P-element lacking the third intron under the control of the Drosophila heat-shock promoter was ligated into the unique BamHI restriction site of pJM314.

expression under non induced conditions, basal transcriptional activity of this promoter (9,14) could produce instability of P element-produced mutations which might complicate the subsequent characterization of the mutants. Therefore, we subdivided the transposon into two components which can be separated by genetic recombination or segregation in the generations following the transposition event. One component, the transposase gene under heat-shock control, was designed to provide heat shock inducible transposase activity but can not transpose because it lacks the 5'*cis*-acting 31 bp inverted repeat that is necessary for transposition (11). The other component has an internal deletion in its coding region, which prevents synthesis of a functional transposase, but it can be mobilized in *trans* by the heat-shock inducible transposase activity. This two component system should permit a transposon-induced mutation to be stabilized by genetic segregation from the active transposase gene that otherwise could be a source of instability.

The general scheme of the plasmid construction is outlined in Figure 1. The final plasmid, pJM647, carries both the heat-shock inducible transposase gene and the defective P-element between both T-DNA borders. It also carries a constitutively expressed neomycin phosphotransferase II gene (NPT-II) as a selectable marker for plant transformation. This plasmid and pBIN19 as a control were introduced into *Nicotiana* cells by *Agrobacterium* mediated transformation (16). In the preliminary experiments, transformed tumor lines were used to analyze the expression of the chimaeric gene. For this purpose we used the *Agrobacterium tumefaciens* strain GV3111 carrying the tumorigenic octopine plasmid pTiB6S3 (17). When this strain carries another plasmid with T-DNA borders, it is capable of mobilizing both T-DNAs to the plant genome. Transformed tumors were selected on hormone-free medium containing kanamycin. Transformed plants were obtained in parallel experiments in which pJM647 and pBIN19 were introduced into the *Agrobacterium* strain LBA4404 which carries a disarmed Ti plasmid in which the entire T-DNA region has been deleted (18).

Expression of the chimaeric transposase gene and the NPT-II gene was analyzed by Northern blots of RNA from transformed tumor tissue. Several tumor lines transformed with either pJM647 or pBIN19, were heat-shocked at 40°C for 1 hour. The results are shown in Figure 2. The transposase transcript induced by the heat-shock promoter should be

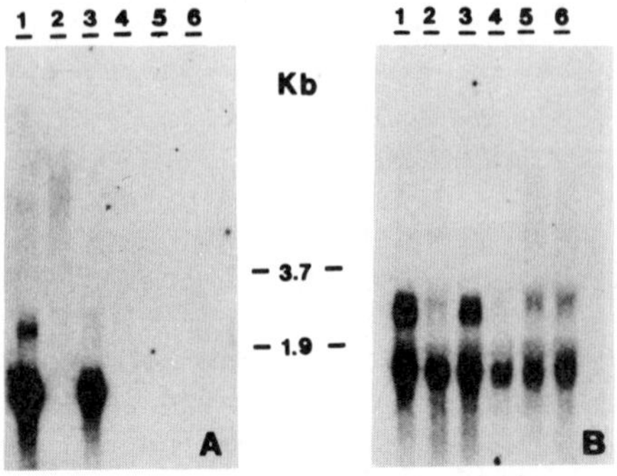

Figure 2. Heat-shock induction of P element and NPT transcripts in transgenic tobacco tumors. Total RNA (10 ug per lane) was analyzed in Northern blots using nick-translated transposase (A) or NPT (B) probes. RNA was isolated from both heat-shocked (1,3,5) and non heat-shocked (2,4,6) tissue. Lanes 1-4 contain RNA from two independent tumor lines transformed with pJM647. Lanes 5-6 contain RNA from a pBIN19 transformed tumor line.

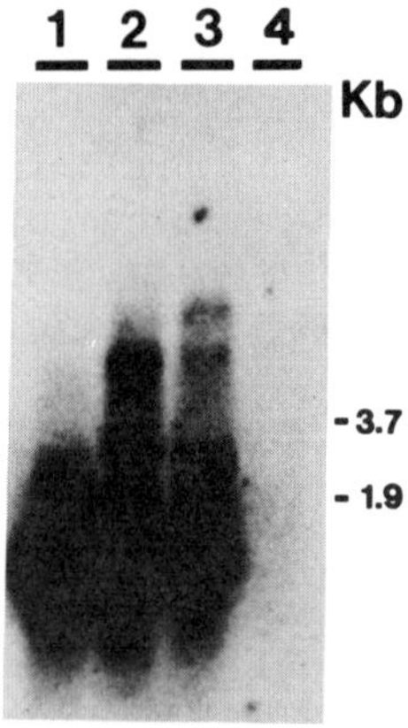

Figure 3. Heat-shock induced P element transcripts in transgenic tobacco. RNA was isolated from heat-shocked tumor transformed with pJM647 (lane 1), and from heat-shocked leaves of two plants transformed with pJM647 (lanes 2 and 3) and one plant transformed with pBIN19 (lane 4).

aproximately 2.5 Kb in length, whereas the NPT-II transcript is 1.6 Kb. Although the major transcript found when hybridizing with the transposase gene is about 1.5 Kb in length, another transcript of approximately 2.5 Kb can also be seen in these samples. Both are induced by heat-shock and are specific to P element transformed tumors. The NPT-II probe hybridizes with a 1.6 Kb transcript that is present in all transformed tumors. A slight induction of this transcript can be observed in the heat-shocked samples of P-element transformed tumors. Also, a longer transcript of more than 2.5 Kb hybridized with the NPT gene in these samples.

In order to see if the same transcript pattern was found in transformed plants, we repeated the experiment with total RNA isolated from heat-shocked leaves of P-element transformed plants. The results are shown in Figure 3. Both transposase transcripts are again found in heat-shocked leaves and low abundance longer transcripts also appeared. No transcript hybridized with the transposase gene in total RNA from a heat-shocked pBIN19 transformed plant.

DISCUSSION

Results presented here indicate that the transposase gene, placed under the control of the *Drosophila* HSP70 promoter, is heat-shock inducible in *Nicotiana* cells. There was also some apparent heat-shock induction of the NPT transcript in those tumor lines transformed with the P-element construct. This could be due to an enhancer effect of the HSP70 promoter on the promoter of the NPT gene. It has recently been shown that this promoter exhibits enhancer activity in mammalian cells (19). However, confirmation of this requires additional work.

Although the transposase gene is heat-inducible, the number and size of the transcripts indicate that they are not properly processed by the tobacco cells. At least three hypotheses can be proposed to explain the origin of the abundant 1.5 Kb transcript. First, there could be premature termination of transcription. Second, incorrect polyadenylation of the transcript could produce a truncated mRNA. For instance, a sequence AATAAA resembling the animal consensus polyadenylation signal is located aproximately 1.3 kb downstream from the promoter of the transposase gene. In general, polyadenylation signals of plant genes are more variable than those of animal genes, but the animal polyadenylation consensus signal is found at the expected position in many plant genes (20). Plants transformed with animal genes do not always correctly and efficiently polyadenylate the foreign transcripts, but polyadenylation generally occurs close to the correct site or at a downstream polyadenylation signal (21). However in one case the transcripts were polyadenylated within the coding region, although close to the 3'end (21).

A third possibility is that the shorter transcript results from incorrect splicing of the two introns in the P-element gene used for these experiments. If the transcript were spliced so that both introns and the internal exon were simultaneously removed, the result could be a transcript of the observed size. Intron splicing sequences in plant genes have recently been reviewed (22). The consensus sequences for the 5' and 3' splice junctions are very similar to those of animals with the exception that a polypyrimidine stretch frequently found upstream of the 3' splice sites in animals, is not as highly conserved in plant intron sequences. Plant pre-mRNAs can be spliced _in vitro_ by a HeLa cell free extract (22), but plant cells could not splice the introns of the human growth hormone pre-mRNA _in vivo_ (23). Recently, Keith and Chua (24) reported that a monocot intron is not efficiently spliced in a dicot species; however when the intron is spliced, it is at the correct site. P-element introns are very short, even smaller than the minimum size necessary for correct intron splicing in animal cells (25). Preliminary analysis of the transcripts from our transgenic tobacco lines suggests that this is not the correct explanation.

The size of the 2.5 Kb transcript can not be precisely resolved by Northern blot analysis to determine whether the two small introns have been correctly processed. If this is a correct transposase transcript, it could result in transposase activity. For this reason we have attempted to detect excision and insertion events at the DNA level. In one experiment, leaf discs of transformed plants were subjected daily to a one hour heat-shock to induce transcription of the transposase gene. Preliminary analyses of the plants regenerated from this material have not shown any evidence of excision or transposition. In another experiment, we are examining the possibility that transposition can be induced during gametogenesis by heat-shocking whole plants during differentiation of the flowering meristems. The progeny of these heat-shocked plants will also be analyzed at the DNA level. Fortunately, the cDNA clone and antibodies against polypeptides encoded by the different open reading frames of the P element are now available (26) and should be very useful in determining the reason for the lack of transposition. Until additional information is available it is not possible to predict whether further modifications of the P element will be required to activate transposition in plant cells.

ACKNOWLEDGMENTS

We are grateful to G. Rubin for providing plasmids pUChs$\pi\Delta$2-3 and pπ25.1 and M. Bevan for pBIN19.

REFERENCES

1. Leutwiler LS, Hough-Evans BR, Meyerowitz EM (1984) Mol Gen Genet 194:15.
2. Meyerowitz EM, Pruitt RE (1985) Science 229:1214.
3. Fedoroff N, Furtek D, Nelson O (1984) Proc Natl Acad Sci USA 81:3825.
4. O'Reilly C, Shepherd NC, Pereira A, Schwarz-Sommer Zs, Bertram I, et al. (1985) EMBO J 4:877.
5. Paz-Ares J, Wienand U, Peterson PA, Saedler H (1986) EMBO J 5:829.
6. Wienand U, Weydemann U, Niesbach-Kloesgen U, Peterson PA, Saedler H (1986) Mol Gen Genet 203:202.
7. Baker B, Schell J, Lorz H, Fedoroff N (1986) Proc Natl Acad Sci USA 83:4844.
8. Laski FA, Rio DC, and Rubin GM (1986) Cell 44:7.
9. Steller H, Pirrotta V (1986) Mol Cell Biol 6:1640.
10. Rubin GM, Spradling AC (1982) Science 218:348.
11. O'Hare K, Rubin GM (1983) Cell 34:25.
12. Karess RE, Rubin GM (1984) Cell 38:135.
13. Spena A, Hain R, Ziervogel U, Saedler H, Schell J (1985) EMBO J 4:2739.
14. Steller H, Pirrotta V (1985) EMBO J 4:3765.
15. Bevan M (1984) Nuc Acid Res 12:8711.
16. Horsch RB, Fry JE, Hoffmann NL, Eichholtz D, Rogers SG, Fraley RT (1985) Science 227:1229.
17. Horsch RB, Fraley RT, Rogers SG, Sanders PR, Lloyd A, Hoffmann N (1984) Science 223:496.

18. Hoekma A, Hirsch PR, Hooykas PJJ, Schilperoort RA (1983) Nature 303:179.
19. Bienz M, Pelham HRB (1986) Cell 45:753.
20. Dean C, Tamaki S, Dunsmuir P, Favreau M, Katayama C, Dooner H, Bedbrook J (1986) Nuc Acid Res 14:2229.
21. Hunt AG, Chu NM, Odell JT, Nagy F, Chua NH (1987) Plant Mol Biol 8:23.
22. Brown JWS, Feix G, Frendewey D (1986) EMBO J 5:2749
23. Barta A, Sommergruber K, Thompson D, Hartmuth K, Matzke MA, Matzke AJM (1986) Plant Mol Biol 6:347.
24. Keith B, Chua NH (1986) EMBO J 5:2419.
25. Green MR (1986) Ann Rev Genet 20:671.
26. Rio DC, Laski FA, Rubin GM (1986) Cell 44:21.

Plant Gene Systems and Their Biology, pages 321–329

VARIABILITY IN 35S PROMOTER EXPRESSION BETWEEN INDEPENDENT TRANSFORMANTS

Joan T. Odell,[1], Ferenc Nagy, and Nam-Hai Chua

Laboratory of Plant Molecular Biology
The Rockefeller University
New York, New York 10021

ABSTRACT The level of expression from the cauliflower mosaic virus 35S promoter was found to vary substantially in independent transgenic plants containing the same 35S-P-CAT chimeric gene construction. The expression level did not correlate with the number of copies of the integrated gene, but did generally correlate with the level of expression of a co-introduced NOS-P-nptII chimeric gene. Progeny of a plant with multiple integrated gene copies and high CAT activity showed a range of CAT activities demonstrating the segregation of 35S-P-CAT gene copies that differ in expression level.

INTRODUCTION

The development of a plant transformation system which makes use of the T-DNA transfer properties of *Agrobacterium tumefaciens* has provided a method for analyzing the expression of foreign genes in plant cells. This system has been used to analyze functional sequences of plant gene promoter regions by comparing the *in vivo* expression of different constructions made *in vitro* (1-4). This type of study requires careful execution and interpretation since the *A. tumefaciens* transformation system itself demonstrates uncontrolled properties, specifically the

[1]Present address: Du Pont Experimental Station, Wilmington, Delaware 19898

number and sites of T-DNA integrations into the plant genome. Different T-DNA vectors and transformation procedures have generated integration numbers of from one to numerous (20 or more) copies per genome of introduced foreign genes. Evidence thus far suggests that the site of integration is random (5,6). Position effects have been proposed to explain the reported variability between transgenic plants in the expression of several promoters including those from pea rbcs-E9, *Nicotiana plumbaginifolia* rbcS-8B, petunia Cab-22R, and nopaline synthase (7-10). The first three promoters are all highly regulated by light, and the fourth has low constitutive activity.

The 35S promoter from cauliflower mosaic virus was shown to be highly active (approximately 30-fold more active than the NOS promoter) in a constitutive manner when samples of pooled transformed callus lines or pooled transgenic plants were assayed (11). By mixing material from independent transformants, any effects of copy number or integration site were minimized. We have now assayed individual transgenic plants separately to assess the extent of variable expression from the 35S promoter. Using chloramphenicol acetyltransferase activity as a marker we have found a wide range in the 35S promoter expression level. However, whether the level is low or high, it retains its constitutive nature. The copy number of the integrated 35S-P-CAT gene was determined for individual plants and was found to have no correlation with expression level. Since this result suggested that strong position effects were acting on the 35S-P-CAT gene, we assessed the activity of a co-introduced gene and also followed CAT activity in the progeny of two plants.

Plasmid Construction and Plant Transformation

As a marker for 35S promoter (35S-P) activity we used the bacterial chloramphenicol acetyltransferase (CAT) coding region. The chimeric 35S-P-CAT-ss 3' (ss 3' representing the polyadenylation signal of the pea rbcS-E9 gene) construction and addition to the pMON145 Ti-plasmid vector (provided by S. Rogers), which contains a nopaline synthase promoter-neomycin phosphotransferase II coding region-nopaline synthase polyadenylation signal (NOS-P-nptII-NOS 3') chimeric gene and a T-DNA homology region (LIH), were described previously (11). This report describes results using the 35S-P fragment that extends to

-941. The constructed pMON145-CAT plasmid is shown in Figure 1. The transfer of this plasmid to *A. tumefaciens* and the generation of transgenic plants was described previously (11).

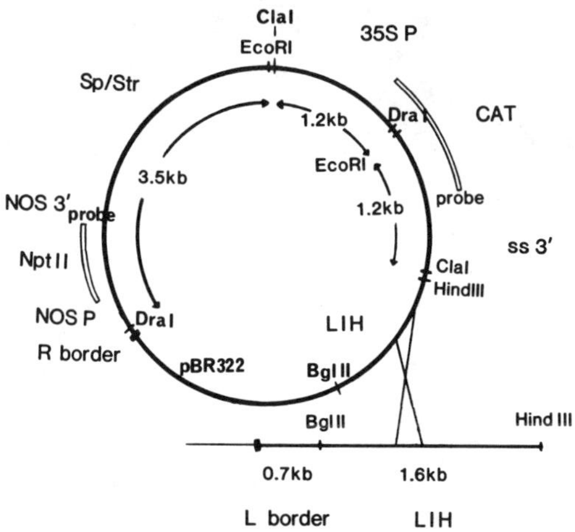

FIGURE 1. Map of pMON145-CAT construction also showing its integration into the T-DNA of the Ti-plasmid. LIH stands for Left Inside Homology and is a BgIII-HindIII fragment inside the left border of T-DNA. Sp/Str is the spectinomycin/streptomycin resistance marker. CAT and nptII Southern blot probes are shown.

35S-P-CAT Gene Analysis

Individual transgenic tobacco plants were assayed for CAT enzyme activity. Figure 2A shows that in leaves of individual plants, the level of CAT activity varies dramatically from slightly above the untransformed control background level in plants 3 and 4, to moderate in plants 7 and 10, higher in plant 5, and highest in plant 15. The range in activities is approximately 30-fold, based on activities of CAT enzyme dilutions (J. Odell, unpublished observation). CAT activities in the roots, stems, and leaves of the low activity plant 3 and the high activity plant 5 are constant for each plant demonstrating no organ-specific differences in expression directed by the 35S promoter. Individual transgenic petunia plants showed similar variability in CAT activity levels (Fig. 2B).

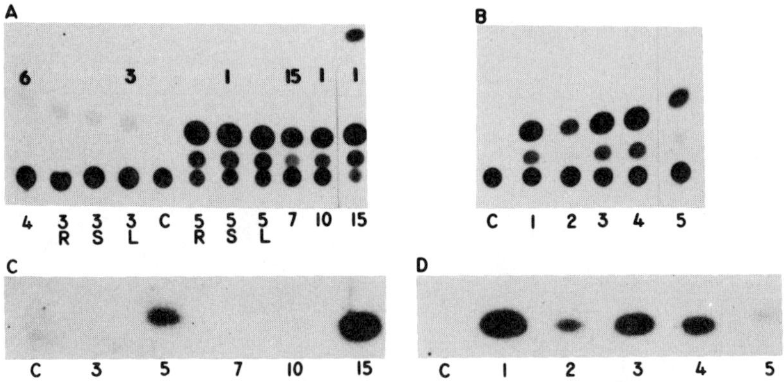

FIGURE 2. A) Cat assays (13) on 2µg of protein from leaf extracts of independent tobacco transformants, as well as root (R) and stem (S) extracts from two plants. Plants are numbered 4,3,5,7,10, and 15. The 35S-P-CAT gene copy number is shown above for each plant. B) CAT assays on leaf extracts from independent petunia transformants. C) NptII assays (14) on 75µg of protein from leaf extracts of the same tobacco plants as in A). D) NptII assays on leaf extracts from the same petunia plants as in B).

To determine whether the differences in CAT activity among transgenic tobacco plants were due to differences in numbers of the 35S-P-CAT gene incorporated into the genomes of the different plants, DNA from each plant was analyzed on a Southern blot. The DNA was digested with EcoRI releasing a 1.2kb fragment that includes the 35S promoter and one third of the CAT coding region (see Fig. 1). When the hybridization of a CAT probe to this 1.2kb fragment is compared to copy number standards, the plants show the following approximate 35S-P-CAT gene copy numbers: plant 3 has 3 copies, plant 4 has 6 copies, plant 5 has 1 copy, plant 7 has 15 copies, and plant 10 has 1 copy (Fig. 3A). Plant 15 also has only 1 copy (data not shown). Clearly the gene copy number does not correlate with CAT enzyme activity. Hybridization of the same filter, after probe elution, to a fragment from the mitochondrial ATPase beta subunit gene (12) gave equal signals from each DNA lane confirming that equal amounts of DNA from each plant were transferred to the Southern blot (data not shown).

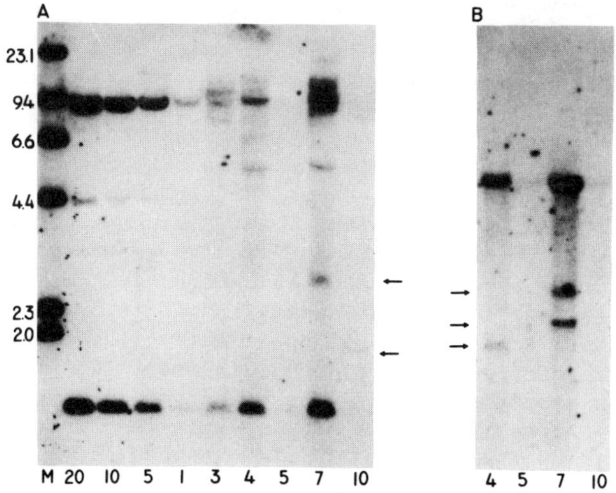

FIGURE 3. A) Southern blot (15) of DNA prepared from independent tobacco transformants 3,4,5,7, and 10. DNA was digested with EcoRI and the blot probed (16) with a nick-translation labeled (17) CAT fragment (see Fig. 1). Lane M is HindIII digested lambda DNA markers. The next four lanes are pMON145-CAT plasmid digested with EcoRI for copy number standards of 20,10,5, and 1. Arrows point to fragments of unexpected sizes (see text). B) Southern blot of DNA from tobacco transformants 4,5,7, and 10 digested with DraI and probed with an nptII fragment (Fig. 1). Arrows are described in A.

Evidence for multiple insertion sites of the 35S-P-CAT gene in the plant genome is also shown in Figure 3A. The hybridizing higher molecular weight bands represent DNA fragments extending from the EcoRI site in the CAT coding region, through the LIH region and T-DNA left border, to the next EcoRI site in the adjacent plant DNA (see Fig. 1). The multiple high molecular weight fragments in lanes with multiple 35S-P-CAT gene copies represent integrations into different sites of the plant genome. DNA from plants 7 and 10 each show a hybridizing band which represents a fragment too small to be a normal border fragment. This data suggests that there is a deletion in the region between the

EcoRI site in the CAT gene and the left border in some copies of integrated foreign DNA.

NOS-P-NPTII Gene Analysis

To determine whether the level of expression of the co-introduced NOS-P-nptII gene corresponds to that of the 35S-P-CAT gene, nptII enzyme activity was assayed in leaf tissue of the same individual transgenic tobacco and petunia plants. Figure 2C shows that tobacco plant 15 has the highest nptII activity and plant 5 has the next highest activity. Activities in plants 7 and 10 are very low and cannot be seen in this particular exposure, while no activity is detected by this assay in plants 3 and 4 (4 not shown). This relationship of nptII activities among the tobacco transformants correlates with the relative CAT activities described above. The same type of corresponding nptII and CAT activities is also seen in the transgenic petunia plants (Fig. 2D).

Southern blot analysis of individual plant DNAs with an nptII probe indicates that this gene is present in approximately the same copy number as the 35S-P-CAT gene (Fig.3B). The DNA was digested with DraI producing a 4.7kb fragment that includes the NOS-P-nptII gene, spec/strep marker and part of the CAT gene (see Fig.1). Plants with low copy number have only the expected 4.7kb fragment while in the high copy number plants 4 and 7, most of the copies are intact but there are also smaller hybridizing fragments. This result indicates that deletions have occurred in some copies of the integrated foreign DNA.

Progeny Analysis

Plants 5 and 7 were each self-fertilized and the seeds were germinated on kanamycin to assay inheritance of the NOS-P-nptII gene. Progeny of plant 5 show a 3:1 ratio of inheritance for kanamycin resistance, as expected for a single dominant trait. When assayed for CAT enzyme activity, all of the kanamycin resistant seedlings show the same level of CAT expression, which is similar to that of the parent plant 5 (data not shown). Seedlings germinated without selection show 3:1 inheritance of CAT expression (Fig. 4A).

Progeny of the high copy number plant 7 show a range of kanamycin resistance levels as determined by their rates of growth on kanamycin medium (data not shown). Unselected seedlings have greatly varying levels of CAT activity (Fig. 4B). We have not yet determined by Southern blot analysis how the various intact and deleted NOS-P-nptII and 35S-P-CAT gene copies are segregating in these progeny.

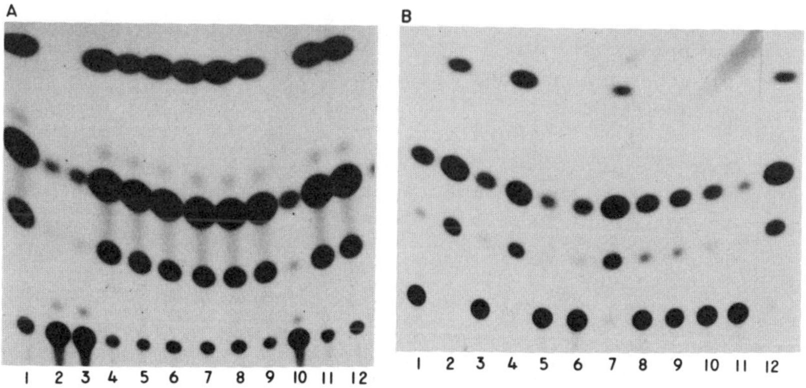

FIGURE 4. A) CAT assays on 2µg of protein from extracts of 12 progeny resulting from selfing of tobacco transformant 5. In this long exposure the activity seen in progeny 2,3 and 10 is the level of control extracts. B) CAT assays on extracts from 12 progeny resulting from selfing of tobacco transformant 7.

DISCUSSION

This study elucidates properties of the _A. tumefaciens_ transformation system which must be taken into account in any analysis of gene expression that utilizes this procedure. Our first observation was that the expression level of the same chimeric gene in individual transgenic plants varies substantially. This phenomenon was true both for the NOS-P-nptII and the 35S-P-CAT gene constructs. The differences in expression levels could not be explained by gene copy number since there was no correlation between these two factors. We have analyzed plants with low copy numbers that have low, medium, and high expression levels. Our high copy number plants have low and medium expression levels. Some seemingly full-size copies of an integrated gene may not be functional due to small rearrangements or mutations that are not detectable on Southern blots. Two plants with one copy of each of the chimeric genes (5 and 15) have the highest levels of expression from both genes, even higher than levels from a plant with 15 copies of each gene. This type of effect can be explained by influences on the promoters of the surrounding plant DNA at the particular site if integration. Whether the 35S promoter is expressed at a low or a high level, it retains its constitutive property.

In our construction, the NOS promoter is adjacent to the right T-DNA border making it very close to plant genomic DNA after integration, while the 35S promoter is approximately 2kb downstream and oriented in the same direction. Thus the influence of the surrounding plant DNA acts over at least several kilobases. Other studies report that the expression levels of two co-transferred genes do not correlate (9,10). In the construction of Jones et al. the two assayed promoters are approximately 8kb apart and oriented in opposite directions. If this positioning is the explanation for the difference in results, each vector construction may have its own characteristics for expression of inserted genes.

Different A. tumefaciens vector systems do have characteristic differences in the numbers of foreign gene copies that are transferred to the plant. The system used in this and our earlier studies (1,11) gives highly variable integrated copy numbers. Other systems give more constant, low levels of integrated copies (9, and F.N. unpublished observation). Therefore characterization of each system is warranted. Since we have found deletions that may inactivate some transferred DNA copies, particularly when there are high copy numbers, a low copy number transfer system is more desirable in terms of stability. Insertions of low copy number also show simple, predictable inheritance, as exemplified in our plant 5. The variable levels of integrated foreign gene expression in the progeny of our high copy number plant 7 indicate that there are unlinked gene copies that have different levels of expression. In conclusion, this study of the highly and constitutively expressed 35S promoter demonstrates that it is highly susceptible to factors that cause variation of expression level in independent transformants.

ACKNOWLEDGMENTS

We thank M. Wong and I. Roberson for technical assistance and C. J. Mauvais for critical reading of the manuscript. This work was supported by a grant from Monsanto Company.

REFERENCES

1. Odell JT, Nagy F, Chua N-H (1985). Nature 313:810.
2. Fluhr R, Kuhlemeier C, Nagy F, Chua, N-H (1986). Science 232:1106.
3. Timko MP, Kausch AP, Castresana C, Fassler J, Herrera-Estrella L, Van den Broeck G, Van Montague M, Schell J, Cashmore AR (1985). Nature 318:579.
4. An G, Ebert PR, Yi B-Y, Choi C-H (1986). Mol Gen Genet 203:245.
5. Zambryski P, Depicker A, Kruger K, Goodman HM (1982). J Mol Appl Genet 1:361.
6. Chyi Y-S, Jorgensen RA, Goldstein D, Tanksley SD, Loaiza-Figueroa F (1986). Mol Gen Genet 204:64.
7. Nagy F, Morelli G, Fraley RT, Rogers SG, Chua N-H (1985). EMBO J 4:3063.
8. Poulsen C, Fluhr R, Kauffman JM, Boutry M, Chua N-H (1986). Mol Gen Genet 205:193.
9. Jones JDG, Dunsmuir P, Bedbrook J (1985). EMBO J 4:2411.
10. An G (1986). Plant Physiol 81:86.
11. Nagy F, Odell JT, Morelli G, Chua N-H (1985). In Zaitlin M, Day P, Hollaender A (eds): "Biotechnology in Plant Science", Academic Press, p 227.
12. Boutry M, Chua N-H (1985). EMBO J 4:2159.
13. Herrera-Estrella L, Van den Broeck G, Maenhaut R, Van Montague M, Schell J (1984). Nature 310:115.
14. Van den Broeck G, Timko MP, Kausch AP, Cashmore AR, Van Montague M, Herrera-Estrella L (1985). Nature 313:358.
15. Southern EM (1975). J Molec Biol 98:503.
16. Thomashow MF, Nutter RC, Montoya AL, Gordon MP, Nester EW (1980). Cell 19:1729.
17. Rigby PWJ, Dieckmann M, Rhodes C, Berg P (1977). J Molec Biol 133:237.

Plant Gene Systems and Their Biology, pages 331–338

WOUND-INDUCED EXPRESSION OF PROTEINASE INHIBITOR II IN POTATO AND TRANSGENIC TOBACCO PLANTS

J. Sanchez-Serrano, M. Keil, H. Pena-Cortes, M. Rocha-Sosa, and L. Willmitzer

Institut für Genbiologische Forschung Berlin, Ihnestr. 63, 1000 Berlin 33, W.-Germany

ABSTRACT The systemic induction of proteinase inhibitor II expression was studied in potato as well as in transgenic tobacco plants. In potato wounding of either tuber or leaf tissue induces its expression in upper and lower leaves as well as in the upper part of the stem. When a potato proteinase inhibitor was transferred into tobacco using Agrobacterium/Ti-plasmid mediated gene transfer, it retained its wound-inducibility. Moreover it was also systemically induced in non-wounded tissues such as leaves, upper part of the stem and petals. Chimeric CAT genes containing different parts of the proteinase inhibitor II gene were wound inducible when both 5'-upstream and 3'-downstream regions of the proteinase inhibitor gene were used in the construction.

INTRODUCTION

Wounding of plants by either mechanical means or by pests results in a variety of changes in the physiological state of the tissue. On the one hand cell wall components such as lignin or suberin are produced at an increasing rate which might be seen as an effort to try to close the wounded surface as rapidly as possible (1). On the other hand the plant produces compounds which are thought to serve as an active defense reaction against the invading pathogen. Besides the formation of low-molecular weight compounds such as phytoalexins certain high molecular weight compounds can be produced. Among them, proteinaceous protease inhibitors represent an important group.

Protease inhibitors are usually present in storage organs of plants such as seeds and tubers. Their activity is in most cases directed against both animal as well as microbial proteases but rarely against plant proteases (2). Tubers of potato are a rich source for a number of different protease inhibitors. Among these are two potent inhibitors of serine endopeptidases called proteinase inhibitor I and II (3,4).

Whereas these inhibitors are normally under developmental control and thus expressed in a tuber-specific fashion, proteinase inhibitor I and II accumulate in leaves when these are wounded by either a chewing insect or by some mechanical means (5,6). The accumulation of both proteinase inhibitors is not confined to the damaged leaf rather wounding leads to their systemic expression in adjacent non-wounded leaves. This systemic response seems to be triggered by a so-called "proteinase inhibitor inducing factor" (PIIF) which most likely is composed of different oligosaccharides probably released from the cell wall as a result of the wounding effect (7).

We have recently isolated and characterized both cDNA's and genomic clones encoding proteinase inhibitor II genes from potato (6,8). These clones have been used in order to answer the following questions:

- Is the systemic induction of the proteinase inhibitor II genes in potato restricted to certain organs of the plant
- Which organs of the plant are able to induce the systemic response
- Is the wound-induced expression kept when the potato gene is transferred into another plant such as tobacco
- Is the systemic induction of the proteinase inhibitor II gene kept in this environment
- Can cis-active regulatory regions be defined which are important for the wound-induction

RESULTS

Wounding of Both Leaves and Tubers Leads to a Systemic Induction in Upper and Lower Leaves As Well As in Upper Parts of the Stem.

In order to analyze the extension of the systemic induction of proteinase inhibitor II in potato plants three leaves of a potato plant were wounded by dialysis clamps (6) and after 20 hours RNA was isolated from upper and lower leaves as well as upper and lower stem segments, roots and tubers. When these RNA's were analyzed for proteinase inhibitor II homologous sequences, a clear hybridization was observed with both upper and lower leaves as well as the upper part of the stem. No hybridization could be detected with the RNA isolated from the lower part of the stem or roots. Tuber RNA gave a prominent signal as to be expected; we did not observe a significant change in the amounts of proteinase inhibitor II RNA in tubers as a result of wounding of leaves. These results therefore show that the systemic induction of the proteinase inhibitor II genes is not restricted to leaf tissue but that activation of the genes can also be observed in upper stem segments of the potato plant.

In order to check whether the systemic induction observed was dependent upon the organ which was wounded experiments were performed wounding tubers of an otherwise unwounded plant. 16 hours later roots, upper and lower stem segments as well as several leaves were analyzed. As a result of wounding tubers proteinase inhibitor II homologous RNA could be detected in both leaves as well as in upper stem segments, whereas no signal was detectable in lower stems as well as roots. The proteinase inhibitor II homologous RNA content of tubers did not change significantly as a result of wounding. These data therefore clearly demonstrate that the signal responsible for systemic induction of the proteinase inhibitor II after wounding can be produced in wounded leaves as well as in wounded tubers. In addition this signal seems to be able to move both upwards as well as downwards in the wounded plant.

A Potato Proteinase Inhibitor II Retains Its Wound-Induced Expression After Transfer into Tobacco Plants.

As a first step in the analysis of the proteinase inhibitor II gene for cis-active regulatory sequences important for its wound-inducibility, we transferred the intact potato gene into tobacco using Agrobacterium/Ti plasmid mediated gene transfer techniques. Tobacco was chosen as a host plant as it does not contain any detectable homology to the proteinase inhibitor II gene family. Intact plants were regenerated and shown by Southern analysis to contain between one and five copies of the gene. Ten independently transformed plants containing at least one copy of the non-rearranged proteinase inhibitor II gene were analyzed for expression of the gene with and without prior wounding of the leaves. Northern gel analysis did not reveal any or only very low levels of expression in non-wounded leaves (cf, lanes 2,4,6 and 8 of Fig. 1). However when the RNA was isolated from leaves which had been exposed to two consecutive woundings, four out of ten plants analyzed showed a prominent

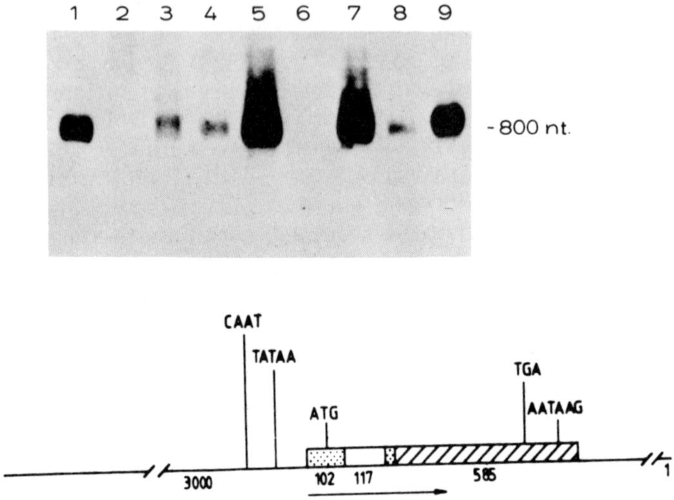

FIGURE 1. Upper part: Influence of wounding on expression of potato proteinase inhibitor II gene in transgenic tobacco plants. Total RNA isolated from leaves of four different transformed tobacco plants either with (lanes 3,5,7 and 9) or without prior wounding (lanes 2,4,6 and 8) were analyzed for proteinase inhibitor II homologous sequences by Northern blot analysis using a proteinase inhibitor II cDNA as probe. Lane 1 shows the hybridization of this probe with RNA isolated from wounded leaves of a potato plant.
Lower part: Schematic structure of the proteinase inhibitor gene used in the upper experiments. Exons are indicated by filled boxes, the intron is shown as an open box. Lengths of exons and introns are given in nucleotides.

band hybridizing with the proteinase inhibitor II cDNA (cf, Fig. 1, lanes 3,5,7 and 9). When the same blots were tested for expression of the co-transferred nopaline synthase gene no significant influence of wounding upon expression of this gene was observable. These results therefore strongly indicate that the wound-induced expression of the potato proteinase inhibitor II gene is controlled by cis-acting regulatory sequences present on the transferred DNA segment shown in the lower part of Figure 1. RNA protection experiments revealed that the transferred proteinase inhibitor II gene did not only retain its wound-inducibility but transcription initiated at the same nucleotide in transgenic tobacco plants as in potato (9).

Systemic Induction of the Potato Proteinase Inhibitor II Gene in Transgenic Tobacco Plants.

As outlined in the first part of the results section a careful analysis of the systemic induction of the proteinase inhibitor II gene in potato has shown it to be systemically induced in upper and lower leaves as well as in the lower part of the stem as result of wounding three middle leaves. We were therefore interested to see whether or not the expression of the potato gene would be induced in non-wounded organs of the plant by wounding some leaves. To this end, a tobacco plant was wounded at the third and fifth leaf and 20 hours after wounding RNA was isolated both from the wounded leaf as well as from non-wounded upper and lower leaves, stem segments, roots and petals. A clear hybridization was observed with RNA isolated from non-wounded upper and lower leaves, stem segments and petals. Thus the potato proteinase inhibitor when transferred into tobacco plants does not only retain its inducibility in leaves after wounding but does also retain its systemic induction in non-wounded organs.

3'-Downstream Sequences Are Necessary For Wound-Induction of Chimeric Genes in Transgenic Tobacco Plants.

In order to determine the location of cis-acting regulatory elements controlling the wound-induction of the potato proteinase inhibitor II gene in transgenic tobacco plants, chimeric genes were constructed consisting of different 5'-upstream sequences of the proteinase inhibitor II gene fused to the coding region of the chloramphenicol-acetyltransferase (CAT) gene from Tn9 and the termination region of gene 7 of the T-DNA of the Agrobacterium tumefaciens plasmid pTi ACH5 (10). The longest construction contained about 1.6 kb of 5'-upstream regions, the shortest construction contained only about 250 bp of upstream sequences derived from the proteinase inhibitor II gene. The fusion to the CAT gene was performed within the 5'-untranslated leader region of the proteinase inhibitor II gene at position + 11. These constructions were transferred into tobacco cells using the Agrobacterium system and whole plants regenerated. For each construct leaves of five independently derived transformed plants were tested for CAT activity with or without prior wounding of the leaf. We were not able to detect any CAT activity in any of more than 60 transformants tested. As the experiments described in the latter paragraph had however shown that the intact gene was wound-inducible in tobacco, these results indicated that sequences contained in the 5'-upstream region were not enough to confer this trait to chimeric genes. Therefore in a next step a 350 nucleotides long fragment of the proteinase inhibitor II gene encompassing its poly-A addition site was introduced into the chimeric genes replacing the terminator region of gene 7 of the T-DNA. When tobacco plants containing a construction comprised of 1.6 kb of 5'-upstream region of the proteinase inhibitor II gene fused at position + 11 to the coding region of the CAT gene and given the 350 nucleotides of the terminator region of the proteinase inhibitor II gene were tested for CAT expression in wounded and non-wounded leaves, a clear activity was visible in the wounded state. This result therefore indicates that the wound-inducible expression is not simply controlled by 5'-upstream sequences of the proteinase inhibitor II gene but that actually both 5'-upstream as well as 3'-downstream sequences are necessary.

DISCUSSION

As a result of wounding leaves of plants from several solanaceous aswell as legume (11) species accumulate proteinaceous serine protease inhibitors, the best studied cases being the proteinase inhibitor I and II of both potato and tomato. Proteinase inhibitor II encoding mRNA becomes detectable in wounded leaves as early as 20 minutes after the wounding event (H. Pena-Cortes, unpublished result), which indicates that the control of wound-induced expression is exerted at the RNA level. As described in the results section the accumulation of proteinase inhibitor II RNA is not confined to the wounded tissue itself but is also observed in several non-wounded parts of the plant such as upper and lower leaves as well as upper part of the stem. The accumulation in non-wounded tissues is probably triggered by cell wall fragments which are released from the plant tissue as a result of wounding. As shown in the results section this signal has to be able to move upwards and downwards in the plant as the proteinase inhibitor is induced in both upper as well as lower leaves. Furthermore leaf as well as tuber tissue have to be able to produce this signal.

Tobacco is devoid of genes homologous to the proteinase inhibitor II gene. Nevertheless when the potato proteinase inhibitor II gene is transferred into tobacco plants inducibility in the wounded tissue itself as well as induction in non-wounded organs is observed. It is reasonable to postulate in tobacco the presence of transacting factors involved in controlling the wound-induced expression of the transferred proteinase inhibitor II gene. Due to the absence of proteinase inhibitor II genes in normal tobacco plants it can furthermore be postulated that these factors normally serve for the wound-induced expression of other genes. Thus it can be assumed that tobacco should contain other wound-induced genes which have at least a partial homology to the proteinase inhibitor II gene. A homology search carried out between the proteinase inhibitor II sequences and other wound-inducible genes yielded a stretch of 34 nucleotides located about 30 nucleotides behind the stop codon in both the proteinase inhibitor II geneand the extensin gene from carrot (12). As seen in Figure 2, 25 nucleotides of these 34 nucleotides are conserved between both genes. As extensin is most likely a ubiquitous proteinone might speculate that the trans-acting factors in-

```
30 GGCCAACTTAAGAAG...ATGAAGTAAAATAATG extensin
   +++++++++++   +     +++++ +++++   +++
26 GGCCAACTTAATTAATGTATGAAATAAAAGGATG proteinase
                                      inhibitor
```

FIGURE 2. Homology between extensin and proteinase inhibitor II genes. Numbers given at the left indicate the position of the nucleotide relative to the respective stop codon.

ducing the proteinase inhibitor II gene in transgenic tobacco would normally be used for induction of the tobacco extensin and most likely other wound-induced genes. The possible biological relevance of this homology is supported by our finding that 5'-upstream sequences of the proteinase inhibitor II gene alone are not sufficient for wound-induced expression of chimeric genes but that indeed both 5'-upstream and 3'-downstream sequences are needed. Run-off transcription experiments performed in nuclei isolated from leaves of potato plants have indicated that the proteinase inhibitor II gene is constitutively transcribed in leaves albeit at a low level (13). If this result is taken together with our observation on the importance of 3'-downstream sequences for wound-induction of chimeric genes, the possibility of wound-inducibility to be mainly due to a differential RNA stability arises. Future tests using different chimeric genes are needed to further elucidate this question.

ACKNOWLEDGEMENT

This work was partly supported by a grant from the Bundesministerium für Forschung und Technologie BCT 035/2. H. Pena-Cortes is a recipient of a fellowship of the DAAD.

REFERENCES

1. Sequeira L (1983). Mechanisms of induced resistance in plants. Annual Rev Microbiol 37:51.
2. Richardson M (1976). The proteinase inhibitors of plants and microorganisms. Phytochemistry 16:159.
3. Bryant J, Green T, Gurusaddaiah T, Ryan CA (1976). Proteinase inhibitor II from potatoes: isolation and characterization of its protomer components. Biochemistry 15:3418.
4. Solyom A, Borsy J, Tolnay P. (1964). Biochem Pharmacol 13:391.
5. Green TR, Ryan CA (1972). Wound-induced proteinase inhibitor in plant leaves: a possible defense mechanism against insects. Science 175:776.
6. Sanchez-Serrano J, Schmidt, R, Schell J, Willmitzer, L (1986). Nucleotide sequence of a proteinase inhibitor II encoding cDNA of potato and its mode of expression. Mol Gen Genetics 203:15.
7. Ryan CA, Bishop PD, Walker-Simmons M, Brown WE, Graham JS (1985). Pectic fragments regulate the expression of proteinase inhibitor genes in plants. In Key J, Kosuge T (eds): "UCLA Symposia on Molecular and Cellular Biology, New Series" 22:319.
8. Keil M, Sanchez-Serrano J, Schell J, Willmitzer L (1986). Primary structure of a proteinase inhibitor II gene from potato (Solanum tuberosum). Nucleic Acids Research 14:5641.
9. Sanchez-Serrano J, Keil M, Schell J, Willmitzer L (1987). Wound-induced expression of a potato proteinase inhibitor II gene in transgenic tobacco plants. EMBO J , in press.
10. Velten J, Schell J (1985). Selection-expression plasmid vectors for use in genetic transformation of higher plants. Nucleic Acid Res 13:6981.
11. Brown W, Takio K, Titani K, Ryan CA (1985). Wound-induced trypsin inhibitor in alfalfa leaves: identity as a member of the Bownan-Birk inhibitor family. Biochemistry 24:2105.
12. Chen J, Varner J. (1985). An extracellular matrix protein in plants: characterization of a genomic clone for carrot extensin. EMBO J 4:2145.
13. Rosahl S, Eckes P, Schell J, Willmitzer L (1986). Organ-specific gene expression in potato: isolation and characterization of tuber-specific cDNA sequences. Mol Gen Genetics 202:368.

Plant Gene Systems and Their Biology, pages 339–349

DISCUSSION:REGENERATION AND TRANSFORMATION OF CEREALS

Thomas K. Hodges and Horst Lorz

Department of Botany and Plant Pathology, Purdue University, West Lafayette, IN, USA
and
Max-Planck Institut für Zuchtungsforschung, D-5000 Koln 30, FRG

During this workshop/discussion the authors presented the current state-of-the-art on regeneration and transformation of cereals. In addition, short platform presentations were given by Sheila Maddock, S. Goldman, and D. Marks and several contributions were made from the auditorium. This summary is presented in two sections: regeneration and transformation.

REGENERATION

Regeneration of cereals (maize, rice, wheat, barley, sorghum, etc.) is of considerable interest because of the potential value of genetically engineering this group of plants. A major breakthrough on cereal regeneration from protoplasts occurred during 1986 with 4 reports that the japonica lines of rice (*Oryza sativa*) could be regenerated from protoplasts (1-4). Not only were these reports of significance in their own right, but they helped to dispell some uncertainty, even with the successful regeneration of small plant-like structures of *Pennisetum americanum* from protoplasts (5), that it might still be impossible to regenerate cereals from protoplasts. Unfortunately, there were no representatives of these laboratories present at this meeting to discuss these important results on rice in detail.

Other than rice, corn has received more

attention regarding regeneration than the other cereals. Green and Phillips (6) first reported that immature embryos of inbred A188 produced a callus which could be induced to produce plants from the callus (referred to as Type 1 callus). Regeneration from Type 1 callus is considered by Springer et al. (10) as an organogenesis phenomenon while others (11,12) consider Type 1 callus as the initial callus produced on the immature embryo, and since this callus produces somatic embryos during the initial stages of growth, regeneration is probably via both organogenesis and somatic embryogenesis. Subsequent studies in various laboratories (7-9) have extended this regeneration from the Type 1 callus of A188 to other inbred and hybrid lines. It has been found that many inbred lines that are recalcitrant for regeneration can be converted to a regenerable hybrid by crossing it to a regenerable inbred e.g. A188 (13). Interestingly, backcrosses of the hybrid to each parent results in a segregation for somatic embryo formation that is consistent with the hypothesis that only two to three genes control this phenomena (9). The quick regeneration from Type 1 callus (within 1 to 2 months plants 3-5 inches tall are produced, 14) is well-suited for selecting and studying somaclonal variation. However, the Type 1 callus is not satisfactory for the development of liquid suspension cultures (the callus is hard and compact) and the regeneration frequency declines rapidly (within 3-6 months) in culture. It is possible, however, to select from the Type 1 callus a more friable and embryogenic callus that grows rapidly (2-fold increase in fresh weight per week). This friable callus was first described by Green (15) and subsequently referred to as a Type 2 callus by Armstrong and Green (16). Others have confirmed the existence of this Type 2 callus (17-19).

Maize Type 2 callus can be maintained and regenerated after 9 to 24 months, depending on the genotype (17,19). Suspension cultures developed from Type 2 callus of B73 x A188 have been shown to be capable of forming somatic embryos and to be regenerable after an additional 9 to 12 months in liquid culture (19). Sheila Maddock of Pioneer Hi-Bred, Inc. reported that their group also have suspension cultures of B73 (and of hybrid combinations involving B73) that were capable of regeneration. Cytological studies revealed chromosome counts of 20 ± 2 (euploid complement 2n=2x=20) in suspension cells; while all the regenerated plants examined so far contained the euploid chromosome number. She also reported that plants regenerated from the hybrid lines were more vigorous than those from inbred (B73) lines. The fine suspension cultures of corn developed from the Type 2 callus would be particularly well-suited for rapidly screening millions of cells (or small cell clusters) for variants or mutants e.g. a herbicide resistant line such as the one selected to be resistant to the class of herbicides known as the imidazolinones (20,21). The embryogenic and regenerable suspension cultures (19) need to be contrasted with the nonembryogenic and nonregenerable suspension culture of Black Mexican Sweet (BMS) corn (22,23).

Protoplasts prepared from the nonregenerable BMS suspension culture are capable of rapidly growing back into a non-embryogenic callus(23). On the other hand, protoplasts prepared from Type 1 callus of inbreds like A188 resynthesize cell walls and initiate cell divisions but stop dividing after microcalli of 0.1 to 0.5 mm diameter are reached (24,25). We have crossed A188 with BMS and prepared a suspension culture from the Type 2 callus of this F1 hybrid (26). Protoplasts obtained from these suspension cultures grow back into an embryogenic callus if cultured for the first two weeks on the BMS protoplast culture system (27) and then transferred to the culture system used for A188 (basically a Murashige and Skoog basal salts medium as described previously, 14). These calli can be induced to form small root- and leaf-like structures of 1-2 cm in length. Experiments are in progress to obtain conditions which will sustain the growth of these small plant-like structures.

Progress on regeneration from protoplasts of other cereals such as wheat, sorghum, etc. is still limited. At the Rothamsted Experimental Station, U.K., suspension cultures of several wheat lines have been obtained from callus derived from immature embryos (Sheila Maddock, this conference). Protoplasts prepared from a suspension culture of line C82d (cv. Copain) would consistently undergo sustained cell divisions, to give rise to non-morphogenetic callus or secondary cell suspensions. The frequency of initial cell divisions was very low. However, this could be improved by using the fastest-growing protoplast-derived colonies to initiate new cell suspensions. Protoplasts isolated subsequently from these secondary suspensions divided at a 10-fold greater frequency (Q. S. Wu, personal communication).

In summary, good progress is being made on developing cell culture systems for most of the cereals. Regeneration from protoplasts of japonica rice into mature plants and of *Pennisetum* and corn into small plant-like structures indicates that it should soon be possible to genetically engineer these cereals by introducing foreign DNA into these regenerable protoplasts (see next section). However, for this to be of real importance for Pennisetum and corn, procedures must still be developed for obtaining mature and normal plants.

TRANSFORMATION

Protoplasts.

Stable genetic transformation of cultured cereal cells has been achieved by direct DNA transfer into protoplasts of wheat. (28), maize (29), rice (30), and *Panicum* (31). In these experiments either kanamycin or hygromycin have been used as selectable marker genes and DNA uptake was achieved by electroporation or by treatment with PEG (polyethylenglycol). In all examples mentioned above protoplasts derived from well growing suspension cultures have been used. In principle direct DNA transfer into protoplasts of cereals is not different from other, e.g. dicotylendonous, protoplast systems. Still at present a limitation remains to the wider application of direct DNA-mediated gene transfer to cereal protoplasts due to the difficulty of regenerating plants from them. The recent success in regenerating plants from protoplasts of rice(1-4) should allow for transformation of rice plants in the near future. The selection of transformed cells with the antibiotic kanamycin is not always satisfying. Additional selectable markers such as methotrexate, hygromycin, or herbicides are actively studied to obtain a more efficient selection scheme.

Transient expression of chloramphenicol acetyltransferase (CAT) or neomycin phosphotransferase (NPT) was detected after PEG treatment or electroporation of protoplasts from several graminaceous species (32-34). Originally only suspension culture derived protoplasts have been used, but it could be shown that dividing as well as non-dividing protoplasts, e.g. root and mesophyll protoplasts, are also suitable for transient expression studies (34,35). Of special interest now is to see to which extent cell type and tissue specific expression is found in such transient expression assays and whether also induction of gene activity can be studied with this technique.

Attempts to Transform Cells and Multicellular Structures.

Due to the difficulty of regenerating plants from graminaceous protoplasts, alternative ways are sought to introduce foreign genes into cereal cells. Plant regeneration is possible from multicellular structures of *in vitro* cultures, specifically from embryogenic callus cultures and cell suspensions. Experiments have been made incubating cells, cell clumps, and immature embryos with DNA and thereafter assaying for CAT- or NPT-activity. Preliminary results showed rather low expression and low reproducibility of the chimeric genes in assays for transient gene activity. Most promising are the approaches where a short enzyme treatment, to obtain partial cell wall digestion, has preceeded the DNA incubation (T. Hodges, this symposium).

Microinjection of DNA directly into cells offers another possibility to achieve DNA transfer without using protoplasts. Experiments with microspore derived small cell colonies of *Brassica napus* have shown that DNA transfer can be successfully introduced by capillary microinjection into multicellular structures, and assaying for transient expression indicated activity of the introduced chimeric genes (G. Spangenberg and G. Neuhaus, pers. communication). It has to be seen whether this method can be applied to cereal cultures and whether genetically stable transformants are obtained by this way.

Agrobacterium-mediated Gene Transfer.

Gene transfer to numerous dicotyledonous plant species is most efficiently performed with the natural vector system of the Ti-plasmid from *Agrobacterium*. The natural host range of *A. tumefaciens* and *A. rhizogenes,* however, is limited and so far only a few monocotyledonous species could be transformed with *A. tumefaciens* , such as *Narcissus* (36), *Asparagus* (37), *Gladiolus* (S. Goldman, this symposium) and *Zea mays* (38,39). The fact that maize is included in this list is very encouraging but so far the evidence for transformation is based only on the detection of opines and no molecular data for the integration of the foreign DNA into the maize genome has been provided yet. In addition, recent reports describing opines in several plant species which have not been infected with Agrobacteria (40), stress the point that additional evidence is needed to demonstrate *Agrobacterium* -mediated gene transfer to *Gramineae* . To circumvent the problem of opines, strains of *A. tumefaciens* carrying NPT as a marker gene have been used for infection of maize, wheat and rice (B. Junker, J. Cabrera, pers. communication). In all species tested NPT-activity could be measured 7-14 days after infection of the seedlings, but in these experiments it could not be absolutely excluded, that the enzyme activity measured was caused by bacterial activity. Further experiments are necessary to give final proof for the transfer of T-DNA into the genomes of wheat, maize, or rice.

A very simple procedure for *Agrobacterium* -mediated gene transfer has been exemplified with *Arabidopsis thaliana* (D. Marks, this symposium). Seeds are incubated with *A. tumefaciens* and in the progeny of the treated material on average 3 out of 1000 seedlings exhibited kanamycin resistance, indicating successful transformation with the chimeric gene which has been used as the selectable marker. It has to be seen whether this method can be applied also to other dicotyledonous and even to graminaceous species.

Transformation of Plants.

The difficulties to regenerate plants from cereal protoplasts and the efforts necessary to handle totipotent *in vitro* cultures made all alternative genetic transformation systems, independent of the vitro culture, very attractive. For many years attempts have been made to introduce DNA into seeds, germinating seedlings, plants, or pollen (41). Pollen mediated transformation has been reported also for maize and formal genetic evidence has been provided (42). So far however, such experiments have not been reproduced by other researchers, and only with the use of cloned DNA and defined marker genes is it possible to follow and to analyze the fate of the introduced genes at the molecular level.

In a different approach plasmid DNA coding for kanamycin resistance has been injected into floral tillers of rye, *Secale cereale* L. (43). Flowering plants were crossed with each other and seeds from DNA-treated plants were germinated in the presence of kanamycin. In 3 plants out of ca. 6000 seeds derived from 350 injected tillers the presence of the gene product (NPT-activity) and the physical presence of the foreign gene (Southern-blotting) could be shown. This injection procedure provides an alternative transformation procedure method which is independent of isolated single cells, an *in vitro* culture system, or specific transformation vectors.

REFERENCES

1. Fujimura T, Sakurai M, Akagi H, Negishi T, Hirose (1986). Regeneration of rice plants from protoplasts. Plant Tissue Culture Lett. 2:74.
2. Abdullah R, Cocking EC, Thompson JA (1986). Efficient plant regeneration from rice protoplasts through somatic embryogenesis. BioTechnology 4,1087-1090.
3. Yamada Y, Yang ZQ,Tang DT (1986). Plant regeneration from protoplast derived callus of rice (*Oryza sativa*). Plant Cell Reports 5:85-88.
4. Coulibaly MY, Demarly Y (1986). Z. Pflanzenzucht 96:79.
5. Vasil V, Vasil IK (1980). Isolation and culture of cereal protoplasts. Part 2: Embryogenesis and plantlet formation from protoplasts of *Pennisetum amrericanum*. Theor Appl Genet 56:97-99.
6. Green CE, Phillips RL (1975). Plant regeneration from tissue cultures of maize. Crop Sci 15:417-421.
7. Lu C, Vasil V, Vasil IK (1983). Improved efficiency of somatic embryogenesis and plant regeneration in tissue cultures of maize (*Zea mays* L.). Theor Appl Genet 66:285-289.
8. Duncan DR, Willams ME, Zehr BE, Widholm JM (1985). The production of callus capable of plant regeneration from immature embryos of numerous *Zea mays* genotypes. Planta 165: 322-332.
9. Hodges TK, Kamo KK, Imbrie CW, Becwar MR (1986). Genotype specificity of somatic embryogenesis and regeneration in maize. Biotechnology 4:219-223.
10. Springer WD, Green CE, Kohn KA (1979). A histological examination of tissue culture initiation from immature embryos of maize. Protoplasma 101:262-281.
11. Vasil, V, Lu C-Y, Vasil IK (1985). Histology of somatic embryogenesis in cultured immature embryos of maize (*Zea mays* L.). Protoplasma 27:1-8.
12. McCain JW, Hodges TK (1986). Anatomy of somatic embryos from maize embryo cultures. Bot Gaz 147:453-460.

13. Hodges TK, Kamo KK, Becwar MR, Schroll S (1985). Regeneration of maize. In: Biotechnology in Plant Science: Relevance to Agriculture in the Nineteen Eighties, eds M Zaitlin, P Day, A Hollaender, Academic Press, Orlando, FL, pp 15-33.
14. Kamo KK, Becwar MR, Hodges TK (1985). Regeneration of *Zea mays* from embryogenic callus. Bot Gaz 146:327-334.
15. Green CE (1982). Somatic embryogenesis and plant regeneration from the friable callus of *Zea mays*. In: Plant tissue culture 1982 (Proc V Int Congr Plant Tissue and Cell Culture), ed. A. Fujiwara, Japanese Association for Plant Tissue Culture, Tokyo, Japan, pp 107-108.
16. Armstrong CL, Green CE (1985). Establishment and maintenance of friable, embryogenic maize callus and the involvement of L-Proline. Planta 164:207-214.
17. Tomes D, Smith OS (1985). The effect of parental genotype on initiation of embryogenesis callus from elite maize (*Zea mays* L.) germplasm. Theor Appl Genet 50:505-510.
18. Vasil V, Vasil IK (1986). Plant regeneration from friable embryogenic callus and cell suspension cultures of *Zea mays* L. J.Plant Physiol. 124:399-408.
19. Kamo KK, Hodges TK (1986). Establishment and characterization of long-term embryogenic maize callus and cell suspension cultures. Plant Sci:45:111-117.
20. Anderson PC, Hibberd KA (1985). Evidence for the interaction of an imidazolinone herbicide with leucine, valine, and isoleucine metabolism. Weed Science 33:479-483.
21. Shaner DL, Anderson PC (1985). Mechanism of action of the imidazolinones and cell culture selection of tolerant maize. In: Biotechnology in Plant Science: Relevance to Agriculture in the Nineteen Eighties, eds. M Zaitlin, P Day, A Hollaender, Academic Press,Orlando, FL, pp 15-33.
22. Sheridan WF (1975). Growth of corn cells in culture. J Cell Biol. 67:396a.
23. Chourey PS, Zurawski DB (1981). Callus formation from protoplasts of a maize cell

culture. Theor Appl Genet 59:341-344.

24. Imbrie-Milligan CW, Hodges TK (1986). Microcallus formation from maize protoplasts prepared from embryogenic callus. Planta: 168: 395-401.
25. Imbrie-Milligan CW, Kamo KK, Hodges TK (1987). Microcallus growth from maize protoplasts. Planta (in press).
26. Kamo KK, Chang KL, Lynn ME, Hodges TK (1987). Embryogenic callus formation and transient transformation of maize protoplasts. Planta, submitted.
27. Ludwig SR, Somers DA, Petersen WL, Pohlman RF, Zarowitz MA, Gengenbach BG, Messing J (1985). High frequency callus formation from maize protoplasts. Theor Appl Genet 71:344-350.
28. Lorz H, Baker B, Schell J (1985). Gene transfer to cereal cells mediated by protoplast transformation. Mol Gen Genet 199: 178-182.
29. Fromm ME, Taylor LP, Walbot V (1986). Stable transformation of maize after gene-transfer by electroporation. Nature 319:791-793.
30. Uchimiya H, Fushimi T, Hashimoto H, Harada H, Syono K, Sugawara (1986). Expression of a foreign gene in callus derived from DNA treated protoplasts of rice (*Oryza sativa* L.). Mol Gen Genet 204:204-207.
31. Ozias-Akins P, Hauptmann R, Vasil V, Tabaeizadeh Z, Horsch R, Rogers S, Fraley R, Vasil I (1987). Stable transformation of *Panicum maximum* and *Triticum monococcum*: more efficient selectable markers for the *Gramineae.* J Cellular Biochem Suppl 11B:28.
32. Fromm M, Taylor LP, Walbot V (1985). Expression of genes transferred into monocot and dicot plant cells by electroporation. Proc Natl Acad Sci USA 82:5824-5828.
33. Werr W, Lorz H (1986). Transient gene expression in a *Gramineae* cell line. A rapid procedure to study plant promoters. Mol Gen Genet 202:368-373.
34. Ou-Lee TM, Turgeon R, Wu R (1986). Expression of a foreign gene linked to either a plant-virus or Drosophila promoter,after electroporation of protoplasts of rice,

wheat,and sorghum. Proc Natl Acad Sci USA 83:6815-6819.

35. Junker B, Zimny J, Luhrs R, Lorz H (1987). Transient expression of chimeric genes in dividing and non-dividing cereal protoplasts after PEG-induced DNA uptake. Plant Cell Rep, submitted.

36. Hooykaas-Van Slogteren GMS, Hooykaas PJJ, Schilperoort RA (1984). Expression of Ti plasmid genes in monocotyledonous plants infected with *Agrobacterium tumefaciens*. Nature 311:763-764.

37. Hernalsteens J-P, Thia-Toong L, Schell J, Van Montagu M (1984). An *Agrobacterium*-transformed cell culture from the monocot *Asparagus officinalis*. EMBO J.3,3039-3041.

38. Graves ACF, Goldman SL (1986). The transformation of Zea Mays seedlings with *Agrobacterium tumefaciens*. Plant Mol Biol 7: 43-50.

39. Grimsley N, Hohn T, Davis JW, Hohn (1987). *Agrobacterium*-mediated delivery of infectious maize streak virus into maize plants. Nature 325:177-179.

40. Christou P, Platt SG, Ackermann MG (1986). Opine synthesis in wild-type plant tissue. Plant Physiol 82:218-221.

41. Ledoux L (1975). "Genetic Manipulation with Plant Material". Plenum Press, New York.

42. Ohta Y (1986). High efficiency genetic transformation of maize by a mixture of pollen and exogenous DNA. Proc Natl Acad Sci USA 83:715-719.

43. De la Pena A, Lorz H, Schell J (1987). Transgenic rye plants obtained by injection of DNA into young floral tillers. Nature 325:274-276.

VI. MOLECULAR MODIFICATION OF THE PHOTOSYNTHETIC APPARATUS

Plant Gene Systems and Their Biology, pages 353–362

THE EFFECT OF CHILLING ON THE PRODUCTION OF PHOTOSYSTEM II LIGHT HARVESTING CHLOROPHYLL A/B BINDING PROTEIN IN TOMATO[1]

Pam Cooper and Donald R. Ort

Department of Plant Biology, USDA/ARS
University of Illinois
Urbana, Illinois 61801

ABSTRACT Nearly two-thirds of the inhibition of photosynthesis caused by dark chilling in tomato is due to impairment of chloroplast function. Yet plants chilled for as long as 45h will eventually recover full photosynthetic capacity if rewarmed in darkness and 100% RH. As a means of determining whether chilling damages any chloroplast proteins, we have monitored protein synthesis in leaves during the recovery period subsequent to 16h chilling at 2C. SDS-PAGE of proteins labelled with [^{35}S]-methionine revealed the synthesis of a protein of M_r = 35kD during the first 2h of rewarming. Two-dimensional PAGE showed the protein exists as a single isoelectric species. The protein was not synthesized by dark controls. A protein of M_r = 27kD abundantly synthesized by control plants declines to undetectable levels in chilled plants. Synthesis resumes when 8h-recovered plants are returned to light. Both polypeptides are synthesized on cytoplasmic ribosomes. Peptide mapping revealed the 27kD protein to be the major light harvesting chlorophyll a/b binding protein of Photosystem II (LHCP-II). We propose the 35kD protein is the unprocessed precursor of LHCP-II, and that chilling causes perturbations in the biosynthetic pathway leading to mature LHCP-II.

INTRODUCTION

Plants are frequently challenged by the environmental stresses sufficient to limit their productivity. Although plant growth relies on the integrated function of many physiological processes, a single

[1]This work was supported in part by a post-doctoral research fellowship awarded to P.C. by Pioneer Hi-bred International.

process can become a limiting factor due to its sensitivity to environmental extremes. In some cases, photosynthesis is that process.

For crop plants of tropical origin, such as tomato and corn, brief exposure to chilling temperatures (0<T<12°C) can have a long-term influence on growth because of the adverse effects on photosynthesis. Tomato plants exhibit a 60% reduction in the maximum rate of net photosynthesis measured at atmospheric CO_2 concentrations following 16h of chilling in darkness at 2°C (1). Two-thirds of this inhibition is due to impairment of the biochemical processes of the chloroplast, and one-third is the consequence of an increase in the stomatal resistance to gas exchange (1).

Our research has focused on identifying sites of chilling damage in the chloroplast. Martin and Ort (2) have observed that tomato plants dark chilled for as long as 45h can recover full light- and CO_2-saturated photosynthetic rates within 12h if allowed to rewarm in darkness and 100% RH, indicating that damage imparted to the chloroplast by chilling can be repaired. Since this repair may involve protein synthesis, we monitored the incorporation of [^{35}S]-methionine into leaf polypeptides synthesized during the recovery phase as a means of identifying photosynthetic proteins altered by chilling. We report the appearance of a 35kD polypeptide and the decline in synthesis of a 27kD polypeptide. Peptide mapping analysis identified the 27kD protein to be the major light harvesting chlorophyll a/b binding protein of Photosystem II.

MATERIALS AND METHODS

Treatment of plant material.

Tomato plants (Lysopersicon esculentum Mill. cv. Floramerica) were grown as previously described (2). Eighteen-day-old seedlings on which the third leaf had emerged were used. Plants were chilled at 2-4°C for 16h in darkness at 100% RH and allowed to rewarm in darkness at room temperature (22-24°C) and 100% RH. Control plants were maintained in darkness at 100% RH and room temperature for 16h plus a length of time equivalent to the recovery period of the chilled plants. Plants from which thylakoids were isolated were maintained in the growth chamber and removed for use during the lighted portion of the diurnal cycle.

Labelling conditions and extractions of proteins.

For plants from which leaf proteins were extracted, the oldest leaflet of the third leaf was abraded lightly in 400 grit carboundum under dim green light and 50 μCi [^{35}S]-methionine (>800 Ci/mmole, Amersham) was applied to the abraded surface. The leaves were harvested and rinsed in ice cold, non-radioactive 1mM L-methionine.

The labelled leaflet was ground in Laemmli sample buffer (3) supplemented with 2mM phenyl methylsulfonyl fluoride and 10mM ascorbate. The homogenate was vortexed with polyvinylpolypyrrolidone (PVPP) and microfuged for 4 min to remove PVPP and insoluble cell debris. The supernatant was boiled for 2 min and stored at -20C.

Plants from which thylakoids were isolated were labelled by painting 300 μCi of [^{35}S]-methionine in 1 ml 1% Tween-20 on lightly abraded upper leaf surfaces. Labelling took place for 3h under room light, after which leaves were rinsed as above and thylakoids isolated by a modified procedure of Kee et al (4). In our experiments, the first resuspension medium contained 50 mM sorbitol to remove contaminating stromal proteins bound to the thylakoids (5). BSA was reduced to 0.2% in the grinding medium and was omitted from successive wash media. Following chlorophyll determination (6) thylakoids were stored at -80C. TCA precipitable radioactivity was determined as described in (7).

Gel electrophoresis and detection of proteins.

Total leaf proteins were separated by SDS-PAGE using the buffer system of Laemmli (3). Two-dimensional gel electrophoresis was by the method of Zurfluh and Guilfoyle (8). The SDS gels in both systems were 12% in acrylamide (w/v). Thylakoid membranes were solubilized in Laemmli sample buffer, boiled 2 min, and separated on SDS gels with a linear acrylamide gradient of 12-17.5% (w/v). For mol. wt. determination, BioRad mol. wt. standards were run alongside samples. Proteins were stained with Coomassie Brilliant Blue. For detection of labelled proteins, Kodak XAR-5 X-ray film was exposed to dried gels. Some gels were impregnated with PPO (9) prior to drying and fluorography at -80C.

Partial proteolytic cleavage of proteins.

Staphylococcus aureus V8 protease was used for limited digestion of proteins according to the method of Cleveland et al (10). Gel slices containing the polypeptides of interest were equilibrated and loaded into slots of a second gel along with 0.5 μg V8 protease. Digestion took place for 30 min in the stacking gel with the current off. The proteolytic fragments were separated on a 12-20% linear gradient polyacrylamide SDS gel containing 4M urea and were detected by fluorography.

RESULTS

Two major changes in protein synthesis were observed in tomato plants after they had been chilled in darkness for 16h. (Fig. 1). A polypeptide of M_r = 35kD was synthesized beginning in the first 30

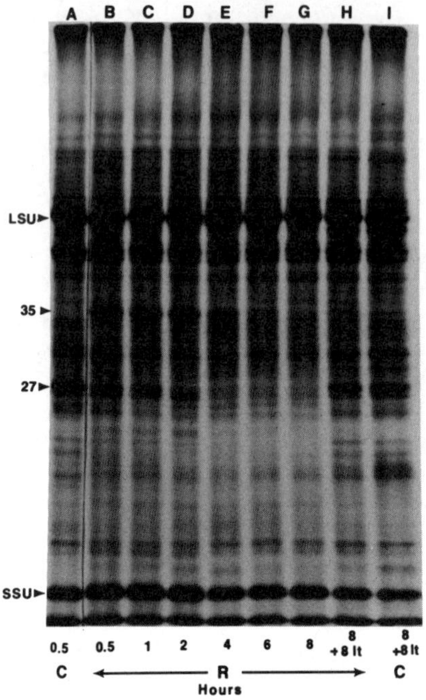

Figure 1. Time course of protein synthesis after 16h of chilling in darkness. Plants were rewarmed in the dark for increasing lengths of time. Labelling with [^{35}S]-methionine took place during the final 30 min of treatment. Lanes: A, control, 16 + 0.5h darkness; B, 0.5; C, 1; D, 2; E, 4; F, 6; G, 8h rewarming after chilling; H 16h chilled + 8h rewarmed; I, control, 16 + 8h darkness. H and I were returned to light for 8h after treatment. LSU, SSU, large and small subunits of Rubisco. Positions of 27 and 35kD proteins are indicated.

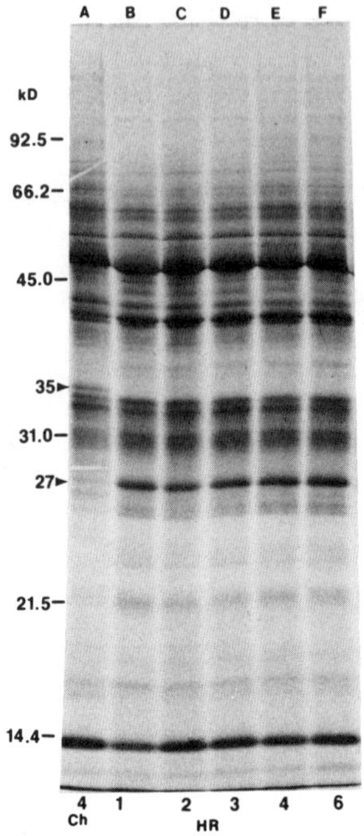

Figure 2. Time course of protein synthesis in plants maintained in continuous darkness. Leaves isotopically labelled as in Figure 1. Lanes: B, 16 + 1h; C, 16 + 2h; D, 16 + 3h; E, 16 + 4h; F, 16 + 6h. Lane A shows proteins from 16h chilled plant dark-recovered 4h. Positions of mol. wt. markers, 27, and 35kD proteins indicated at left.

min after the plants had been rewarmed (Fig. 1, Lane B). This polypeptide was not synthesized by unchilled controls maintained in darkness for the same total length of time (Fig. 1, Lane A). The synthesis of the 35kD polypeptide increased through the first two hours of recovery (Lane B-D) and thereafter its synthesis declined (Lanes E-G). Two dimensional analysis of newly synthesized proteins from chilled recovering leaves revealed the 35kD protein exists as a single isoelectric species (Fig. 3).

A polypeptide of M_r = 27kD normally synthesized in abundance by control leaves (Fig. 1, Lane A, and Figure 2), exhibited a decline in synthesis in chilled leaves as recovery progressed (Fig. 1, Lanes B-G). Beyond 4h recovery, its synthesis was barely detectable. Prolonged dark treatments alone did not cause a loss of 27kD protein synthesis or induce the 35kD protein (Fig. 2, Lanes B-E). However, if chilled plants recovered 8h were returned to a lighted growth chamber, synthesis of the 27kD protein was quickly restored to control levels (Fig. 1, Lanes H and I).

Protein synthesis inhibitors were employed to determine whether either of the proteins were encoded by the chloroplast genome. Plants were dark chilled or maintained in darkness at room temperature for 16h, then during the first 30 min of recovery, a slightly abraded leaflet was submerged in 20 µg/ml cycloheximide or 200 µg/ml chloramphenicol dissolved in 1% Tween-20. Controls were submerged in 1% Tween-20 alone. The leaves were rinsed and allowed to incorporate label for 1h prior to extractions of proteins.

Since the large subunit of ribulose bisphosphate carboxylase/oxygenase (Rubisco) is synthesized on chloroplast ribosomes, whereas the small subunit is synthesized on cytoplasmic ribosomes, the relative levels of Rubisco subunit synthesis served as internal controls for the efficacy and selectivity of the inhibitors. Large, but not small, subunit synthesis occurred in cycloheximide-treated leaves (Fig. 4, B and E), whereas the converse was true for chloramphenicol-treated leaves (Fig. 4, C and F). Cycloheximide, but not chloramphenicol, inhibited synthesis of the 35kD polypeptide in chilled leaves (Fig. 4, E and F). The same was true for the synthesis of the 27kD protein in control leaves (Fig. 4, B and C). Neither protein as detectable even when the gel was over-exposed (data not shown).

That these polypeptides were synthesized in the cytoplasm considerably increased the range of their possible identities and sites of function. Because we were concerned with possible relationships the polypeptides had with photosynthesis, we isolated thylakoid membranes as an attempt to determine whether either polypeptide had its function in the photosynthetic membrane. The thylakoid polypeptides from chilled and unchilled plants were qualitatively identical (Fig. 5). The newly synthesized 35kD protein found in chilled leaves was not associated with thylakoids. However, several quantitative changes were apparent on both the stained gel and on its autoradiograph. The levels of newly synthesized β-subunit of the coupling factor, the major Photosystem II light-harvesting complex protein (LHCP-II), and two unidentified polypeptides were significantly reduced in thylakoids from chilled leaves (Fig. 5, Lanes F and H) as compared to those from control leaves (Fig. 5, Lanes E and G). This difference is reflected in the levels of accumulated protein for the β-

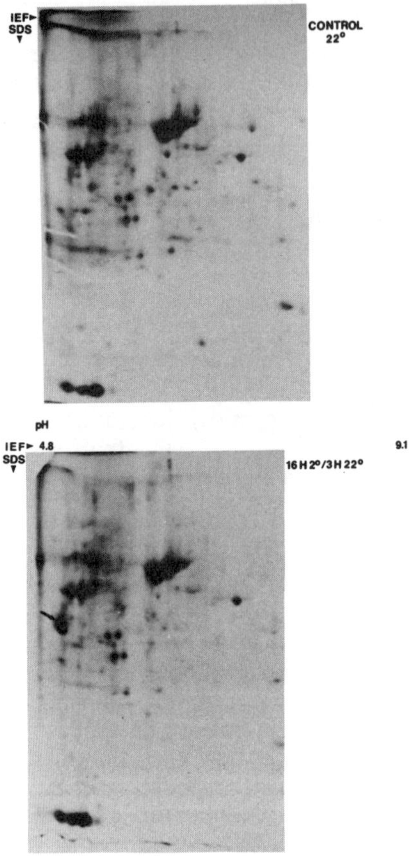

Figure 3. Two-dimensional analysis of proteins synthesized in control and chilled plants after 3h dark recovery. Arrows indicates 35kD protein.

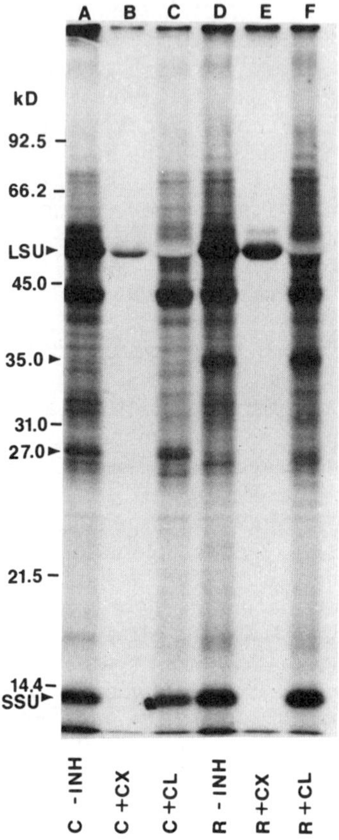

Figure 4. Effect of inhibitors on protein synthesis in chilled and unchilled leaves. Lanes: A-C, unchilled; D-F, chilled. A, D, 1% Tween-20; B, E, 20 μg/ml cycloheximide; C, F, 200 μg/ml chloramphenicol. Ten μg protein loaded in each lane. Positions of LSU, SSU, 27, and 35 kD proteins indicated by arrowheads.

subunit and unknown polypeptides, but not for LHCP-II (Fig. 5, Lanes A-D).

Based upon its molecular weight and relative abundance, we suspected the 27kD polypeptide was the major light harvesting complex

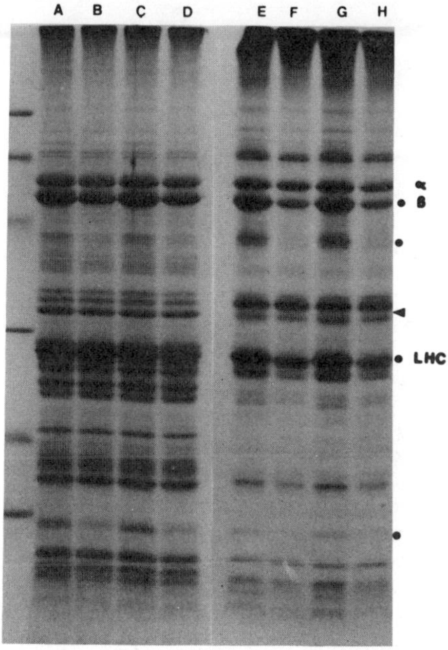

Figure 5. Effect of chilling on protein composition of tomato thylakoid membranes. Plants were labelled for 3h in darkness subsequent to 16h in darkness at 22 or 2°C. Sample volumes equivalent to 20 µmol chlorophyll were loaded in each lane. Lanes A-D, Coomassie blue stained proteins; E-H fluorograph of newly synthesized proteins. Controls, lanes A, C, E, G; chilled, lanes B, D, F, H. Dots indicate proteins reduced in chilled leaves. Arrow denotes theoretical position of 35kD protein as calculated from a mol. wt. standard curve. Mol. wts. of standards at far left, in kD top to bottom: 92.5, 66.2, 45.0, 31.0, 21.5, 14.4

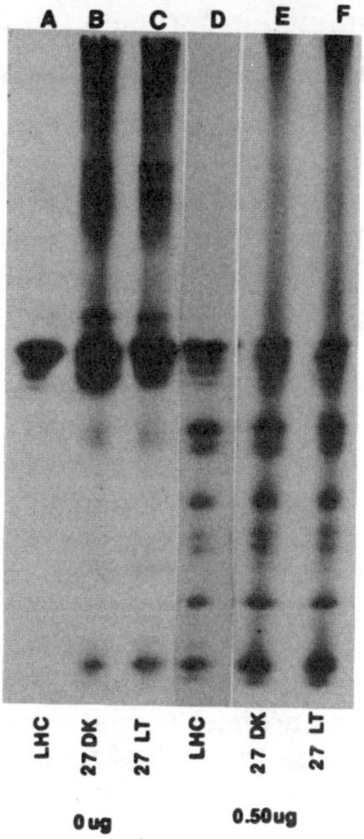

Figure 6. Partial proteolytic digestion of the 27kD protein compared to authentic tomato LHCP-II. Radioactive LHCP-II obtained from thylakoids isolated as described in Materials and Methods. The 27kD protein was obtained from light grown plants on those kept in darkness 20h. Lanes: A-C, undigested controls, D-F proteins digested with 0.50 µg V8 protease. A and D, LHCP-II; B and E, 27kD protein from dark-treated plants; C and F, 27kD protein from light-grown plants.

protein of Photosystem II. This hypothesis was reinforced by the apparent reduction of labelled LHCP-II present in the thylakoids of chilled plants. Comparison of the partial proteolytic digestion patterns of the 27kD protein with authentic LHCP-II supported our hypothesis (Fig. 6). The patterns generated by the proteolytic fragments of 27kD protein from light-grown plants or plants maintained in darkness for 20h were identical to the digestion pattern of LHCP-II.

DISCUSSION

Chilled tomato plants allowed to rewarm in darkness exhibit altered leaf protein synthesis compared to unchilled dark controls. In chilled leaves a 35kD protein is synthesized within the first 30 min of rewarming. Its synthesis increases for two hours, then declines to a barely detectable level. Inhibitor studies showed this protein to be synthesized on cytoplasmic ribosomes, and 2D-PAGE showed it to exist as a single isoelectric species. Upon rewarming there is also the progressive decline in synthesis of a 27kD protein, also shown to be synthesized on cytoplasmic ribosomes. Peptide mapping identified this protein as the major light-harvesting chlorophyll a/b-binding protein of Photosystem II (LHCP-II).

Although light could induce LHCP-II synthesis in leaves rewarmed for 8h, it was not darkness alone that caused the loss of synthesis. Synthesis of LHCP-II can continue even following prolonged darkness in mature green leaves (11). Control tomato leaves were able to synthesize LHCP-II even after 22h of continuous darkness. One explanation is that chilling may accelerate the degradation of messages coding for LHCP-II. Light is known to regulate LHCP-II mRNA synthesis (12), so returning plants to light may have restored LHCP-II synthesis by inducing transcription of new message.

Loss of message is perhaps the simplest explanation for the decline in LHCP-II synthesis. However, there is evidence that chilling may affect other components of the LHCP-II biosynthetic pathway. LHCP's are encoded by the nucleus and synthesized on cytoplasmic ribosomes as soluble high molecular weight precursors (pLHCP) (13, 14). The import of the pLHCP into the chloroplast is a post-translational event. Presumably the first step is recognition of the precursor by a putative receptor on the envelope (15), followed by uptake and cleavage of an amino-terminal transit peptide. A cleavage step can occur after the insertion of pLHCP into the thylakoid membrane (14).

Hayden, et al (16) have shown that a 31kD polypeptide accumulates in the thylakoids of maize plants chilled in high light. This protein is immunochemically related to LHCP-II. They suggest the 31kD protein is a pLHCP which accumulates in stressed thylakoids because the proteolytic processing step is damaged. We did not observe any unique proteins in tomato thylakoids as a result of dark chilling. However, preliminary peptide mapping data from our laboratory indicates from primary structure homology between LHCP-II and the 35kD protein found in leaf extracts (not shown). It may be that chilling also interrupts post-translational processing of pLHCP in tomato, perhaps at the recognition step. This would certainly explain the increase in 35kD protein synthesis concomitant to the loss of LHCP-II synthesis. Inability of pLHCP to enter the chloroplast would also explain the reduced levels of mature LHCP-II in thylakoids from chilled leaves. Using immuno-blot analysis, we are presently conducting experiments to determine whether the 27 and 35kD polypeptides are in fact related.

In conclusion, the experiments described here point to the LHCP-II biosynthetic pathway as a chill-labile process in tomato, with possibly more than one component of the pathway being susceptible to injury.

REFERENCES

1. Martin B, Ort DR, Boyer JS (1981). Impairment of photosynthesis by chilling temperatures in tomato. Plant Physiol 68:329.
2. Martin B, Ort DR (1985). The recovery of photosynthesis in tomato subsequent to chilling exposure. Photosyn Res 6:121.
3. Laemmli UK (1970). Cleavage of structural proteins during the assembly of the head of bacteriophage T4. Nature 227:680.
4. Kee SC, Martin B, Ort DR (1986). The effects of chilling in the dark and in the light on photosynthesis of tomato: Electron transfer reactions. Photosyn Res 8:41.
5. Strotmann H, Hesse H, Edelmann K (1973). Quantitative determination of coupling factor CF_1 of chloroplasts. Biochim Biophys Acta 314:202.
6. Graan T, Ort DR (1984). Quantitation of the rapid electron donors to P_{700}, the functional plastoquinone pool, and the ratio of the photo systems in spinach chloroplasts. J Biol Chem 259:14003.
7. Mans RJ, Novelli DG (1960). A convenient, rapid and sensitive method for measuring the incorporation of radioactive amino acids into protein. Biochem Biophys Res Comm 3:540.
8. Zurfluh LL, Guilfoyle TJ (1982). Auxin-induced changes in the population of translatable messenger RNA elongating sections of soybean hypocotyl. Plant Physiol 69:332.
9. Jen G, Thach R (1982). Inhibitors of host translation in encephalomyocarditis virus-infected L cells: a novel mechanism. J Virol 43:250.
10. Cleveland DW, Fischer SG, Kirschner MW, Laemmli UK (1977). Peptide mapping by limited proteolysis in sodium dodecyl sulfate and analysis by gel electrophoresis. J Biol Chem 252:1102.
11. Bennett J (1981). Biosynthesis of the light-harvesting chlorophyll a/b protein: polypeptide turnover in darkness. Eur J Biochem 118:61.
12. Tobin EM, Silverthorne J (1985). Light regulation of gene

expression in higher plants. Ann Rev Plant Physiol 36:569.

13. Schmidt GW, Bartlett SG, Grossman AR, Cashmore AR, Chua N-H (1981). Biosynthetic pathways of two polypeptide subunits of the light harvesting chlorophyll a/b protein complex. J Cell Biol 91:468.
14. Chitnis RR, Harel E, Kohorn BD, Tobin EM, Thornber JP (1986). Assembly of the precursor and processed light-harvesting chlorophyll a/b protein of Lemna into light-harvesting complex II of barley etiochloroplasts. J Cell Biol 102:982.
15. Cline K, Werner-Washburne M, Lubben TH, Keegstra K (1985). Precursors to two nuclear-encoded chloroplast proteins bind to the outer envelope membrane before being imported into chloroplasts. J Biol Chem 260:3691.
16. Hayden DB, Baker NR, Percival MP, Beckwith PB (1986). Modification of the photosystem II light harvesting chlorophyll a/b protein complex in maize during chill-induced photoinhibition. Biochim Biophys Acta 851:86.

Plant Gene Systems and Their Biology, pages 363–370

TARGETING OF PROTEINS INTO CHLOROPLASTS[1]

Kenneth Keegstra*, Cynthia Bauerle*, Thomas Lubben*, Sjef Smeekens+ and Peter Weisbeek+

*Department of Botany, University of Wisconsin
Madison, WI 53706
and
+Department of Molecular Cell Biology
State University of Utrecht
Utrecht, The Netherlands

INTRODUCTION

A typical plant cell contains many different compartments, each with its own set of metabolic processes. Consequently each compartment has its own set of enzymes, including many which are uniquely found in a single compartment and are therefore useful as marker enzymes in fractionation studies. Most of the proteins found in these various compartments are encoded in nuclear genes and synthesized on cytoplasmic ribosomes. The mechanisms responsible for targeting proteins to the proper compartment can be divided into at least two different types. The first is a cotranslation mechanism whereby proteins destined for compartments of the endomembrane system are transported across the endoplasmic reticulum membrane as they are being synthesized on membrane bound ribosomes (1). These proteins are subsequently segregated to the proper location via sorting mechanisms and membrane flow (vesicle traffic) through the endomembrane system (2). A great deal of work has been done on this targeting mechanism in animal systems; considerably less has been done in plant systems.

The mechanism responsible for targeting of proteins to

[1]This work was supported by the US National Science Foundation, the US Department of Energy, The Netherlands Foundation for Chemical Research and by a NATO Collaborative Research grant.

chloroplasts and mitochondria is significantly different. Proteins destined for these locations are synthesized on soluble cytoplasmic ribosomes and are transported into the respective organelles after translation is complete. Translocation of proteins into mitochondria has received a lot of attention in yeast and animal systems (3). Mitochondrial transport in plant cells presumably operates in a similar manner, although there is yet no confirmation of this point. Translocation of proteins into chloroplasts has received considerable study in plant cells (4); a brief summary of the major conclusions are described below.

Recognition of these principles of protein targeting poses additional complexity for those interested in manipulating the properties of plants via recombinant DNA techniques. Vectors are available for introducing new genes into the nuclear genome; specific promoters are becoming available for expressing the introduced genes at the proper time and place. However, it is also necessary to have signals which can direct the gene product to the proper intracellular location.

TARGETING OF PROTEINS TO THE STROMA

Chloroplasts are structurally the most complex organelle within a plant cell. They have three different membranes, i.e. the inner and outer envelope membranes and the thylakoid membrane. These three membranes enclose three aqueous compartments, i.e. the thylakoid lumen, the stromal space and the intermembrane space between the two envelope membranes. For five of these six compartments, at least some of the proteins are encoded in nuclear genes and synthesized on cytoplasmic ribosomes. We know very little about the proteins located in the space between the two envelope membranes.

The stromal compartment contains the largest amount of chloroplast protein. Transport of proteins to the stromal space is now understood better than transport to other compartments (4). It was first demonstrated almost ten years ago that the small subunit of ribulose-1, 5-bisphosphate carboxylase/oxygenase (Rubisco SS), a major stromal protein, is synthesized in the cytoplasm as a higher molecular weight precursor (4). The extra peptide is present at the amino terminus and because of its role in transport of the precursor protein into chloroplasts is called a transit sequence (4). The transit sequence is removed by proteolysis either during or shortly after transport. In addition to Rubisco SS, many different chloroplast proteins are synthesized as a higher molecular weight precursors (4).

Recent experiments from several laboratories have demonstrated that a transit peptide is necessary and sufficient for transport to the stroma. The necessity of a transit peptide for transport was first demonstrated by Mishkind et al. (5) who showed that a precursor with part or all of its transit peptide removed could not be transported into chloroplasts. This has subsequently been confirmed using modified precursors generated by recombinant DNA techniques. Precursors lacking a transit peptide can not be imported into chloroplasts (6). Precursors with a deletion in the transit peptide have either lost the ability to be imported or have impaired ability to be imported, depending upon the extent and location of the deletion (Lubben, Moore and Keegstra, unpublished observations; Smeekens and Weisbeek, unpublished observations). Efforts to define essential regions of the primary sequence have generally not been successful and it has been suggested that the secondary structure of the transit peptide is the essential feature (7).

The first demonstration that a transit peptide is sufficient for targeting a protein to the stroma came from the work of Schreier et al. (8) and Van den Broeck et al. (9). Using both in vivo and in vitro approaches, they demonstrated that the Rubisco SS transit peptide from pea was capable of transporting a bacterial protein, neomycin phosphotransferase (NPTII), into tobacco chloroplasts. More recently Lubben and Keegstra used a quantitative in vitro import assay to demonstrate that a Rubisco SS transit peptide from soybean could efficiently transport a cytoplasmic heat shock protein into pea chloroplasts (11). All of the studies described above have employed Rubisco SS transit peptide, however, Smeekens et al. have demonstrated that ferredoxin and plastocyanin transit sequences are capable of directing passenger proteins into pea chloroplasts (11; Smeekens, Van Steeg, Bauerle, Bettenbroek, Keegstra and Weisbeek, submitted for publication). The general conclusion derived from these studies is that it is possible to direct a number of different passenger proteins into the chloroplast stroma simply by adding a transit sequence onto the amino terminus of the passenger. A similar conclusion has been reached in studies of mitochondrial targeting (3).

It should be noted that this conclusion is not always valid. Van Steeg et al. have demonstrated that the efficiency of import into mitochondria is dependent upon the passenger protein and that some proteins are imported poorly or not at all (12). For chloroplast import, Lubben et al. have observed that certain passenger proteins are imported very poorly or not at all (Lubben, Gatenby, Ahlquist and Keegstra, submitted for publication). For example, they found that soybean SS transit sequence caused only limited import of the brome mosaic virus coat protein. Wassman et al. observed inefficient in vitro import of NPTII when only the SS transit sequence was used for targeting, but found that import improved when 23 amino acids from the mature region of SS was also included (13). In contrast, Lubben et al. observed that addition of 11 amino acids of mature SS gave only limited improvement of import and the addition of almost the entire mature SS abolished import completely (Lubben, Gatenby, Alhquist and Keegstra, submitted for publication). Thus, the addition of segments of the mature protein does not always improve import efficiency.

Although in vitro import studies have several advantages, i.e. they are relatively easy to conduct and allow one to quantitatively measure import directly, in vivo studies also have advangages. Kuntz et al. (14) measured the in vivo accumulation of two NPTII constructs similar to these studied in vitro by Wassman et al. (13). Kuntz et al. observed very little NPTII accumulation in chloroplasts when the transit peptide plus 23 amino acids of mature SS were used for targeting, whereas a substantial amount of NPTII accumulated in chloroplasts when only the transit peptide of SS was used for targeting. Several explanations could account for the differences between their results and those of Wassman et al. For example, the two proteins could be differentially sensitive to protease degradation, either in the cytosol before import or in the plastid after transport. However, these results do serve as an indicator that factors in addition to import efficiency serve to control the extent of accumulation of foreign proteins within chloroplasts.

TARGETING OF PROTEINS WITHIN CHLOROPLASTS

Relatively little is know regarding the mechanisms responsible for targeting proteins to locations other than the stromal compartment. One important practical question is whether transit sequences from precursors destined for a specific location would target other passenger proteins to this location. Cline has provided evidence that LHCP, an integral protein of the thylakoid membrane, is first imported into the stromal space and then is inserted into the thylakoid membrane (15). It has not been determined whether the insertion step is determined by the transit sequence, the mature protein, or a combination of the two.

Smeekens et al. (11) have examined the targeting of plastocyanin (PC). The targeting of this protein is particularly interesting because it must cross three membranes between the site of precursor synthesis in the cytosol and the location of PC function in the thylakoid lumen. Smeekens et al. demonstrated that proper targeting and processing of PC could be reconstituted in vitro (11). They also provided evidence that PC targeting occurs in two steps. The first is the import of the precursor across the two envelope membranes. This process is presumably via a mechanism similar to the import of prSS and other stromal proteins. In the stromal space, part of the transit peptide is removed by a processing protease to generate an intermediate-sized molecule. The second step is hypothesized to involve translocation of the PC intermediate across the thylakoid membrane (11). Inside the thylakoid lumen the intermediate is cleaved to the mature size by a second processing protein (16). Although the broad outlines of this two step model have considerable experimental support, many details for each of the steps need to be studied further.

Smeekens et al. attempted to use the transit sequence from plastocyanin to direct a non lumenal protein to the thylakoid lumen. The first efforts employed ferredoxin as the passenger protein. Although the PC transit sequence could direct ferredoxin into chloroplasts, the ferredoxin ended up in the stroma as a form slightly larger than authentic ferredoxin (11). From this result they concluded that the first step in the plastocyanin import pathway had occurred, but that the second step and its associated processing event was blocked. Several possible explanations of the block in the second step were considered, including the possibility that ferredoxin could not be transported out of the stroma because of its many strong associations with other stromal components. To test this idea, a chimeric precursor containing a non chloroplastic passenger protein was constructed. For this purpose, a yeast mitochondrial superoxide dismutase (SOD) was used. Again the same results were obtained, i.e. the plastocyanin transit peptide was able to direct the import of SOD into chloroplasts, but the protein accumulated in the stroma as a form smaller than the precursor but larger than authentic SOD (Smeekens, Van Steeg, Bauerle, Bettenbroek, Keegstra and Weisbeek, submitted for publication). Further studies with other passenger proteins are underway, but up until this point, it has not been possible to target a non lumenal protein to the thylakoid lumen.

Although the features required for a protein to be a successful passenger are not understood for either step, it can be concluded that the requirements for acceptable passenger proteins are less stringent for the first step than for the second. This observation points to the possibility that protein transport across the envelope membranes is fundamentally different than transport across the thylakoid. The possibility that chloroplasts may contain two distinct protein transport mechanisms makes this organelle very attractive for further studies on protein targeting.

SUMMARY

Significant advances have been made in understanding how proteins are targeted to chloroplasts (4). It is clear that a transit sequence is necessary and sufficient (in many cases) to target a passenger protein into chloroplasts (8-11). Evidence has been presented that a transit peptide is also important in targeting a protein to the proper location within a chloroplast (11). However, it has not yet been possible to use transit sequences to target foreign passenger proteins to non-stromal locations. All foreign passenger proteins that have been imported into chloroplasts have been found in the stromal space, regardless of the transit sequence used for targeting.

These results have important implications for those interested in modifying plants via recombinant DNA techniques. At least for certain proteins, it is possible to target a protein to the stromal space of a chloroplast simply by addition of a transit sequence to the amino terminus. However, several cautionary notes must be added. First, this strategy apparently does not work for all proteins and we currently do not understand enough about the mechanism of translocation to predict the efficiency of import. Secondly, it is not yet possible to target a protein of interest to any location within chloroplasts except the stromal space. Finally, factors other than targeting efficiency may also be important in vivo in determining the final amount of a foreign protein that accumulates inside chloroplasts.

Despite these significant advances, it is also clear that much still needs to be learned before we understand the details of the targeting mechanisms. Some of the major unanswered questions are: What translocation mechanism is responsible for transporting proteins across the two bilayer membranes of the chloroplast envelope? What other types of topogenic sequences are necessary to target proteins to the proper location within chloroplasts? What mechanisms are responsible for intraorganellar targeting?

REFERENCES

1. Walter P, Lingappa VR (1986). Mechanism of protein translocation across the endoplasmic-reticulum membrane. Ann Rev Cell Biol 2:499.
2. Farquhar MG (1985). Progress in unraveling pathways of golgi traffic. Ann Rev Cell Biol 1:447.
3. Douglas MG, McCammon MT, Vassarotti A (1986). Targeting proteins into mitochrondria. Microbiol Rev 50:166.
4. Schmidt GW, Mishkind M (1986). The transport of proteins into chloroplasts. Ann Rev Biochem 55:879.
5. Mishkind ML, Wesler SR, Schmidt GR (1985). Functional determinants in transit sequences: import and partial maturation by vascular plant chloroplasts of the ribulose-1,5-bisphosphate carboxylase small subunit of Chlamydomonas. J Cell Biol 100:226.
6. Anderson S, Smith SM (1986). Synthesis of the small subunit of ribulose-bisphosphate carboxylase from genes cloned into plasmids containing the SP6 promoter. Biochem J 240:709.
7. von Heijne G (1986). Mitochondrial targeting sequences may form amphiphilic helices. EMBO J 5:1335.
8. Schreier PH, Seftor EA, Schell J, Bohnert HJ (1985). The use of nuclear-encoded sequences to direct the light-regulated synthesis and transport of a foreign protein into plant chloroplasts. EMBO J 4:25.
9. Van den Broeck G, Timko MP, Kausch AP, Cashmore AR, Van Montagu M, Herrera-Estrella L (1985). Targeting of a foreign protein to chloroplasts by fusion to the transit peptide from the small subunit of

ribulose-1,5-bisphosphate carboxylase. Nature 313:358.
10. Lubben TH, Keegstra K (1986). Efficient in vitro import of a cytosolic heat shock protein into pea chloroplasts. Proc Natl Acad Sci USA 83:5502.
11. Smeekens S, Bauerle C, Hageman J, Keegstra K, Weisbeek P (1986). The role of the transit peptide in the routing of precursors toward different chloroplast compartments. Cell 46:365.
12. Van Steeg H, Oudshoorn P, Van Hell B, Polman JEM, Grivell LA (1986). Targeting efficiency of a mitochondrial pre-sequence is dependent on the passenger protein. EMBO J 5:3643.
13. Wasmann CC, Reiss B, Bartlett SG, Bohnert HJ (1986). The importance of the transit peptide and the translocated protein for protein import into chloroplasts. Mol Gen Genet 205:446.
14. Kuntz M, Simons A, Schell J, Schreier PH (1986). Targeting of protein to chloroplasts in transgenic tobacco by fusion to mutated transit peptide. Mol Gen Genet 205:454.
15. Cline K (1986). Import of proteins into chloroplasts. Membrane integration of a thylakoid precursor protein reconstituted in chloroplast lysates. J Biol Chem 261:14804.
16. Hageman J, Robinson C, Smeekens S, Weisbeek P (1986). A thylakoid processing protease is required for complete maturation of the lumen protein plastocyanin. Nature 324:567.

Plant Gene Systems and Their Biology, pages 371–381

GENETIC MODIFICATION OF psbA GENES IN THE CYANOBACTERIUM SYNECHOCYSTIS 6803[1]

Christer Jansson,[2] Richard J. Debus,[3]
Heinz D. Osiewacz, Michael Gurevitz,
and Lee McIntosh

Plant Research Laboratory and Department of Biochemistry,
Michigan State University,
East Lansing, MI 48824, U.S.A.

ABSTRACT psbA in Synechocystis 6803 was found to belong to a small multigene family with three copies. The reading frames for the three psbA genes were interrupted by insertion of marker genes conferring antibiotic resistance. The transformant inactivated in all three psbA genes cannot grow photosynthetically, but can be cultured photo-heterotrophically. This transformant, lacking the D-1 polypeptide, has no oxygen evolving capacity, and displays a high yield of chlorophyll a fluorescence.

INTRODUCTION

The reaction center core of photosystem II (PS II) contains at least six integral membrane polypeptides; CP-47, CP-43, D-1, D-2, and the two subunits of cytochrome b-559 (1-4). Cofactors include the reaction center chlorophyll a (chl a) P680, the intermediate acceptor pheophytin, the primary quinone acceptor Q-A, the secondary quinone acceptor Q-B, and a non-heme iron (5).

[1]This work was supported in part by Department of Energy contract DE-AC02-76ER01338.
[2]Supported by the McKnight Foundation and a fellowship from the Swedish Natural Science Research Council.
[3]Supported by National Science Foundation Postdoctoral Fellowship grant DMB-8608566 and the McKnight Foundation.

Where and how the different cofactors are bound in the protein matrix of PS II is still unclarified. It has been inferred that D-1 provides a binding site for Q-B (6). CP-47 and CP-43 have both been identified as chl-binding polypeptides (7,8), and it has been suggested from spectroscopic analyses that CP-47 is the site for the primary charge separation between P680 and Q-A (9,10). However, based on the recent determination of the crystal structure of reaction centers in photosynthetic bacteria (11,12) and on the regional amino acid homology between the L and M subunits of bacterial reaction centers and D-1 and D-2, it has been proposed that the charge separation in PS II takes place in a D-1/D-2 unit (12,13). The recent isolation of a photochemically active complex of D-1, D-2 and cytochrome b559 (14,15) gives weight to the latter proposal.

To obtain more information on the role of D-1 and its interaction with other PS II polypeptides, we have genetically engineered the facultative heterotrophic cyanobacterium *Synechocystis* 6803 to obtain a transformant inactivated in *psbA*, the gene encoding the D-1 polypeptide. Some properties of this transformant are described.

MATERIALS AND METHODS

Growth conditions. *Synechocystis* 6803 was grown as described (16). For selection and propagation of transformants, 10 µM 3-(3,4-dichlorophenyl)-1,1-dimethyl-urea (DCMU; Sigma), 5 mM glucose and appropriate antibiotics (Sigma) were added to the medium. The concentrations of antibiotics used were; 5 µg/ml kanamycin, 10 µg /ml streptomycin, 20 µg/ml spectinomycin and 2.5 µg/ml chloramphenicol.

DNA isolation and hybridization. Genomic DNA from *Synechocystis* 6803 was isolated from CsCl gradients. Southern blot analyses were performed according to standard procedures (17).

Western blot analyses. The protocol of Blake et al. (18) was followed, using an antibody against the D-1 polypeptide from *Amaranthus hybridus* (19).

Functional analyses. All steps were carried out at room temperature. Cells were harvested and washed three times in assay medium (30 mM Hepes pH 7.4, 20 mM NaCl) containing 200 mM sorbitol. The washed cells were suspended in assay medium and kept in the dark. Oxygen evolution was measured in a Rank Brothers oxygen electrode with saturating white light. The reaction mixture is given in the legend to Fig. 3. Chl *a* fluorescence induction was measured in a laboratory-constructed fluorimeter (20) with a 690 nm (10 nm bandwidth) interference filter (Baird-Atomic) in front of the photodiode and a blue (Corning 4-96) excitation filter. Further details are given in the legend to Fig. 4. Chl *a* was determind according to Mackinney (21).

RESULTS

Cloning of the psbA gene family. In the cyanobacterium *Synechocystis* 6803, *psbA* exists as a small multigene family with three copies, *psbA*-1, *psbA*-2 and *psbA*-3 (Figs. 1 and 2). *PsbA*-1 was cloned by screening a recombinant library of *Synechocystis* 6803 DNA fragments in pUC 18 with the *psbA* gene from *Amaranthus hybridus* (22) as a probe. *psbA*-2 was cloned by screening a pUC 19 library with *psbA*-1 as a probe. *psbA*-3 was cloned from a EMBL 3 Lambda library and subsequently subcloned as a smaller fragment in pUC 19.

Inactivation of the psbA genes. The procedure used to inactivate the *psbA* gene family is shown in Fig. 1. *psbA*-1 was inactivated by insertion of transposon Tn 903 (23), conferring kanamycin resistance. The insertion was made into a unique Hpa I site close to the 5' end of the *psbA*-1 gene. To inactivate *psbA*-2, a cassette conferring streptomycin and spectinomycin resistance (24) was inserted into a blunt-ended unique Nco I site close to the 3' end. *psbA*-3 was inactivated by insertion of a chloramphenicol resistance gene (25; obtained in pRL 171, J. Elhaj and P. Wolk, unpublished results) into a blunt-ended unique Nco I site close to the 3' end. Constructs containing the inactivated *psbA* genes were cloned in pUC 19, which does not replicate in *Synechocystis*. After transformation (16) the samples were spread on membrane filters (0.45 μm, Fisher Scientific) resting on plates with non-selective agar medium. The plates were incubated for 2 days to allow for maximum phenotypic expression. Transformants were selected by transferring the filters to solid media containing glucose, DCMU and appropriate antibiotics. The location of the insertions in the chromosomal *psbA* genes was confirmed by Southern blot analyses (Fig. 2).

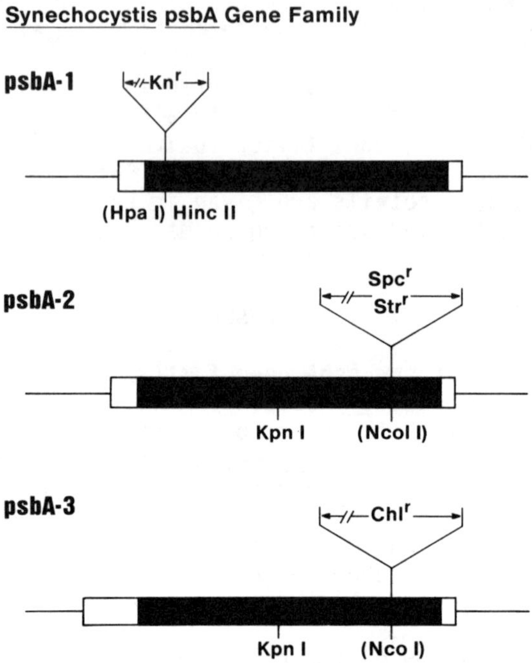

Fig. 1. The psbA genes in Synechocystis 6803 were inactivated by insertion of marker genes, conferring antibiotic resistance. The kanamycin resistance gene (Km^r) was inserted as a Hinc II fragment, the streptomycin and spectinomycin resistance cassette (Str^r/Spc^r) as a Sma I fragment, and the chloramphenicol resistance gene (Chl^r) as a Hinc II fragment.

PS II properties. Inactivation of psbA-1 alone or of psbA-1 and psbA-2 did not affect the viability of the cells growing under photoautotrophic conditions. Inactivation of all three psbA genes, however, resulted in a transformant incapable of photosynthetic growth. This triple-transformant was grown photoheterotrophically on glucose.

Western blot analyses confirmed that the D-1 polypeptide is absent in thylakoid membranes from the triple-transformant (not shown). By examining the properties of the transformant we could then begin to characterize the functional perturbations resulting from the loss of the D-1 polypeptide. Fig. 3 shows the PS II

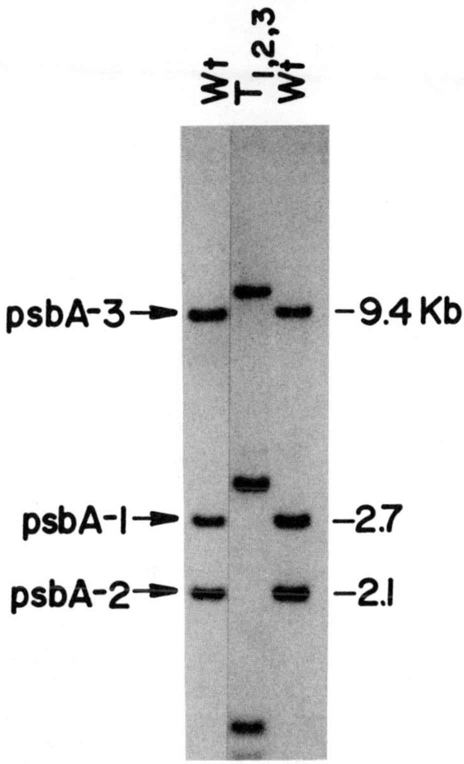

Fig. 2. Southern blot analysis of genomic DNA from *Synechocystis* 6803, showing the three native *psbA* genes in the wild type (Wt) and the inactivated genes in the triple-transformant ($T_{1,2,3}$). The DNA was digested with Hind III and probed with a combination of *psbA*-1, *psbA*-2, and *psbA*-3. The inactivated *psbA*-1 and *psbA*-3 genes gave rise to larger Hind III fragments as compared to the native genes, due to the Km^r and Chl^r inserts. The inactivated *psbA*-2 gene showed two smaller fragments because of two Hind III sites in the Str^r/Spc^r cassette.

electron transport activity, assayed as oxygen evolving capacity. The wild type showed high rates of oxygen evolution. The activity was inhibited by addition of DCMU, which blocks electron transfer, probably by binding to the D-1 polypeptide (2,26). In contrast, the transformant did not evolve oxygen. These results indicate that the absence of the D-1 polypeptide causes an

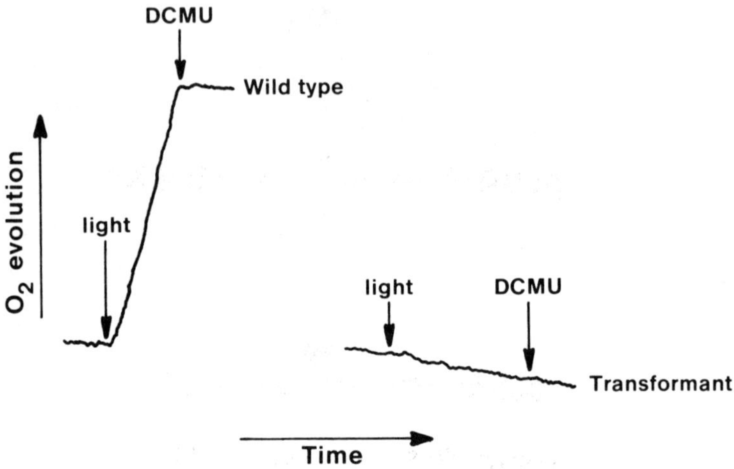

Fig. 3. Oxygen evolution measured with 1 mM 2,5-dimethyl-benzoquinone and 3 mM ferricyanide as a combined electron acceptor, with or without 10 µM DCMU. The reaction mixture contained 25 mM Hepes pH 7.4, 20 mM NaCl, and 15 µg chl _a_ /ml. The rate for the wild type was 201 µmole O_2/mg chl _a_ /h. No activity could be detected in the transformant.

inhibition in electron transport between the water splitting system of PS II and the plastoquinone (PQ) pool.

To obtain more information about the inhibitory site(s), room temperature measurements of chl _a_ fluorescence induction were performed (Fig. 4). A 690 nm narrow band emission filter (see Materials and Methods) was used to eliminate most of the phycobilisome fluorescence (27). The wild type showed a transient in fluorescence yield similar to what has been reported for other cyanobacteria (27). Addition of DCMU abolished the transient and caused a rapid rise in fluorescence to a higher steady state level. The transformant lacked the transient feature observed for the wild type and displayed rapid kinetics and a high maximum fluorescence yield. Addition of DCMU had no effect.

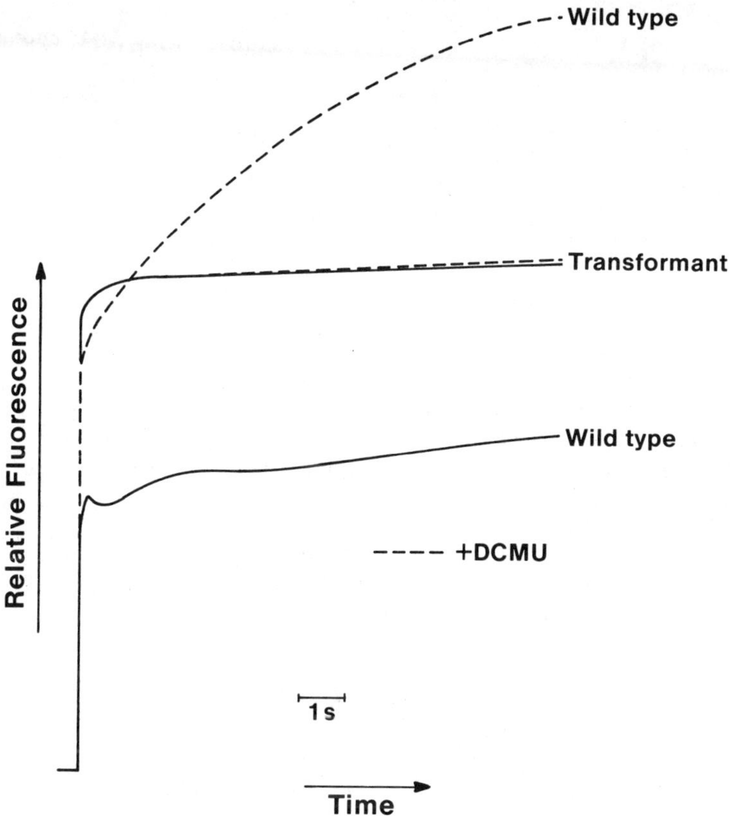

Fig. 4. Room temperature chl a fluorescence induction measured at 690 nm (10 nm bandwidth) using a broadband blue excitation filter. The reaction mixture contained 30 mM Hepes pH 7.4, 20 mM NaCl, and 5 μg chl a/ml, with or without 10 μM DCMU. Samples were dark adapted for 10 min. Note that the (+) DCMU and (-) DCMU traces for the transformant overlap.

DISCUSSION

The presence of a psbA gene family in Synechocystis 6803 (Figs. 1 and 2) is consistent with results from other cyanobacteria (28-31). In higher plants and algae there is generally one psbA gene per plastid genome. One exception is Chlamydomonas with two identical copies (32).

Why there are multiple psbA genes in cyanobacteria is not known. One hypothesis is that the D-1 polypeptide might have capacities in addition to its role in PS II (28,31). Thus, the polypeptide products from different psbA genes could serve different functions in cyanobacterial cells.

From the measurements of oxygen evolution and fluorescence induction (Figs. 3 and 4) it is clear that the loss of the D-1 polypeptide shuts off the electron transfer in PS II. The extent and exact nature of the inhibitory effects, however, remains to be elucidated. The high fluorescence yield of the transformant observed at 690 nm suggests the presence of PS II chl a. Possibly, those chl a molecules that normally funnel excitation energy to P 680 have been disconnected and the excitation decays as fluorescence. Alternatively, the high fluorescence could result from a block in electron transfer on the reducing side of PS II, between Q-A and the PQ pool. If the high fluorescence yield results from a block on the reducing side this would imply that the water splitting system is still able to feed electrons to P680. A more complete understanding of the fluorescence data will require that the integrity of the oxidizing side of PS II be carefully examined.

ACKNOWLEDGEMENTS

We thank Kristin Rorrer for skillful technical assistance and Dr. J.G.K. Williams for initial collaboration.

REFERENCES

1. Satoh K (1985). Protein-pigment photosystem II reaction center. Photochem Photobiol 42:845.
2. Arntzen CJ, Pakrasi HB (1986). Photosystem II Reaction Center: Polypeptide Subunits and Functional Cofactors. In Staehelin LA, Arntzen CJ (eds): "Encyclopedia of Plant Physiology." Berlin: Springer, vol 19, p 457.
3. Ghanotakis DF, Yocum CF (1985). Polypeptides of photosystem II and their role in oxygen evolution. Photosynth Res 7:97.
4. Widger WR, Cramer WA, Hermodson, M, Herrmann RG (1985). Evidence for a hetero-oligomeric structure of the chloroplast cytochrome b559. FEBS Lett 191:186.

5. Diner BA (1986). The reaction center of photosystem II. In Staehelin LA, Arntzen CJ (eds): "Encyclopedia of Plant Physiology." Berlin: Springer, vol 19, p 422.
6. Arntzen CJ, Darr SC, Mullet JE, Steinback KE, Pfister K (1982). Polypeptide determinants of plastoquinone function. In Trumpower BL (ed): "Function of Quinones in Energy Conserving Systems." New York: Academic Press, p 443.
7. Delepelaire P, Chua N-H (1979) Lithium dodecyl sulfate/polyacrylamide gel electrophoresis of thylakoid membranes at 4 °C: Characterization of two additional chlorophyll-protein complexes. Proc Natl Acad Sci USA 76:111
8. Camm EL, Green BR (1980). Fractionation of thylakoid membranes with the nonionic detergent Octyl-β-D-glucopyranoside: Resolution of chlorophyll-protein complex II into two chlorophyll-protein complexes. Plant Physiol 66:428.
9. Nakatani HY, Ke B, Dolan E, Arntzen CJ (1984). Identity of the photosytem II reaction center polypeptide. Biochim Biophys Acta 765:347.
10. Yamagishi A, Katoh S (1985). Further characterization of the two photosystem II reaction center complex preparations from the thermophilic cyanobacterium _Synechococcus_ sp. Biochim Biophys Acta 807:74.
11. Deisenhofer J, Epp O, Miki K, Huber R, Michel H (1985). Structure of the protein subunits in the photosynthetic reaction center of _Rhodopseudomonas viridis_ at 3 Å resolution. Nature 318:618.
12. Michel H, Deisenhofer J (1986). X-ray diffraction studies on a crystalline bacterial photosynthetic reaction center: A progress report and conclusions on the structure of photosystem II reaction centers. In Staehelin LA, Arntzen CJ (eds): "Encyclopedia of Plant Physiology." Berlin: Springer, vol 19, p 371.
13. Trebst A (1986). The topology of the plastoquinone and herbicide binding peptides of photosystem II in the thylakoid membrane. Z Naturforsch C 41:240.
14. Nanba O, Satoh K (1987). Isolation of a photosystem II reaction center consisting of D-1 and D-2 polypeptides and cytochrome _b-559_. Proc Natl Acad Sci USA 84:109.

15. Okamura MY, Satoh K, Isaacson RA, Feher G (1987). Evidence of the primary charge separation in the D_1D_2 complex of photosystem II from spinach: EPR of the triplet state. In Biggins J (ed): "Progress in Photosynthesis Research." Dordrecht: Martinus Nijhofff Publishers, vol 1, p 379.
16. Williams JGK (1987). Construction of specific mutations in the PS II photosynthetic reaction center by genetic engineering methods in the cyanobacterium Synechocystis 6803. In Glaser A, Packer L (eds): "Methods in Enzymology." In press.
17. Maniatis T, Fritsch EF, Sambrook J (1982). "Molecular Cloning: A Laboratory Manual." Cold Spring Harbor, New York: Cold Spring Harbor Laboratory Press.
18. Blake MS, Johnston, KH, Russel-Jones GJ, Gotschlich EC (1984). A rapid, sensitive method for detection of alkaline phosphatase-conjugated anti-antibody on Western blots. Anal Biochem 136:175.
19. Ohad I, Kyle, DJ, Hirschberg J (1985). Light-dependent degradation of the Q_B-protein in isolated pea thylakoids. EMBO J 4:1655.
20. Paterson DR, Arntzen CJ (1982). Detection of altered inhibition of photosystem II reactions in herbicide-resistant plants. In Edelman M, Hallick RB, Chua N-H (eds): "Methods in Chloroplast Molecular Biology." Amsterdam: Elsevier Biomedical Press, p 109.
21. Mackinney G (1941). Absorption of light by chlorophyll solutions. J Biol Chem 140:315.
22. Hirschberg J, McIntosh L (1983). Molecular basis of herbicide resistance in _Amaranthus hybridus_. Science 222:1346.
23. Oka A, Sugisaki H, Takanami M (1981). Nucleotide sequence of the kanamycin resistance transposon Tn 903. J Mol Biol 147:217.
24. Prentki P, Krisch HM (1984). In vitro insertional mutagenesis with a selectable DNA fragment. Gene 29:303.
25. Bagdasarian M, Bagdasarian MM, Coleman S, Timmis KN (1979). New vector plasmids for gene cloning in _Pseudomonas_. In Timmis KN, Puhler A (eds): "Plasmids of Medical, Environmental and Commercial Importance." Amsterdam: Elsevier, p 411.
26. Pfister K, Steinback KE, Gardner G, Arntzen CJ (1981). Photoaffinity labeling of an herbicide receptor protein in chloroplast membranes. Proc Natl Acad Sci USA 78:981.

27. Fork DC, Mohanty P (1986). Fluorescence and other characteristics of blue-green algae (cyanobacteria), red algae, and cryptomonads. In Govindjee, Amesz J, Fork DC (eds): "Light Emission by Plants and Bacteria." Orlando: Academic Press, p 451.
28. Golden SS, Brusslan J, Haselkorn R (1986). Expression of a family of psbA genes encoding a photosystem II polypeptide in the cyanobacterium Anacystis nidulans R2. EMBO J 5:2789.
29. Curtis SE, Haselkorn R (1984). Isolation, sequence and expression of two members of the 32 kd thylakoid membrane protein gene family from the cyanobacterium Anabaena 7120. Plant Mol Biol 3:249.
30. Mulligan B, Schultes N, Chen L, Bogorad L (1984). Nucleotide sequence of a multiple-copy gene for the B protein of photosystem II of a cyanobacterium. Proc Natl Acad Sci USA 81:2693.
31. Nierzwicki-Bauer SA, Haselkorn R (1986). Differences in mRNA levels in Anabaena living freely or in symbiotic association with Azolla. EMBO J 5:29.
32. Erickson JM, Rahire M, Rochaix J-D (1984). Chlamydomonas reinhardii gene for the 32000 mol. wt. protein of photosystem II contains four large introns and is located entirely within the chloroplast inverted repeat. EMBO J 3:2753.

Plant Gene Systems and Their Biology, pages 383–395

Biosynthesis and function of carotenoids in plants

Pablo A. Scolnik

E. I. DuPont de Nemours, Plant Science Group, Experimental Station, 402/4252, Wilmington, DE 19898

ABSTRACT Carotenoids are plastid pigments that are synthesized by nuclear-encoded enzymes. In chloroplasts, their main function is photoprotection. They are also responsible for the synthesis of the phytohormone abscisic and quantitatively different in proplastids, chloroplasts, chromoplasts and amyloplasts, suggesting a role for these compounds in plastid differentiation. In the absence of colored carotenoids chloroplasts do not develop and the expression of several nuclear genes is affected. We summarize our current knowledge of the biochemistry and the physiological roles of carotenoids, and we discuss the use of carotenoid biosynthesis as a model system for studying nuclear-cytoplasmic interactions.

INTRODUCTION

In plants there are three main classes of pigments: chlorophyll, carotenoids and anthocyanins. The first two are localized in the chloroplast whereas anthocyanins reside in the cytoplasm. Carotenoids are present in all photosynthetic and many non-photosynthetic organisms.

Carotenoids derive their name from the fact that they are the major pigments of the carrot root. Chemically, these pigments are aliphatic,

aliphatic-alicyclic or aromatic structures composed of five-carbon isoprene groups. Color derives from the presence of a chromophore, a set of conjugated C-C double bonds. The major carotenoid subgroups are carotenes and oxycarotenoids (xanthophylls). The former include the hydrocarbon carotenoids, and the latter the hydroxy, epoxy, furanoxy, and oxy derivatives of the carotenes. Genetic evidence indicates that in plants the enzymes of the carotenoid pathway are nuclear-encoded. Carotenoids are essential components of chloroplasts, thus regulation of the synthesis of these compounds is one way by which the nucleus can control chloroplast differentiation. Nuclear control of organelle biogenesis has been extensively documented in plants. However, more recent information suggests that chloroplast differentiation may influence nuclear gene expression (1,2,3). This article will summarize carotenoid biosynthesis and function, in the context of nuclear-cytoplasmic interactions in higher plants.

Functions of carotenoids in photosynthetic organisms

Although numerous functions have been assigned to carotenoids, the two most relevant ones seem to be photoprotection (4) and abscisic acid (ABA) biosynthesis (5). The photoprotection provided by carotenoids is essential to all photosynthetic organisms because chlorophyll, as well as many other porphyrins, can mediate the conversion of oxygen triplets (ground state) into oxygen singlets. Oxygen singlets are highly reactive species that can kill an organism by oxidizing its macromolecules. Plants devoid of colored carotenoids (Fig. 1) sustain extensive damage to their chloroplasts when grown at high light. This damage, however, seems to have little effect on other cellular compartments. Photooxidative damage to the chloroplast membranes has been extensively documented in several carotenoid-deficient plants. As we discussed below, photooxidation of the chloroplast has profound effects on the expression of certain nuclear genes.

The second area in which carotenoids are

essential for plants is ABA biosyntesis. This phytohormone is probably present in all higher plants and it is involved in processes such as bud and seed dormancy, elongation growth, stomatal opening, root growth, geotropism, fruit ripening and senescence. According to a postulated "indirect pathway" model, ABA is made from the carotenoid violaxanthin (Fig. 1). Although at the present time the enzymes for ABA biosynthesis have not been isolated, the fact that in plant leaves carotenoids are localized in the chloroplast membranes strongly suggest that at least the initial steps in the conversion of violaxanthin into ABA occurs in the chloroplast.

Biochemical-genetic approaches for studying the biosynthesis of carotenoids and ABA.

Although a variety of prokaryotes and eukaryotes contain carotenoids, a detailed knowledge of the molecular aspects of the biosynthetic process is available only in the photosynthetic bacterium *Rhodobacter capsulatus.* Genes affecting pigment biosynthesis in this organism are clustered in a 40 kb region of the genome (6). Using transposon mutagenesis, Zsebo and Hearst (7) identified eight genes that affect carotenoid biosynthesis. Through biochemical analyses of carotenoid mutants, functions have been assigned to five of these genes. Due to the strategic position that phytoene dehydrogenase (the enzyme that converts phytoene into phytofluene) has in the pathway (Fig. 1), we used genetic methods to localize *crtI,* the gene that is most likely to code for this enzyme (8).

Chloroplast carotenoids have been the subject of extensive genetic and biochemical research, although not at the molecular level. Qualitatively, β-carotene is the most important chloroplast carotenoid. Among the xanthophylls, lutein, neoxanthin and violaxanthin predominate. The biosynthetic pathway is depicted in Fig. 1. The sequence of reactions that leads to the formation of geranyl-geranyl pyrophosphate (GGPP) is common to the biosynthesis of all isoprenoids. The first carotenoid formed is the colorless compound phytoene. Formation of colored carotenoids requires a series of dehydrogenation reactions that extend the set of conjugated double bonds that constitutes

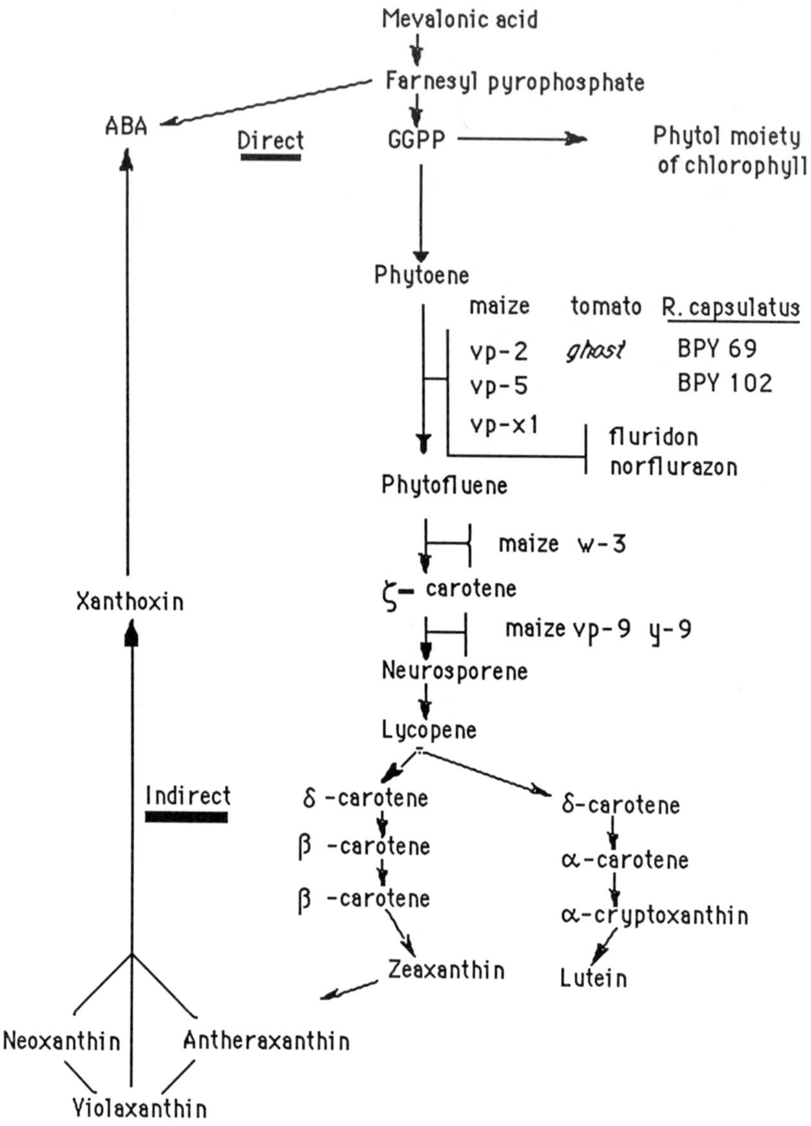

Figure 1: biochemical-genetic pathway of carotenoid biosynthesis in plants. Compounds before ζ-carotene are colorless. Postulated pathways for ABA biosynthesis are underlined.

the chromophore. In addition, phytoene in higher plants appears to be a 15-*cis* -isomer whereas the colored carotenoids are all-*trans*. Thus, isomerization must occur at some point in the biosynthetic pathway. Cyclization and the introduction of oxygen functions occurs later in the pathway (Fig. 1).

Except for phytoene synthetase, enzymes for carotenoid biosynthesis have proven to be difficult to purify in an active form. According to currently accepted models, the enzymes for desaturations, cyclizations and hydroxylations would form a highly organized multienzyme complex in the membranes of the chloroplast envelope (9). The solubilization schemes used so far would disrupt the organization of this complex, rendering the enzymes inactive.

In analyzing the process of carotenoid biosynthesis in the chloroplast one must keep in mind that these compounds are present as complexes with pigment-binding proteins and chlorophyll. Thus, the accumulation of carotenoids in the chloroplast is the result of the coordinated synthesis and assembly of these three components. Unfortunately, little is known about the sites of interaction of carotenoids with chlorophyll a/b binding proteins. Also, chloroplast carotenoids are present mostly in the thylakoids. If the site of synthesis is indeed the envelope, a translocation mechanism must exist. We suggested that binding of carotenoids to the chlorophyll a/b binding proteins (CAB) could be coupled to the process of transport of these cytoplasmic-synthesized proteins into the chloroplast, with CABs binding carotenoids as they cross the chloroplast membrane.

An interesting system for *in vitro* reconstitution of light harvesting complexes for photosystem II has been recently published (10). In addition to apoproteins and chlorophyll a and b, the reconstitution requires certain xanthophylls. These results confirm the role of carotenoids in the organization of photosynthetic membrane complexes.

Two possible pathways, "direct" and "indirect", have been suggested for the biosynthesis of ABA (5). Labeling experiments indicate that the indirect pathway is responsible for stress-induced ABA synthesis (11). Inhibition of carotenoid biosynthesis with norflurazon (Fig. 1) in barley also results in a drop in ABA, confirming the connection between carotenoids and this phytohormone. Evidence supporting the direct pathway comes from experiments in which ^{3}H-mevalonic acid and ^{14}C-phytoene were simultaneously applied to leaves (5). ABA became labelled only with ^{3}H. However, this experiment is highly questionable because it is unlikely that phytoene can be incorporated under these conditions. Thus, the results currently available indicate that ABA is made from a cyclic carotenoid, probably violaxanthin (Fig. 1). This xanthophyll is particularly abundant in the chloroplast envelope. It is possible that this could be connected to the first stages of ABA biosynthesis in the envelope.

Inhibition of carotenoid biosynthesis with norflurazon in bean has effects that resemble those of red light: shortening of internodes, leaf expansion, development of leaves at the second and third internodes and root development. The herbicide does not act directly on phytochrome since its effects are not reversible by far-red light. Thus, it is likely that the effects observed are due to the inhibition of ABA biosynthesis.

Mutants affecting carotenoid biosynthesis in plants.

Several mutants affecting carotenoid biosynthesis have been described in maize (Fig. 1) (12). In all cases, the genes responsible are nuclear recessive. The *cl1* mutation results in a yellow or white seed endosperm and albino seedlings. No C_{40} carotenoids were detected in this mutant, thus the block probably occurs before phytoene. *cl1* can be specifically suppressed by a dominant modifier nuclear gene Cl_M. Several alleles of this locus partially or completely suppressed the albinism of the seedling but not of the endosperm. Another maize mutation that affects the formation of colored carotenoids is *w*-3. It is listed in Fig. 1 as accumulating phytofluene due to early reports of this compound accumulating at low levels. However, in many alleles the only compound that accumulates is phytoene. Thus, it is possible that the block is after phytoene and that the accumulation of phytofluene in some alleles results from leakiness. An interesting feature of *w*-3 is that the mutation appears to be somewhat temperature-sensitive. Depending on the genetic background, *w*-3 can make as much as 14% colored carotenoids as wild-type at 37°C, but only 2.2% at 22°C. Plastid differentiation in *w*-3 has been extensively studied. At light levels that are too low to have significant destruction of the chlorophyll by photooxidation, *w*-3 plastids contain long single thylakoids. Under the same conditions, wild-type plastids contain similar thylakoids but a few grana were observed. When switched to high light, *w*-3 thylakoids disintegrate and only vesicles and membrane fragments can be observed after 24 h.

At least four other maize mutations that affect carotenoid biosynthesis and that result in photosensitation of chlorophyll have been described. One of these mutants accumulates ξ-carotene, the other one, lycopene. The albescent (*alb*) mutant has a phenotype that suggests that the regulation of carotenoid biosynthesis in the dark is altered (13). This mutant accumulates phytoene and phytofluene in the dark. Upon illumination at high light the chlorophyll is photooxidized. However, if the seedlings are incubated in low light for 3 days, colored carotenoids form. Furthermore, when the light is increased an apparently normal plant develops. Thus, it would seem that the synthesis of carotenoids may be regulated by different pathways in the light or in the dark, and that the dark regulation is affected in *alb*.

In addition to the maize mutants described, which affect mostly chloroplast carotenoids, mutants of other plant systems have been described which affect chromoplasts in various organs. The most extensively studied system is the chromoplast of the tomato fruit. The tomato carotenoid mutants described so far, with the exception of *ghost* and "high pigment", affect chromoplasts but not chloroplasts. As in the maize mutants, the genes responsible for these mutations show normal Mendelian inheritance.

We have recently analyzed the *ghost* mutation that had been originally described by Rick et al. (1959)

(14, 15) This mutant shows a variable pattern of pigment synthesis during development. Cotyledons are green but true leaves are white. During further growth green clonal sectors appear on the white leaves. White *ghost* leaves accumulate phytoene and no chlorophyll. Green *ghost* leaves accumulate wild-type carotenoids and chlorophyll. The plastids formed in white leaves show little internal membrane structures, with only occasional single thylakoids observed. In contrast, normal chloroplasts develop in green leaves. Thus, it seems that the *ghost* mutation affects the activity of phytoene dehydrogenase. The lack of colored carotenoids results in the destruction of chlorophyll and chlorophyll-containing membranes of the chloroplast. In the green leaves the mutation is transiently suppressed. This results in the restoration of carotenoid biosynthesis, and consequently, in the accumulation of chlorophyll and normal chloroplast differentiation. The green phenotype is not inheritable and the molecular basis for this epigenetic phenomenum is not currently understood.

The other mutation that affects both chloroplasts and chromoplasts is "high pigment" (*hp*). This mutant shows higher levels of both colored and colorless carotenoids in leaves and fruits. However, it not known whether this mutation affects carotenoid biosynthesis, pigment biosynthesis in general and/or plastid formation.

An interesting locus affecting chromoplast carotenoids is *at*. In *atat* fruits lycopene accumulation is almost abolished but β-carotene is still present. It is likely that there are two pathways for carotenoid biosynthesis in tomato fruits. Evidence for this comes from the fact that at temperatures higher than 30°C lycopene biosynthesis is inhibited, but β–carotene accumulates normally. Since in leaves the latter pigment predominates, it is possible that there are "leaf" and "fruit" pathways, which would tend to accumulate β–carotene and lycopene respectively. The "fruit" pathway would be temperature-sensitive. The "leaf" pathway would be active in early stages of fruit maturation, before chloroplasts become chromoplasts. Thus, the observation that fruits of homozygous *at* plants accumulate β-carotene at levels similar to leaf chloroplasts could be explained by postulating that *at* affects only the "fruit" pathway. However, *ghost* causes the accumulation of phytoene in both leaves and fruits. Thus, if there are indeed two pathways, the branching must occur after phytoene dehydrogenase (16).

Other loci affecting carotenoid biosynthesis in the fruit have been described; *og* may control enzymes for the conversion of phytoene into lycopene, *t* may control an isomerase affecting phytofluene, *vo* may control the enzyme for the conversion of ξ-carotene into neurosporene, *B* controls the ring closure of lycopene, and *Del* may control the conversion of lycopene into δ-carotene (16).

Carotenoids, chloroplast differentiation and nuclear gene expression.

As we explained above, evidence obtained from maize, tomato and tobacco indicates that the lack of colored carotenoids causes the internal membranes of the chloroplast to be destroyed by photooxidation. Thus, formation of mature chloroplasts requires the simultaneous synthesis and assembly of chlorophyll, carotenoids and pigment-binding proteins. Also, as noted above, ABA is made from carotenoids. Thus, biosynthesis of colored carotenoids, which is controlled by nuclear genes, is necessary for both normal chloroplast development and the production of an essential phytohormone.

Work conducted in several plant systems indicates that the expression of nuclear, photosynthesis-related genes is decreased in plants in which carotenoid biosynthesis is blocked either by herbicide treatment or by mutation. The genes studied were the ones for chlorophyll a/b binding (*CAB)* protein and, in some cases, also the genes for the small subunit of RubisCO (*SSU*). In tobacco, barley and tomato *ghost,* the effect was shown to take place at the transcriptional level. In the case of some maize mutants, expression of *CAB* genes was restored when the seedlings were placed under low light (17). Under these conditions, photooxidation does not occur and chloroplasts from wild-type and carotenoid-deficient plants are similar.

It has been proposed that nuclei and chloroplasts communicate through a bidirectional regulatory mechanism. The nuclei would control chloroplast development by the well-documented production of most chloroplast proteins. Chloroplasts would control nuclear gene expression by generating a signal that would act as a positive activator for the expression of at least some of the nuclear genes that code for chloroplast proteins. So far, the existance of this signal has not been demonstrated.

CONCLUSIONS

The regulation of plastid differentiation is a central problem in our understanding of plant development. Proplastids, chloroplasts, amyloplasts and chromoplasts are present in different stages of the development of a plant. Each plastid form contains a specific set of proteins. It is reasonable to expect that a tight regulatory interaction between nuclear and plastid genomes is responsible for the timely expression of specific plastid proteins.

Evidence obtained from two different experimental lines implicate chloroplasts in the regulation of expression of nuclear-encoded, photosynthesis-related plastid proteins. First, a correlation has been observed between the presence of chloroplasts in plant tissues and the expression of some nuclear genes that code for chloroplast proteins. For instance, *CAB* genes are expressed in leaves and stems but not in roots. More direct involvement of chloroplast in the regulation of nuclear gene expression comes from experiments conducted in potato. Chloroplast formation can be induced under special conditions in potato roots, which otherwise contain amyloplasts. Expression of three leaf-specific cDNAs was observed in chloroplast-containing but not in regular roots (18). Second, the results with carotenoid-deficient plants described above would suggest that active chloroplasts are needed to activate the expression of some nuclear genes.

The hypothesis that mature chloroplasts generate a signal that travels to the nucleus and activates transcription is very attractive because it would explain many aspects of the tissue-specific regulation of genes for photosynthesis. The challenge at the present time is to design experiments to demonstrate that this signal exists and then, to explore its nature. One possible experimental approach would be to introduce mature chloroplasts or fraction thereof into carotenoid-deficient cells, looking for activation of nuclear genes. This approach is becoming technically feasible but protocols for obtaining expression in protoplasts of genes such as *CAB* and *SSU* have first to be developed.

Our knowledge on the genetic and physiological aspects of carotenoid biosynthesis has been summarized. Clearly, carotenoid and ABA biosynthesis could be used as model systems for exploring nuclear-cytoplasmic interactions. However this would require identification at the molecular level of both structural and regulatory genes for both pathways. Of particular interest are the enzymes that convert violaxanthin into ABA. As we explained above, the enzymes for carotenoid biosynthesis are probably organized as a membrane-bound complex that is difficult to purify in an active form. It also possible that at least the enzymes for the early steps in the conversion of violaxanthin into ABA are also part of this complex. The field of purification of membrane proteins is advancing rapidly due to new detergents and separation techniques. New breakthroughs could occur by applying these new techniques to the field of carotenoid biosynthesis. Another interesting possibility is the use of microbial models. As exemplified by the work on *Rhodobacter capsulatus*, it is considerably easier to clone genes for carotenoid biosynthesis in microbes than in plants. If these microbial genes contain regions of conservation with plant genes, they could be used as hybridization probes.

Regardless of the approach used, we can expect developments in the field during the next few years, and a sense of excitement can be felt as this information starts filling the gaps in our knowledge of gene expression during plant development.

REFERENCES

1. Mayfield SP, Taylor WC (1984). Carotenoid-deficient maize seedlings fail to accumulate light-harvesting chlorophyll a/b binding protein (LHCP) mRNA. Eur J Biochem 144:79.
2. Batschauer A, Mösinger E, Kreuz K, Dörr I, Apel K (1986). The implication of a plastid-derived factor in the transcriptional control of nuclear genes encoding the light-harvesting chlorophyll a/b protein. Eur J Biochem 154:625.
3. Simpson J, Van Montagu M, Herrera-Estrella L (1986). Photosynthesis-associated gene families: differences in response to tissue-specific and environmental factors. Science 233:34.
4. Anderson IC, Robertson DS (1960). Role of carotenoids in protecting chlorophyll from photodestruction. Plant Physiol 35:531.
5. Milborrow BV (1983). Pathways to and from abscisic acid. In Addicott FT, (ed): "Abscisic Acid", New York: Prager Scientific, p 79.
6. Taylor DP, Cohen SN, Clark GW, Marrs BL (1983). Aligment of genetic and restriction maps of the photosynthesis region of the *Rhodopseudomonas capsulata* chromosome by a conjugation-mediated marker rescue technique. J Bact 154:580.
7. Zsebo KM, Hearst JE (1984). Genetic-physical mapping of a photosynthetic gene cluster from *R. capsulata*. Cell 37:937.
8. Giuliano G, Pollock D, Scolnik PA (1986). The gene *crtI* mediates the conversion of phytoene into colored carotenoids in *Rhodopseudomonas capsulata*. J Biol Chem 261:12925.
9. Bramley P (1985). The *in vitro* biosynthesis of carotenoids. In Paoletti R, Kritchensky D (eds): "Advances in Lipid Research", New York: Academic Press, p 243.
10. Plumley FG, Schmidt GW (1987). Reconstitution of chlorophyll a/b light-harvesting complexes:

xanthophyll-dependent assembly and energy transfer. Proc Natl Acad Sci 84:146.

11. Creelman RA, Zeevart JAD (1984). Incorporation of oxygen into abscisic acid and phaseic acid from molecular oxygen. Plant Physiol 75:166.

12. Kirk JTO (1978). Genetic control and plastid biochemistry. In Kirk JTO, Tilney-Bassett RAE (eds): "The Plastids", New York: Elsevier, p 525.

13. Troxler RF, Lester R, Craft FO, Albright JT (1969). Plastid development in albescent maize. Plant Physiol 44:1609.

14. Rick CM, Thompson AE, Bauer O (1959) Genetics and development of an unstable chlorophyll deficiency in *Lycopersicon esculentum*. Am J Bot 46:1.

15. Scolnik PA, Hinton P, Greenblatt IM, Giuliano G, Delanoy MR, Spector DL, Pollock, D. (1987). Somatic instability of carotenoid biosynthesis in the tomato *ghost* mutant and its effect on plastid development. Planta 171:1.

16. Scolnik PA, Giuliano G, Pollock D, Hinton P (1986). Molecular genetics of carotenoid biosynthesis: the tomato *ghost* mutant and mutants of the photosynthetic bacterium *Rhodopseudomonas capsulata*. In Randall DD, Miles CD, Nelson CJ, Blevins DG, Miernyk JA (eds): "Current Topics in Plant Biochemistry and Physiology" vol 5, University of Missouri-Columbia, p 142.

17. Taylor WC, Burgess DG, Mayfield SP (1986) ibid. p 117.

18. Eckes P, Schell J, Willmitzer L (1985) Organ-specific expression of three leaf/stem specific cDNAs from potato is regulated by light and correlated with chloroplast development. Mol Gen Genet 199:216.

Index